James

To celebrate the
last main day
of our collaboration!
Dear Trevor Fuerher
Jean
15 January
2006

The Diary of Nicholas Peacock, 1740–1751

The Diary of Nicholas Peacock, 1740–1751

The worlds of a County Limerick farmer and agent

EDITED BY
Marie-Louise Legg

FOUR COURTS PRESS

Set in 9.5 pt on 12.5 Sabon for
FOUR COURTS PRESS LTD
7 Malpas Street, Dublin 8, Ireland
e-mail: info@four-courts-press.ie
http://www.four-courts-press.ie
and in North America
FOUR COURTS PRESS
c/o ISBS, 920 N. E. 58th Avenue, Suite 300, Portland, OR 97213

© Marie-Louise Legg 2005

A catalogue record for this title
is available from the British Library.

ISBN 1–85182–899–0

All rights reserved. No part of this publication may be
reproduced, stored in or introduced into a retrieval
system, or transmitted, in any form or by any means
(electronic, mechanical, photocopying, recording or
otherwise), without the prior written permission of
both the copyright owner and the publisher of this book.

Printed in England
by Antony Rowe Ltd, Chippenham, Wilts.

CONTENTS

List of illustrations	6
Family tree	7
Map	8
Acknowledgments	10
INTRODUCTION	11
Glossary	33
Bibliography	41
THE DIARY	45
Index	229

ILLUSTRATIONS

(appear between pages 128 and 129)

1 James Henry Brocas, *Old Baal's Bridge, Limerick*, National Gallery of Ireland
2 British School, *Mrs Valentine Quin, née Mary Widenham (1682–1776)* (private collection)
3 British School, *Valentine Quin (d. 1744)* (private collection)
4 George Petrie, *Kilmallock, Co. Limerick 1826*. Visited by Nicholas Peacock in November 1745. From J.N. Brewer, *The Beauties of Ireland* (London 1825)
5 *Lohort Castle, near Mallow, Co. Cork c.1738*. Visited by Nicholas Peacock in August 1747. Engraved by J. Toms
6 *Itinerant spademen holding 'loys', used particularly for potato cultivation c.1820*. Ulster Folk and Transport Museum
7 George Edward Pakenham, *Irish road scene c.1737*. From the Journal of George Edward Pakenham, 1737–90
8 John George O'Brien (Oben), *A View of Adare, Co. Limerick 1793* (George Stacpoole collection)

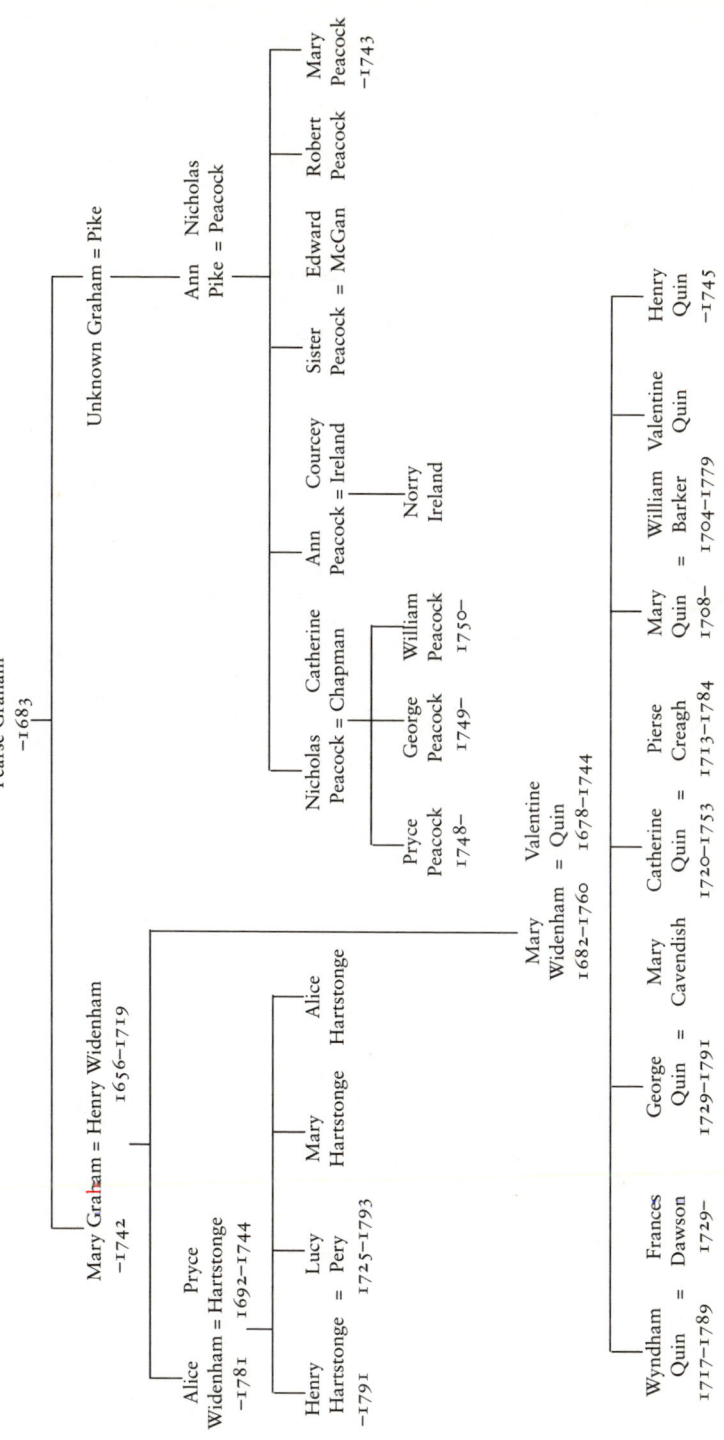

Conjectured family tree of Nicholas Peacock

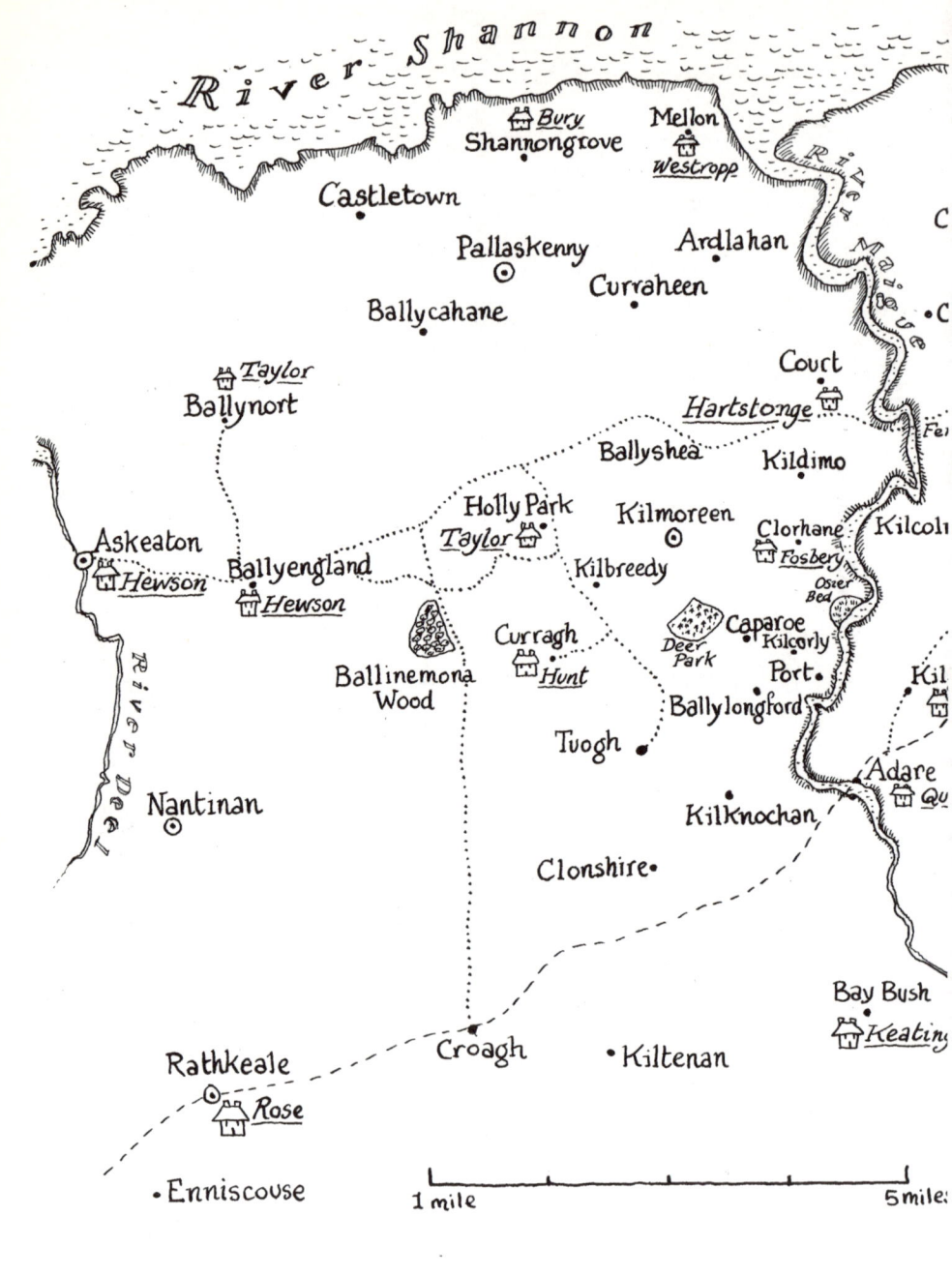

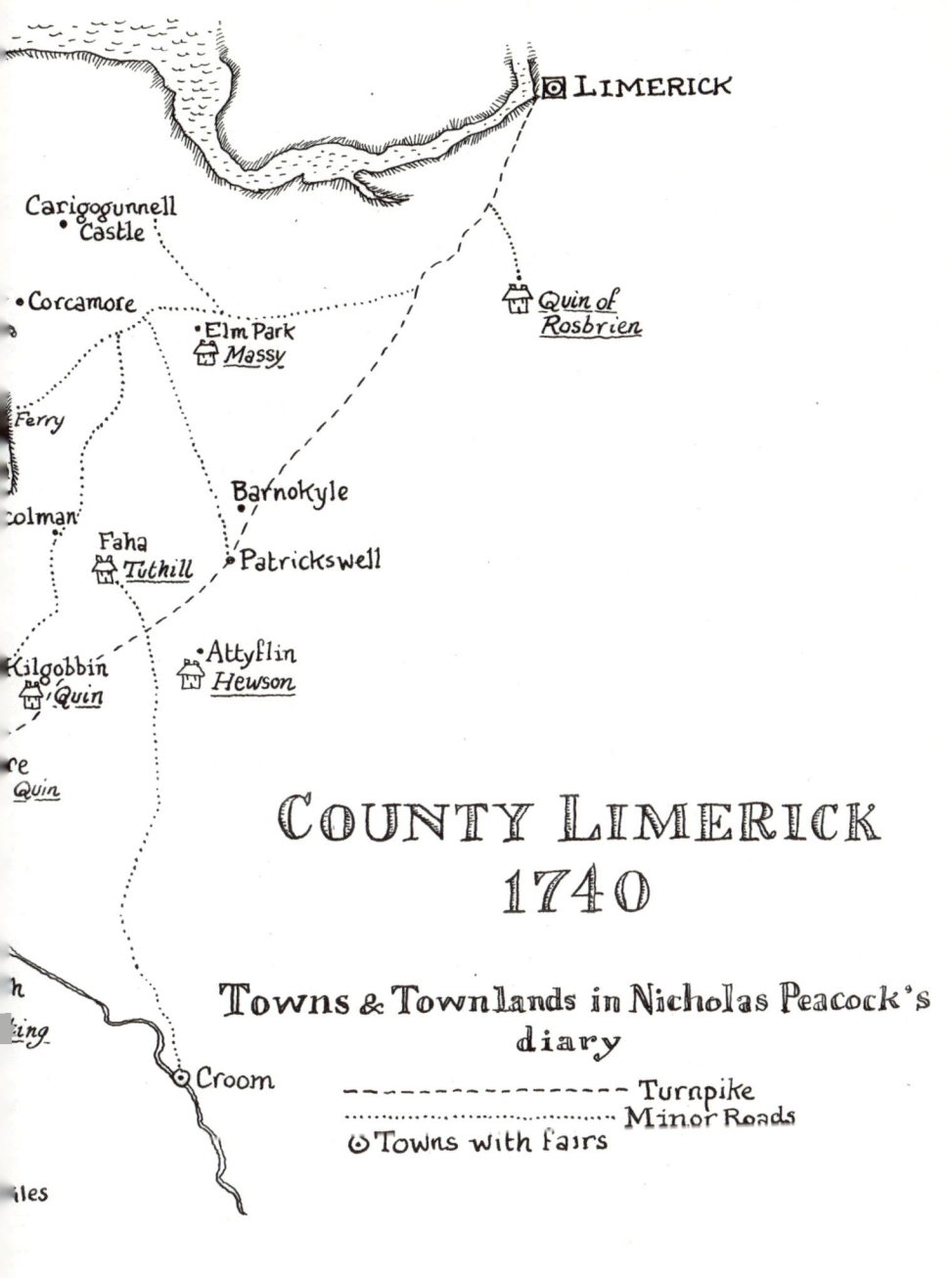

ACKNOWLEDGMENTS

I am deeply grateful to Toby Barnard and David Dickson for suggesting that I might edit Nicholas Peacock's diary. Their intimate knowledge of the period and of daily life in eighteenth century rural Ireland has been of tremendous value to me and a great source of support. I am indebted to John Logan and the University of Limerick for their hospitality while I was working there, and to the staff of the University Library Special Collections for their help with the Dunraven Papers. I am very grateful to Gerard Lyne, keeper of manuscripts at the National Library, for permission to publish this diary. I should particularly like to thank Tom Desmond and the staff of the National Library of Ireland for making the uncatalogued Limerick Papers available to me and for much useful advice. Many archives, libraries and librarians have given invaluable support, including Ray Refaussé at the Representative Church Body, Robert Mills, Librarian of the Royal College of Physicians of Ireland, the National Archives, the Society of Genealogists, London, the London Library, the British Library. Tom Jago advised me on brewing and Ms Mary Hartstonge of Christchurch, New Zealand, Gerry Dobson of Dallas, Texas and Mr William Peacocke of Limerick were most helpful in providing details of their family histories.

The following have given permission to reproduce illustrations: George Stacpoole, Lady Dunraven, the National Gallery of Ireland, the Courtauld Institute of Art. I am grateful to Desmond Fitzgerald, the Knight of Glin for advice on illustrations.

The map of Peacock's countryside and neighbours was drawn by Martin Rowson, to whom, as ever, I am very grateful.

This publication has been supported with grants from the Marc Fitch Fund and the Irish Georgian Society, to whom I am extremely grateful.

Marie-Louise Legg
London, November 2004

INTRODUCTION

The undulating fields in eastern Co. Limerick that border the river Maigue between Adare and the sea are some of the most fertile lands in Ireland. The Maigue with its tributaries drains the area; the rich alluvium is formed into large meadows which line the banks up to its junction with the Shannon which is tidal above Adare.[1] This is the corcas land, which has been favourably compared to that of the Golden Vale in Tipperary.[2] These fields have been reclaimed from the river by a system of artificial embankments; the resulting topsoil is a rich black mould deposited on top of yellow or blue clay.[3] The land and its villages are marked by numbers of ruined tower houses and priories which are a testament to its value.

This was the land farmed by Nicholas Peacock, a man from the middling gentry living in the middle of the eighteenth century. We can deduce from the diary that he lived in Kilcorly and Kilmoreen townlands, on a bend in the right bank of the Maigue north of Adare. He was fortunate that it was of such exceptionally good quality. From 1740 to 1751, Peacock kept a detailed diary of his work and life, and this has left a valuable record of the day to day life of a man living at a level of Irish society of which we know little or nothing. This diary is one of the most complex and detailed accounts existing of life in rural Ireland in the mid-eighteenth century. Peacock was not a grandee, although he was invited to the houses of grand neighbours, nor was he a simply a small farmer, although he both employed and entertained his farmer neighbours to whom he lets land for grazing. He was a son of a cadet branch of the Widenham family of Court, the ruins of whose house still stand north of Adare. He was a freeholder, land agent to his relations the Hartstonges. His diary records his rise in status through marriage and how his life changes from being a bachelor sharing his house with his servants to his married life with a small family, who exchanges entertainment with his neighbours on more equal terms.

NICHOLAS PEACOCK AND HIS FAMILY

In the diary, the writer gives no hint about his identity. It is the date of his wedding to Catherine Chapman in 1747 that makes it possible to trace him in the

---

[1] Lamplugh et al., p. 51. The full citation of sources (abbreviated in the footnotes) can be found in the Bibliography at pp 41–3. [2] Willes, p. 52 [3] Lamplugh et al., p. 57

published marriage bonds of the Cloyne diocese, now destroyed.[4] Despite exhaustive searching, there is no other record of his existence, except for a reference in his great aunt Widenham's will to her 'nephew Nicholas Peacocke'.[5]

Nicholas Peacock collected rents, tithes and church rates, built two houses, got married and had at least three children. On 31 August 1751 his diary stops. It is just possible that then or soon after he became ill and died.

The Peacock family had originally acquired land in Ireland through William Peacock, a London 'painter steyner' who, as an Adventurer, subscribed £200 to support Cromwell's army in Ireland. William's subscription entitled him to a grant of 840 statute acres in Munster, including land in Graigue, in the barony of Pubblebrien, Co. Limerick.

A census of 1659 listed a 'Mr Peacock and his two brothers', who may have included William Peacock and a George Peacock of York who was granted 880 statute acres in Pubblebrien. This is probably the same land as William Peacock's grant, which he may have passed to George.[6] George Peacock, who also bought land worth £600 in Pubblebrien in 1654 from a Yorkshire widow, Susan Hoyle, married Aphra Maunsell of Ballyvorine, Co. Limerick, and died childless in 1678, leaving his land to his nephew, Richard Peacock of Graigue. Richard died in 1691, also childless, and the estate passed to his brother, James Peacock of Graige and Cahir Etna, now Fort Etna.[7]

We can also trace Nicholas Peacock through his employers, Price and Alice Hartstonge. The Hartstonges were originally from Norfolk, and their Irish connection had begun in the late sixteenth century when Ralph Hartstonge of South Repps, Norfolk, married Thomasina Curzon of Bealaha, Co. Clare. He thereby acquired estates in Co. Limerick, including that of the Lacy family in Bruff. Ralph and Thomasina's son Standish Hartstonge (1627–1701), became an Irish justice of assize until 1686, when he temporarily left Ireland for his Radnorshire estates, and on his return in 1691 he was made a baron of the Exchequer. He married Elizabeth Jermy of Gunton, Norfolk and had three sons and four daughters. The eldest son, Francis (1651–88), married Mary Brettridge of Co. Cork and the youngest, John, became archdeacon of Limerick and bishop of Ossory and Derry. Francis died during his father's lifetime, and the next son, Standish, was MP between 1695 and 1727 for Kilmallock, for Ratoath, and for St Canice, Kilkenny. He married Anne Price, daughter of a Welsh judge, Mr Justice Price of Presteigne. Standish died in 1751[8]. His eldest son was Price Hartstonge (1692–1743), who was portreeve of St Canice's Kilkenny and MP for Charleville from 1727 until he died.

Price Hartstonge married Alice Widenham.[9] Alice and her sister Mary (1682–1763) were the daughters and co-heiresses of Henry Widenham of Court

---

4 I am grateful to Toby Barnard for his help in identifying Nicholas Peacock from H.W. Gillman (1900), p. 23. 5 Will of Mary Widenham of Court, Co. Limerick, 13 June 1736. Welply. 6 Beggan, pp 101–2. 7 *Mungret/Crecora Parish Journal*, p. 26. 8 23 July 1751. 9 Oliver, 'The Hartstonges and Radnorshire: Part 1', p. 32. Alice Widenham executed a marriage agreement to marry Pryce

Introduction

who married a Mary Graham.[10] Henry settled £6000 on Alice on her marriage, and in his will, he left Court, Kilcolm, Dromore and other lands in the parish of Kildimo.[11] Mary Widenham referred in her will of 1736 to her 'nephew Nicholas Peacocke'.[12] The Widenhams had been granted the lands of Tough, Corra, Ballygowan, Ballymelick and a ferry over the river Maigue in 1684, and had purchased further land in Limerick from the Commissioners for Forfeitures in 1703.[13] It is with this land that Peacock was concerned in his diary. Mary, Henry and Mary Widenham's other daughter, married Valentine Quin, son and heir of Thady Quin of Adare, Co. Limerick who was called to the Bar at the Middle Temple in April 1698 and admitted to the King's Inns in 1709. Valentine Quin converted late to protestantism in 1734. In her marriage settlement, Mary brought several townlands, originally purchased by her father from the trustees of the Forfeited Estates.[14]

It is now thought the Mary Widenham, née Graham, had a sister whose Christian name is unknown and who married a Mr Pike. Their daughter Ann married Nicholas Peacock senior and they had four or possibly five children, one of whom as Nicholas Peacock junior. The other children were Robert, who became a hearth money collector, and who may have been 'my brother' who lived at Enniscouse. In April 1743, Nicholas recorded that he stood godfather to an unnamed brother's child who shortly afterwards died.[15] There were three sisters. One was Anne, the wife and then widow of Coursey Ireland. (Valentine Quin left £20 to 'Anne Peacock', but his will may have been made before her marriage.)[16] Another sister, 'sister McGan', was married to Edward McGan and lived in Limerick. He also had a sister Molly, who died in March 1743.[17] There is one further clue which ties Nicholas Peacock to the land near Court. That is in 1744 Nicholas took a lease of land at Kilmoreen on the west bank of the river Maigue, and two years later he built a house there, completed in 1749. His eldest son Price was born in 1748, and a Price Peacock of Kilmoreen was listed as a freeholder in 1766.[18]

From the beginning of the diary it is clear that Peacock is anxious to get married; he is open about the fact that he needed someone to share his life and take care of his house.[19] In 1740 he was involved in negotiations with a neigh-

---

Hartstonge on 15 September 1720. Registry of Deeds 32 531 30813. **10** I am extremely grateful to Gerry Dobson of Dallas, Texas, USA, who is descended from Daniel Widenham and knows the family history in detail. **11** Will of Henry Widenham of Court, Co. Limerick, 2 August 1719, D3196/K/1/2. **12** Will of Mary Widenham of Court, Co. Limerick 13 June 1736. Welply. **13** Widenham pedigree NLI mic 8303; Grove White, vol. 2, pp 166–7. **14** Alice Widenham needed to raise £3000 quickly from the land left to her by her father, and £6000 had to be paid to Sir Standish Hartstong to maintain his son Price, and Alice Widenham. Registry of Deeds, Widenham and Widenham, 2 August 1719; will of Henry Widenham, proved 19 December 1719; W. Betham, Genealogical Abstracts, series i, vol. 71, p. 67. **15** 7 April 1743. **16** Valentine Quin's will was witnessed by Alice Craven, Vere Hunt, John Martin and George Evans, and was proved in 1744. Mary Quin's will was witnessed by Jane Evans, John Martin senior and John Martin junior. Welply Will abstracts RCB 80/B2/1/f.191; Vicars. **17** 18 March 1743. **18** Freeholders of the County of Limerick 1776 NA M.1322. **19** 17 February 1746.

bour's daughter for which another neighbour, John Fosbery, acted as the intermediary. An intermediary charged with negotiating the terms of the marriage settlements was also used in 1747, when Peacock successfully courted Catherine Chapman of Lisdogan, Co. Cork. A neighbour, Mr Bastable, was the intermediary with her father. Peacock himself acted as intermediary in negotiations for the marriage of one of his relations, Elizabeth Ireland, possibly his niece. It sounds as though Elizabeth Ireland had no expectations, and would have to be helped by her richer landed relatives. Peacock discussed her situation with his aunt, Mary Quin, and was charged with negotiating the size of the settlement and with the organization of obtaining the money from another relation. George Quin, Mary Quin's brother-in-law, provided £20, which Peacock gave to Betty Ireland, now 'alias Attkins'.[20]

In January 1740, Peacock was trying to marry a Miss Blunt, daughter of John Blunt of Kilcorly. Despite having 'got all consent but her own', he came away from a supper with the Blunts dissatisfied with what had been proposed as her dowry for, as he says, '35l would not maintain her'.[21] Thus balked at this attempt, he let the matter alone for five years. In September 1745, he planned to visit Co. Cork, and was given money for his expenses by Mrs Hartstonge. While staying in the Charleville area, he had supper with the Chapman family at Lisdogan.[22]

Nicholas met Catherine when he travelled to Liscarroll to view land acquired by the Hartstonges after the death of Brettridge Badham, who was for a time agent to the earl of Orrery.[23] Land in the area had been granted to Roger Bretridge in 1657, and in 1659 Bretridge held 12 English and 51 Irish acres in Castlemagner.[24] The castle and lands granted to Roger Bretridge eventually came to the Hartstonges, probably through the marriage settlement between Mary Brettridge and Francis Hartstonge, the uncle of Price Hartstonge, Peacock's employer. While visiting Castlemagner, Peacock stayed with the Thornhills of Rockfield, Ballyheen. Sophia Thornhill, the wife of Richard Thornhill, was Brettridge Badham's daughter.[25] One of the townlands in the ownership of the Brettridges/Hartstonges was Lisraduggan – possibly Lisdogan or Lisdougan, the home of Catherine Chapman. After 1747 his closest relation was, of course, his wife Catherine Chapman. It is possible that she was a daughter of John Chapman of Gortnagrass and Summerville, Co. Cork, who in 1712

20 3, 10, 14, 17, 23, 24 February 1750. 21 23 July 1740; 30 July 1740; 4 August 1740. 22 16 September 1745; 24 September 1745. 23 Badham was the brother-in-law of Henry Boyle, later Speaker of the Irish House of Commons. Through these connections he became collector of Youghal and a commissioner of the revenue in Cork. A Tory, he became involved in dishonest dealings, was condemned as 'obnoxious' and dismissed. See Barnard, *A new anatomy*. 24 Grove White, vol. 4, p. 25; vol. 2, p. 120. 25 Grove White, vol. 1, pp 211–13; vol. 2, p. 104, 189; Registry of Deeds 15 68 6788, Memorial of a marriage settlement between Brettridge Badham and Sophia daughter of the rt. Hon. the Lord Baron John Kingston and James King, Sophia King's brother. Lands to the value of £7000, 12 October 1715; 128 382 87295. Badham and Purcell, 16 November 1767, refers to Sophia, relict of Richard Thornhill and administratrix of Brettridge Badham's estate.

Introduction   15

married Catherine daughter of Arthur Hyde of Castle Hyde, near Fermoy. They had one daughter and two sons, John and Arthur.[26]

The following January he returned and, although Mr Chapman was away, he was invited to stay the night by the women of the house. The next day, he visited Bastable, the intermediary. They must have discussed the possibility of Nicholas marrying Catherine, because three weeks later he heard from Bastable that Mr Chapman was not prepared or able to settle even half the money on his daughter that had been expected.[28] This was disappointing news, but Nicholas was not entirely discouraged, and spent a night 'museing on ye want of one yt would take care of my house'.[29] Probably he was invited to renew his suit the following April, because he immediately prepares to return to Lisdogan when he has news of Miss Chapman from another Cork neighbour. When he arrived, at first all does not go smoothly. On speaking directly to Mr Chapman about Catherine, they still cannot agree, and once again he prepares to go away. But this time he 'wod not be let'.[30] The next day, he and Mr Chapman talked again and, with Catherine's consent, the match was agreed.[31]

The arrangements for the marriage went ahead at once. Peacock returned home and ordered his clothes for the wedding: a blue coat, scarlet trousers, 2 fustian waistcoats, a hat, and a wedding ring for which he pays 7s. 7d. Sensibly he also buys some dishes, plates and a saucepan for the house.[32] On 6 May, as if to reassure himself as to the size of his fortune, he lists the number of his cattle, sheep, and horses.[33] Having thus settled his house he packed his things and went to Co. Cork 'wth an intent to go to be happy or miserable for ever'. Arriving the next day, he finds that he cannot marry immediately but has to have a licence, supported by a bond, which he gets from Cloyne for 18s. 11d. The following day he returns to Lisdogan, drunk.[34] His fears about his future happiness were groundless. On 9 May 1747 he and Catherine Chapman were married and he wrote that he 'from thence take ye begining of ye date of my happyness'. Calling her 'my comfort', he asked God 'to give us his blessings of love constancey and happyness',[35] and called her by a pet name, 'Catty'.

26 Grove White, vol. 4, p. 236. 28 26 January 1746. 29 17 February 1746. 30 11–12 April 1747. 31 13 April 1747. 32 18 April 1747. 33 'I have this day 138 sheep 57 cows two bull 3 three yr olds 6 yearlings 16 lambs and 9 horses and mares': 6 May 1747. 34 5–9 May 1747. 35 9 May 1747.

THE LAND AGENT

Peacock kept his journal partly as an account book, to keep track of the complex money affairs involved in farming, and of the transactions that were part of his duties as agent in buying and selling and collecting rents and taxes. Agents were frequently cadet members of a landlord's family, as was Peacock. It was a part-time job, for which they were paid between £20 and £40 a year, and their tasks were multiple and difficult.[36] They had to collect rents on the gale days in May and November, and they charged their employer a proportion of the rent they managed to get. They took legal action against tenants and acted as bankers to their employers. In doing so, they had to be able to deal with complex financial transactions. When their employers were away, in Dublin or perhaps in England, any failure by their agent to make regular returns of the state of their estates would be met with unease.[37] Agents would find themselves in the centre of family disputes, and family members desperate to pay bills could exert great pressure. In the 1750s, Richard Bourke of Dromsallagh, Co. Limerick, was agent to the French family who lived in Dublin. Henry French gambled on horses and cards, and frequently attempted to draw in advance on the estate, which was itself in debt.[38] After losing 'all my money' on a horse at Lurgan which 'had at Hollow all the World to Nothing he fell Down in a slough … so was beat', French wrote a bill for £50 which he expected Bourke to settle by putting pressure on the tenants to pay their May rent. The bill arrived before French's letter, and Bourke paid it 'to give it the greater honour'.[39]

One of the major responsibilities of a land agent was to take action in cases where tenants failed to pay rent. Agents had to have some knowledge of the law on landlord and tenant, and the procedure to recover monies due for rent was long drawn out and complex. An affidavit had to be filed in court by the landlord (or his agent), stating that one year's rent was due. As a form of pressure, it was open to the landlord to distrain property on the tenant's land as a pledge for rent due. The goods distrained were limited by law: ploughing animals, tradesmens' tools and freehold fixtures could not be distrained.[40] The distraint of cattle was a mixed blessing, because cattle had to be put in a pound, guarded by a pound keeper and fed. The landlord could, as one nineteenth-century hand-

---

36 On agents, cf. T. Barnard, *A new anatomy of Ireland*, pp 208–38.   37 Henry French to Richard Bourke, 23 May 1749, NLI MS 8475 (1) Bourke papers.   38 Henry French to Richard Bourke, 23 May 1749; same to same Gort, 15 July 1752; same to same, 20 May 1753. NLI MS 8475 (1) Bourke papers.   39 Henry French to Richard Bourke, 6 September 1753; Bourke to French, 12 October 1753. NLI MS 8475 (1) Bourke papers.   40 Dutton (1726), p. 536.

book put it, be left 'in temporary charge of a ménagerie'.[41] If, by a week after the action for distress, the rent was still not paid, the landlord posted a notice of sale. Until the sale, the goods distrained still belonged to the tenant, and any profits from the sale had to be returned to him.[42] If a tenant's goods were taken without justification, then the tenant could take out a *replevin* against the landlord and have them returned. The landlord could not take goods in distraint that could not be returned to the tenant in good condition, e.g. meat, growing crops and fruit.[43]

Peacock is constantly engaged in the service of writs and subpoenas, in arranging to seize and drive stock, in taking action for trespass and, in the ultimate, in taking tenants to court to recover outstanding rents. He takes legal advice in difficult cases, like that of Henry and Thomas Studdert where repeated applications for payment had been unsuccessful. The Studderts were sons of the Revd George Studdert (1685–1738), who had been rector of Kilpeacon and Rathkeale, and they rented land from the Hartstonges. In April 1745, Peacock began to put pressure on them for rent that must have been due from the previous November gale at least. By October that year he is repeatedly trying to meet one of them in Rathkeale or in Limerick, but the brothers either fail to turn up or make promises to pay which they fail to keep. One of the Studderts does say he will sell some sheep in January 1746 and he pays £9 2s. to Peacock on 28 January 1746. But by March that year, Peacock is forced to ask his lawyer, Bryen McMahon, to execute a warrant against one of the Studderts. While he still attempts to meet them and get some money – he extracts a further £3 on 6 June 1746 – he gets McMahon to put in hand the distraining of Henry Studdert's effects. On 24 November he gets a *justicies* – a writ empowering a sheriff to take action normally taken by a higher court.and then on 26 January 1747, he asks his lawyers for a *fieri facies* – which would enable the sheriff to break into the Studderts' property to distrain on their goods in order to raise the sum that is owing to the Hartstonges.[44] This is followed up with an execution – an order to carry out the terms of a judgment of the court.[45] In dealing with the Studdert brothers, Peacock talked or wrote to his lawyers John Wallis and Bryen McMahon on at least fifteen occasions, and John Wallis sent Peacock a bill for costs on 12 March 1747 for £8 19s. 0d.

In 1744, the Hartstonges themselves faced process of the law, and Peacock had to deal with the consequences. Price Hartstonge died in Dublin on 4 February 1744. The Hartstonge land at Bruff had been mortgaged in 1706, and Price Hartstonge had borrowed £1,150 secured on land in Kilkenny. A private act of parliament had enabled him also to raise £2,000 through a mortgage on the lands of Court and Kilcollum in 1726.[46] Probably because the estate was

---

41 De Moleyns, pp 368–9. 42 De Moleyns, pp 376, 387, 834. 43 Dutton (1726), pp 537, 539. 44 Dutton (1726), p. 600. 45 15 April 1745 and 25 March 1747. 46 Registry of Deeds, 15 274 7481, Badham and Lysaght, 10 and 11 October 1715; 18 192 8821, Hartstonge and North, 18 March 1716; An Act for enabling Pryce Hartstonge of Bruffe in the County of Limerick esq. to rise

loaded with debt, the creditors closed in. What is perhaps surprising is Peacock's reaction. From his own experience as an agent, he must have known that it was unlawful to conceal goods that were to be distrained.[47] But on receiving the news from Dublin of Hartstonge's death, Peacock immediately made contact with Standish Grady, who was married to Price Hartstonge's niece Catherine Quin, to secure and hide the Hartstonges' moveable possessions at Court.[48] (Peacock and Grady went to Limerick to secure property there, only to find goods already seized and sold.)[49] Grady sealed the doors and a chest of drawers at Court, and Peacock tried to evade the sheriff's drivers by sending the (more valuable) black cattle to Standish Grady's at Elton, and he dispersed the furniture to the Burys at Shannongrove and to other neighbours. The rest of the stock was hidden by dividing it between the tenants.[50] He remained on guard to protect what remained, but on 1 March he and Grady decided not to take any further action without legal advice. Four days later, the advice came that the stock should be given up to the sheriff. With an alarm of war in Europe and general atmosphere of fear of uprising at home – there had been a proclamation against the Roman Catholic clergy – it was a bad time.[51] (Peacock had already received orders from Dublin not to cultivate the Court lands because they would be let.)[52] Despite attempts to buy them off, eventually the drivers were successful, and appraisers came to value the stock, and seized further stock and produce.[53] Fees had to be paid to everyone.[54] On 20 April, 'the malincoly cant' of all the stock and chattels began. Peacock made an attempt to save the stock that belonged to the Hartstonge children, but without success.[55] The estate must have been substantial, as the sale lasted from 20 to 26 April. Peacock himself benefited from the cant: he bought timber, slates, plates, dishes, and barrels for himself. Even then, the work has not been complete. Peacock travelled to Dublin to report to Mrs Hartstonge what had happened, only to find 'all in confusion' on his return, when further goods had had to be surrendered to the sheriff.[56]

The preparation for a funeral after the death of one of his family devolved heavily on the land agent. Mary Widenham died on 5 October 1742. Immediately Peacock went to Limerick and ordered a coffin and 'all other preparasions'. Over the next three days he wrote out lists of those to be invited to the funeral, sent out a hearse and 'all other things belonging to it' from Limerick, and ordered a suit of mourning and stockings for himself.[57] At about 10 o'clock on the day of the funeral, he dressed himself and helped to give out

£2000 by mortgage of all or any part of the land of Court and Colum, 12 Geo I, chap. 3 (private acts). I am very grateful to Dr James Kelly for tracing this act in his record of private acts of the Irish Parliament. 47 15 Geo. II, c. i, ss. 3 and 4 provided that if goods were fraudulently removed or concealed from distraint the debt could be increased by a factor of 2. Goods removed could be traced and distrained within 20 days. de Moleyns, pp 374–5. 48 7–8 February 1744. 49 9 February 1744. 50 8, 9, 10, 11 February 1744. 51 6, 8 March 1744. 52 26 February 1744; 4, 5 March 1744. 53 Appraisers had to be three or more 'honest and discreet persons' from the barony. Dutton (1726), p. 628. 54 16, 27, 30, 31 March 1743/4; 1 April 1744. 55 20 April 1744. 56 6 May 1744. 57 5 October 1742; 6 October 1742.

scarves and gloves to the mourners, of which he kept a strict account. Afterwards he accounted to Mrs Widenham's sister for all the monies spent by him, which totalled 17s. 6½d.[58] Of the ceremony he writes briefly but movingly, 'About one she was taken out and carryed to her long home.'[59]

THE FARMER

Peacock rented land himself, probably for grazing, as others did from him. In January 1740, he was paying rent to a Henry Page, and in April he agreed to lease 26 acres of land, valued at 12s. an acre for 21 years at Portacacha, near his own property at Kilcorly, from John Bryen of Ballylongford.[60] He also paid rent to Colonel Edward Taylor of Ballynort.[61] The terms of his leases seem favourable. When he lets part of the north east park at Kilmoreen he asks for £17 10s. 0d. for the first year and £18 5s. 0d. a year for every year after that. On his part, he was to give his two tenants straw and thatch for one cabin and help to build the walls of two other cabins.[62] This corcas land was occasionally flooded by high tides, and sometimes animals were drowned.[63]

After the terrible winters of 1739/40 and 1740/41, it was observed that the state of the land was such that subsidies were needed to encourage liming and marling to increase productivity in corn, flax, hemp and hops. It was said that 'Moory sowr' land could be reclaimed by plowing and burning and if 'well tylled' and drained over four years would be three to four times greater in value.[64] Although his land was of good quality, Peacock improved it in various ways by liming it, spreading dung and burning. Limestone was burned to make it suitable for spreading, often on fallow land. A rough kiln would be constructed in a field where limestone gravel would be laid in alternate layers with peat. A fire would be lit and allowed to burn for 48 hours before the lime was raked out.[65] Burning was done by stripping off the surface sod, drying it, then burning it and spreading the ash back on the soil to fertilise it. (An act of parliament was passed in 1743 forbidding this process, but it proved ineffective.)[66] All these methods were particularly suitable for bringing marginal land into cultivation.[67]

There is no doubt that Peacock rotated his crops, but it is difficult to trace his system because he uses field names irregularly, if at all. However, his references to ploughing fallow land certainly indicate that a rotation system was used. Occasionally land was left fallow to prepare it for grain crops the following year. 'Green-fallowing' meant the cultivation of a root crop to break and cleanse the land. 'Summer' fallowing meant repeated ploughing (cross plough-

---

58 8 October 1742. 59 7 October 1742. 60 19 January 1740; 1 April 1740. 61 2 December 1741. 62 4 February 1748. 63 Lamplugh et al., p. 57; the tide flooded the corcas land in 1746, 1747/8 and 1748/9. Animals were lost in three successive years. See 1 February 1747; 15 May 1746; 29 February 1748; 1 February 1749. 64 William Jessop to Charles Smyth 4 December 1741; Lenihan, pp 337–8. 65 Bell and Watson, p. 35. 66 Bell and Watson, p. 27. 67 Bell and Watson, p. 37.

ing) over the course of the year. Ploughs would have been made of wood, with an iron sole, or sull. Ploughing needed three men: one to guide the plough, a driver to walk in front and lead the animals and, sometimes on heavy land, a boy to lean on the beam of the plough to keep it steady.[68] Peacock writes of Irish spades. Loys, one-sided Irish spades, were common substitutes for ploughs in Munster as they were efficient in preparing the ground for potatoes.[69]

After ploughing, the land had to be harrowed to further break up the soil and make it ready for sowing. In 1741, Peacock's harrowing was held up by the late and severe frost.[70] Sowing was done by hand, with the seed often carried in sowing sheets. Grain crops and hay were reaped by scything. Hay was brought into the hay yard (haggard) and constructed into a stack which was then thatched with straw.[71] When it was needed it was cut with a hay knife.[72]

Peacock grew flax and had it spun and woven. Flax is an extremely labour-intensive crop which requires well-prepared land with a fine tilth. The seed was threshed from a previous crop and picked over by hand to eliminate weeds. Peacock sowed flax seed in April and pulled it in August. In 1742 he refers to selling flax seed to the Linen Board in Limerick, for which he receives a receipt, 'wch being a new thing I [need] to know how to wright one on Ocation'.[73] Usually sown between late March and early April, flax was pulled about fourteen weeks after sowing. Flax intended for fine linen was pulled early and there is evidence that much of Peacock's flax was of this type.[74] Pulled flax then had to be retted, that is, soaked for one or two weeks in water to separate out the woody central part of the stalk. Although Peacock had water from the Maigue nearby, he probably would not have used the river because flax made the water poisonous to fish.[75]

Once the process of harvesting and preparing the flax was complete, it was sent to spinners. Rather than selling the raw flax, growers preferred to sell it after it was spun into yarn, so that local women were able to earn money from spinning.[76] The flax and hemp grown by Peacock was spun for thread, string, and rope by local men and women who were paid in kind (bushels of oats) or in cash. Peacock writes of both linen and of tow thread, which was spun from the shorter fibres of flax and used for rope and string. These fibres had been separated from the longer and finer threads intended for linen. The linen thread was sent to weavers where it was woven into bandles of cloth, the size of the bandles being regulated by law. The Linen Board increasingly attempted to control the quality of linen thread and of cloth produced; there were regulations for the width of cloth and the size of hanks of thread.[77] If it was to be bleached, the cloth was sent to a bleaching green. Peacock refers to a well-known bleaching

---

68 Bell and Watson, p. 69. 69 Bell and Watson, p. 49. 70 17 February 1741. 71 4 October 1742. 72 16 December 1740. 73 6 February 1742. 74 Flax sown on 27 April 1741 was pulled 13 weeks later on 27 July; 82 men and woman were engaged in pulling flax on 15 August 1744. 75 Bell and Watson, pp 160–71. 76 Gill, p. 38. 77 Bandle cloth, which was produced by Peacock, had to be 12" to 14" wide. Hanks of thread had to be bundled in 12 cuts and not contain more than 120 threads. Flaxen yarn and tow yarn must not be mixed and should be of the same quantity; Gill, p. 6; Dutton (1718), pp 205–6.

green near him at Pallaskenry and another at Court.[78] Other fabrics that he makes are flannel and frieze and canvas, rope and string from spun hemp. From his linen and wool cloth he has and his servants' clothes made by a tailor in Limerick, and another tailor visits the house. His worsted and thread stockings are either knitted by local women or bought in Limerick.

Peacock's main arable crops were oats, wheat and barley (English and Dutch and bear barley). He also grew rape, used in the finishing process in the textile industry and as fodder, and hemp, used for rope and string; rape was suitable for boggy land but was tricky to harvest.[79] He also grew peas and beans for fodder and for the house, and in his gardens he grew onions, parsnips, cucumbers, saffron and garlic.[80] He had a 'hot bed' to force vegetables.[81] Oats were the most important crop, being thought particularly suitable for the Irish climate and soil and used both as a cereal and for animal fodder. Wheat, although less suitable to Irish conditions, was ground for bread flour and produced a hard straw, suitable for thatching. Thatch has always been a roofing material in Co. Limerick, and Peacock thatches buildings frequently. He also mentions cutting reeds, and it is possible that in some cases he thatched with reed, not straw.[82] Peacock's grain was sown by hand from a canvas sheet, probably attached to the labourer's waist.[83] Winter wheat was thought to be suitable for Irish conditions, and between 1740 and 1749 Peacock sows his wheat from 22 September to 19 November.[84] Barley was used to brew beer and porter and to make whiskey. Choosing the most suitable moment to harvest barley posed difficulties: too early, the grain shrivelled, and too late, the grain heads broke off.[85] Grain crops were harvested with scythes, a blade attached to a long straight shaft, or sned. During use, scythes need to sharpened frequently, and Peacock bought a scythe stone during harvest in 1745 for 5d.[86]

The final process in grain production is threshing and winnowing. Peacock threshes oats for seed between 10 December and 17 February, and sows seed correspondingly later.[87] It is interesting that he grew 'naked' (rhealba) oats, which would have been easier to thresh as they had an atrophied husk which was easily separated from the seed.[88] Threshing was done with flails. The sheaves of grain would be laid on the ground, either in the open air or in a barn (Peacock does not specify), and beaten. The most common type of flail in Ireland was made

---

78 15 March 1742; 10 June 1743. 79 Thirsk, pp 72–3, 78. 80 Saffron, 26 May 1746. 81 21 March 1746. 82 15 September 1740. 83 6 January 1741. See Bell and Watson, p. 103. 84 12 November 1740, 22 September 1741, 17 November 1743, 19 November 1747, 25 October 1748, 11 November 1749. 85 Interesting that Peacock sowed English barley on the same date, 28 February 1741, as a member of the Lucas family in Co. Clare: cf. Diary of a member of the Lucas family, NLI MS 14,101. 86 18 July 1745. 87 Oats threshed, 21 January 1740, sowed 27 February 1740; threshed, 24 January 1741, sowed, threshed 12 January 1741; threshed 2 January 1742, sowed, 22 January 1742; threshed 9 April 1743, sowed 30 May 1743; threshed, 17 January 1745, sowed, 21 February 1745; threshed, 10 June 1747, sowed, 8 March 1747; threshed, 17 February 1748, sowed, 14 March 1748; threshed, 25 October 1749, sowed, 14 March and 23 October 1749; threshed, 10 December 1750, sowed 1 March 1750. 88 24 March 1742.

from two sticks, 3–4 feet in length and tied together. The thresher holds one stick, wielding the other stick to beat the grain from the chaff. The last step to produce grain for seed or for milling was winnowing, separating out the grain from the chaff. Peacock carried out winnowing by spreading a sheet on the ground and pouring the grain on to it from a height.[89] The wind would blow away the chaff and the heavier seeds would remain on the sheet.

Peacock's farming operations were, of course, conditioned by the weather. The diary begins during the severe cold weather and drought which persisted from December 1739 to September 1741.[90] The frost which had begun in December 1739 continued throughout January, and the coldest night was observed on 27 January 1740.[91] Although there was a thaw in February and early March, the weather was extremely dry during March and April and this continued throughout May and June. Even in July there was frost. The autumn brought further frosts in October and cold winds and in November there were frequent and sharp frosts. In 1740 he could not sow beans until 5 March and could not plant potatoes until 26 March. At the end of that year he tried to set out cabbage plants but was unable to do so because the ground was frozen so hard, and 'several suffered by the frost'.[92] In January 1741 there was a thaw, and the banks of the Maigue flooded over, but fortunately the flood did not affect him.[93] The hard weather continued into the next year when the ground was so hard they could not plough for flax.[94] The persistent cold weather brought great hardship to many all over Ireland, but nowhere does Peacock mention the distress, or the relief that was being given out very near to him. In Rathkeale, the dowager Lady Southwell and her daughter were reported to have distributed money and medicines against the 'mortal flux now rife among the Poor'. They gave beef and oats and lowered the price of corn.[95] The only direct reference that Peacock makes to the great frost and famine is in 1740, when he mentions receiving a pamphlet, the 'Elogey on ye Petatoe', published that year in Dublin.[96] But Table A reveals the abnormally high price of oats in 1741. Peacock got 14s. a barrel that year, and 21s. a barrel in 1742. This can be compared with prices ranging from 1s. to 6s. a barrel for oats in the years between 1744 and 1746.

Apart from the great horse fair at Cahirmee, Co. Cork, the local fairs seem to have taken all goods and crops. Peacock sold oats at Askeaton, and sheep at Ballingarry and the fair at Croom took cattle.[97] He sold oats, butter, fowl, barley, wheat, peas, wool sheep, and beans at Limerick. Occasionally he would send goods to a fair, but fail to make a sale. In February 1745 he was unable to

---

89 26 May 1743. 90 Dickson, p. 11. 91 Rutty, p. 101. 92 12 December 1740. 93 1 January 1741/2. 94 27 April 1741. 95 *Pue's Occurrences*, 24–27 January 1740/1. 96 *An Elegy on the much lamented death of those excellent patriots and lovers of the country, the family of the potatoes in the kingdom of Ireland, who fell by a general massacre, being confined and starved alive by cold and hunger ... between the 26th day of December, and the 18th day of January in the year 1739* (Dublin 1740); reprinted in Carpenter (1998), pp 248–52. 97 15 June 1743; 1 April 1747; 22 April 1747.

Introduction

sell his wheat, and blamed it on a fall in the market.⁹⁸ He was unable to sell his butter on three occasions in 1749.⁹⁹ What is noticeable from an analysis of the price of oats and butter between 1740 and 1751, in Table A, is that he gradually transfers from the production of grain crops to dairying. This may be due in part to an agreement he made in 1746 with Martin fitzGerald, a dairy farmer, to farm his dairy herd in return for a proportion of the production of butter. Peacock and fitzGerald contracted that fitzGerald should have a hundredweight of butter and 7s. for every cow he cared for. The herd was to be limited to 40 cows and 20 calves reared to adulthood.¹

Table A: Comparison of prices in oats and butter, 1740–50

The measures are both of capacity and of weight. The definitions of bushels and quarters varied across Ireland, so these must be regarded as approximate.

barrel: 14 stones weight, or 196 lbs
bushel: 4 pecks or 8 gallons
peck: 2 gallons
firkin: quarter of a barrel

stone: 14 lbs weight
qr. (quarter): 8 bushels or 28 lbs
cwt. (hundred weight): 112 lbs

Oats

1740	7s. 8d. a barrel	
	2s. 8½d. a barrel	
	4s. 0d. a barrel	
1741	14s. 0d. a barrel	
1742	11d. a stone	
	£1 1s. 0d. a barrel and a half	
	6d. a peck	
	6d. half a peck	
	13s. 0d. a barrel	
	£1 1s. 0d. a barrel	
1743	5d. a peck	
	10d. a bushel	
	3s. 5d. a peck	
	3d. a peck	
1744	7d. a bushel	
	4s. 0d. a bushel	
1745	1s. 0d. a barrel	
	2s. 6d. a bushel	

Butter

1740: 4s. 5d. a lb.
£1 3s. 1d. a firkin

1743: 14s. 0d. a firkin
18s. 3½d. for 2 qr. 26 lbs
£1 8s. 6d. for 1 cwt. 1 qr. 2 lbs
3s. 5d. a peck

1744: 18s. 0d. a cwt.
17s. 0d. a cwt.
18s. 0d. a cwt.
£1 1s. 6d. a cwt.

1745: £1 0s. 4d. for 1 cwt. 17 qr. 2 lbs
£16 9s. 9d. for 2 cwt. 0 qr. 2 lbs
£1 17s. 6d. a cwt.
£1 15s. 0d. for 1 cwt. 0 qr. 12 lbs

98 2 February 1745.   99 12 February; 18 February 1749; 28 July 1749.   1 18 November 1746.

		£2 0s. 0d. for 1 cwt. 1 qr. 5 lbs
		£9 16s. 1d. for 6 cwt. 0 qr. 26 lbs
		£4 16s. 5d. for 3 cwt. 0 qr. 2lbs
1746	7s. 0d. a barrel	£3 9s. 6½d. 2 cwt. 2 qr. 2 lbs
	6s. 0d. a barrel	£2 19s. 6½d. for 2 cwt. 1 qr. 26 lbs
		£10 12s. 1½d. for 8 cwt. 3 qr. 10 lbs
		£4 16s. 8d. for 4 cwt. 0 qr. 10 lbs
		£6 3s. 9½d. for 4cwt. 3 qr. 2 lbs
		£2 15s. 0d. for 2cwt. 0 qr. 13 lbs
		£1 4s. 0d. for 1 cwt.
		£1 5s. 10½d. for 1 cwt. 0 qr. 14 lbs
1747		£1 8s. 6½d. for 1 cwt. 0 qr. 16 lbs
		£10 1s. 3d. for 7 cwt. 2 qr. 27 lbs
		£6 13s. 5d. for 5 cwt. 0 qr. 15 lbs
		£6 19s. 0½d. for 5 cwt. 0 qr. 17 lbs
		£2 17s. 7½d. for 2 cwt. 0 qr. 15 lbs
		£6 8s. 11d. for 5 cwt. 2 qr. 12 lbs
		16s. 6½d. for 0 cwt. 2 qr. 16 lbs
1748	6s. 0d. a barrel	£6 9s. 1d. for 4 cwt. 2 qr. 26 lbs
	9s. 0d. a barrel	£7 0s. 7½d. for 1 cwt. 3 qr. 22 lbs plus
	8s. 6d. a barrel	2 cwt 2 qr 27 lbs
		£7 19s. 1½d. for 4 cwt. 0qr. 15 lbs
		£2 1s. 2½d. for 1 cwt. 1 qr. 9 lbs
		£3 15s. 0d. for 2 cwt 1 qr. 22 lbs
		£6 2s. 7½d. for 3 cwt. 3 qr. 23 lbs
		£3 13s. 2d. for 2 cwt. 2 qr. 13 lbs
		£3 6s. 5½d. for 2 cwt. 2 qr. 1 lb.
		£2 18s. 1½d. for 2 cwt. 9 qr. 22 lbs
1749		£5 15s. 8d. for 3 cwt. 2 qr. 26lbs
		£2 5s. 1d. for 1 cwt. 1 qr. 23 lbs
		£2 1s. 7d. for 1 cwt. 1 qr. 6 lbs
		£8 2s. 0d. for 5 cwt. 0 qr. 7 lbs
		£7 16s. 11d. for 4 cwt 3 qr. 4 lbs
		£2 1s. 3d. for 1 cwt. 1 qr. 1 lb.
		£8 3s. 4d. for 5 cwt. 1 qr. 22 lbs
		£1 17s. 3d. for 1 cwt. 1 qr. 4 lbs
		£1 14s. 9d. for 1 cwt. 1 qr. 7 lbs
		£3 14s. 8d. for 2cwt. 3 qr. 2lbs
1750		£1 0s. 11d. for 0 cwt. 3 qr. 3 lbs
		19s. 10½d. for 0 cwt. 3 qr. 2 lbs
		£2 2s. 6d. for 1 cwt. 1 qr. 10lbs
		£5 0s. 7d. for 2 cwt. 2 qr. 2 lbs
		17s. 3d. for 0 cwt. 3 qr. 4 lbs

Introduction

He had men and women apprenticed to him, and these were supplemented by day labourers at times of sowing and harvesting. If his labourers did not turn up to work, he impounded their cattle.[2] He contracted out dairying to a neighbour under an agreement to pay him pro rata in butter and money for cows reared.[3]

Peacock's dealings with his neighbours, his landlords and with the payment of tithes, and through his servants with fairs and markets were complex. He paid wages and accounts partly in specie and partly in goods: barley, corn, beans, grazing. When he paid in money, this too was complex, because the currency was mixed. Quadruple pistoles, moidores and French crowns were all in circulation, and some coin was suspect.[4] This cannot be attributed only to Limerick's position as a important port, trading with Nantes, Bordeaux, Cadiz and the West Indies, because the wide variety of currencies used was common in eighteenth century Ireland. Another agent, Benjamin Pratt in Co. Westmeath, received rent from one of his tenants in moidores, four-pound pieces, a 'gold cob', pistoles and guineas.[5] Like Peacock, Pratt's farm accounts were complex. They too were reckoned in money, skins, bushels of bear barley, corn, and grazing for two cows.[6] Ready cash must have been essential to pay the numerous small disbursements for tolls, wayage and customs which figure whenever Peacock goes to Limerick. Pratt has entries for 'publick money' which are probably the same.[7] The system of payments was complicated by the custom of paying 'earnest', a small deposit on a larger transaction, and it is clear that an account book or diary such as these kept by Peacock and Pratt were essential to keep track of the progress of credit and debit. Peacock crosses out debts when they have been paid.

As a member of the local gentry, Peacock was himself expected to assist with county administration and the judicial process. These were carried out by the grand jury and the petty jury. From what Peacock writes, although it is not entirely clear, he served on a grand jury and a petty jury. Grand jurors had administrative and judicial functions. They raised monies to build and repair roads and buildings, and they held preliminary hearings of cases to come before the assize courts, to decide whether there was a 'true bill' against the defendant.

---

2 23 May 1746. 3 18 November 1746. 4 Examples are: he pays quadruple pistoles for butter, 13 July 1745; 2 four pound peices and a giney, 31 August 1745; moidores and French crowns, November 1741 'light' money not accepted on two occasions. 6 and 20 April 1750. 5 O'Connor, p. 41; farming account book of Benjamin Pratt, 1726-42, NLI MS 5248. 6 18 March 1737. Pratt calculated that in Irish money 8 moidores were worth £11 14s. 0d.; the 'gold cob' £3 13s. 0d.; 5 four pound pieces £19 08s. 04d. and 3 forty shilling pieces £5 16s. 06d. Pratt, ibid. 7There were three types of tolls: 'tolls thorough' imposed by corporate bodies to pay for repairs to walls, streets and bridges. They were paid at the entrance to the town. Tolls which derived from the original grants for fairs or markets and tolls authorized by parliament for particular purposes. They were charged on articles sold on market day. Wayage was charged by the weighmaster who weighed goods for buyers and sellers. See report of the Commissioners appointed to inquire into the state of fairs and markets in Ireland, PP 1852-3 [1674] XLI, 74, p. 11.

They were selected by convention, not statute, and their sole qualification was that they should be freeholders. From his description of serving on a jury at the assizes in 1747 and 1748, it sounds as though he was then serving on a petty jury.[8] Petty jurors did not have to deal with roads and buildings as did grand jurors. Their duty was to hear a case and return a verdict of guilty or not guilty. It was possible for leaseholders as well as freeholders to be empanelled on a petty jury, and a lease for several lives was regarded as equivalent to a freehold. To serve on a petty jury Peacock would in any case have had to have been a protestant, to have had an income of at least £15 a year and a lease for a number of lives.[9] To serve on a jury of whatever kind in Limerick meant both inconvenience and expense. He had to stay overnight in the city, and in August 1747 he noted his expenses 'during the Syzes' were 6s. 10d. As well as the assizes, which were held four times a year,[10] there were also lower courts which were available to Peacock for civil cases of debt. Sheriffs, while having various specific duties, also had their own 'Turn-court', which they convened to hear minor cases.[11] He refers to the Sheriff's court in dealing with some business of Mrs Hartstonge. There was also the Tholsel court, which Peacock may have used to take action against a tenant defaulting on his rent, as he consults with a 'Mr Jackson' in 1745/6 about a legal matter, and Jackson may have been an attorney appearing at the Limerick Tholsel Court.[12] Tholsel courts were presided over by local magistracy and the lord mayor of the city.

The high spot in Peacock's service as a juror was the hearing of the case of Hugh Massy, charged with the abduction of Frances Ingoldsby.[13] Sadly he fails to supply the lascivious details of the case. The bill of indictment against Massy was heard in Cork in 1745, but was thrown out by the grand jury. More charges were brought against Massy and against his brother-in-law, John Bourcher, at the Limerick assizes in Trinity term 1745, and Peacock was sworn in as a member of the jury.[14] The following year he again served on a jury at the assizes.[15]

The river Maigue bordered and watered Peacock's land, and transport by river was used for turf and cider.[16] However, the river does not seem to have been used by people to journey up to Limerick. In 1745 Peacock constructed a weir across the Maigue in order to catch fish, and he caught salmon there soon after its construction.[17] The weir was probably made of sticks, since a month after its construction, it 'went away', only to float back the next day on the tide, after which it was reconstructed.[18] The most convenient link for goods, cattle and people between Court and Limerick was by road, crossing the Maigue by a ferry established in the seventeenth century.[19] The river was navigable when the roads

---

8 10, 11 and 12 August 1747; 30 March 1748. 9 Garnham, pp 135–6. 10 Garnham, p. 76. 11 18 January 1746; Garnham, p. 94–5. 12 20, 12 November, 31 December 1745; 20 January 1745/6. 13 The history of the abduction of Frances Ingoldsby and the circumstances that surrounded it, is recounted in fascinating detail in T.C. Barnard, *The abduction of a Limerick heiress*. 14 Barnard, op. cit., pp 23, 27; diary, 17 August 1745. 15 2 April 1746. 16 10 December 1743; 1 December 1744. 17 20–23, 26 March 1745; 28 March, 8 and 21 April 1745. 18 5–9 May 1745.

Introduction

might be deep in snow, but if the ferry was not working, a traveller would have to go south to the bridge at Adare. When a boat went aground in the winter of 1744/5, oats could not be delivered to Court.[20] Peacock of course rode everywhere. He twice falls off, once quite badly when he was rushing to prepare for his wedding and fell off on the road to Limerick. He had to spend the next day in bed.[21] One of his first purchases for his wife was a side saddle.[22] The Hartstonge family had a coach.[23] When Peacock brought a bale of wool back from Limerick, he paid 8s. for hire of a pillion.[24] Both roads and the ferry across the Maigue cost money in tolls. Every time Peacock visited Limerick he noted sums of around 2d. to pay for the turnpike, and sometimes quite a lot of money had to be paid out. In 1745 Peacock had to pay £1 1s. to use a road through Ballycraheen.[25]

He went to Dublin on business for the Hartstonges twice between 1740 and 1750. His ride there took three days, through Co. Tipperary, Queen's County and King's County. Regrettably he says nothing of where he stayed or what he did and saw in Dublin. The Hartstonges had a house in Dawson Street, where he probably stayed.[26]

DOMESTICITY

The land occupied by Peacock was a group of three townlands situated north of Adare on a bend of the river Maigue at Kilcorly, Portacacha and Kilmoreen. In April 1740 he leased farmland at Portacacha from his neighbour, John Bryen, and took possession on 1 May. The next year he built a barn, and marked its foundation by putting $6^{1}/_{2}d.$ under the south east corner stone, 'the first stone laid'.[27] In 1742, he built himself a house and stables there.[28] In 1744, he leased land from Brettridge Badham close by at Kilmoreen, and by 1746 he was building a new house there also. He had the plans of the house and orchard marked out, and the foundation was laid in May and June 1746.[29] After his marriage in 1747, he enlarged his bedroom where he threw out a new window.[30] He took advantage of local sales to buy timber and slates to use in his building work. During the acquisition of land and his building work he was advised by a surveyor, one of the Guerins of Limerick, and his building was carried out by Cyprian Purcell, a jack of all trades, one of which was pulling teeth.

Peacock is tantalizing about the furnishing of his house. After his marriage he had a new bed made, with headboard and tester.[31] Peacock writes about buying pewter dishes and flatware, tumblers, and candle snuffers. He spends what seems to be quite a lot of money –18s. – on 'wooden ware' for the house.[32]

---

19 Henry Widenham had been granted the fees from the ferry by Charles II. Peacock collected fees for it. Cf. 19 June 1745; 30 June 1745; 7 July 1745. 20 7 March 1745. 21 26 January 1743. 22 18 April 1747. 23 25 November 1745. 24 18 July 1744. 25 The ferry boat cost 2d. on 21 November 1743: see 28 October 1745. 26 *Pue's Occurrences*, 19–23 July 1751. 27 30 June 1741 28 28 June 1742; 16 July 1742. 29 10 May 1746; 14 May 1746; 10 June 1746. 30 30 September 1747; 10 October 1747. 31 1 October 1747. 32 22 November 1743.

Particularly after his marriage, he adds to the household necessities in saucepans, dishes, plates and a grater. In 1748 he puts aside 12½ pounds of linen yarn for curtains.³³ He buys prints at a cant in Rathkeale in 1743.³⁴ At the beginning of almost every year he bought an almanac.³⁵ He had much trouble with his clock which was probably a simple brass wall clock without a case. In 1743 he had it mended and hung it, but immediately one of his servants broke the 'pendely spring'.³⁶ The clock was still broken the following April, and he asks the versatile Cyprian Purcell to mend it, but he could 'do no good'.³⁷

His house must have been furnished with books, as throughout the diary he often refers to reading at home, and he bought books for Mrs Hartstonge from Welch, a printer in Limerick from whom he bought newspapers.³⁸ The only books he definitely identifies are the pamphlet on the 'Elogey on ye Petatoe' and a book of 'Persian Tails', probably François Petit de la Croix, *Thousand and One Days*, a popular version of the Arabian Nights, with which he diverted himself when he was snowed in.³⁹ He occasionally reports events in Dublin and elsewhere. He is daily involved with his neighbours, either in business or in the pursuit of pleasure, and he stands godfather to his neighbours' children, as they for his children.⁴⁰

He spent his leisure before his marriage in heavy drinking (he writes almost naively, 'when I have drink I find People come to drink it'). Once, after he got drunk with his brother-in-law he slept for a day and a night. The following day he was 'craw sick'.⁴¹ He entertained his neighbours, often at some cost to himself. John Johnston's son ate 'all my chees and wod eat ye Davill if chees'. Young Hartstonge went hunting and brought neighbours to dinner, and Peacock gave them a 'bely full of punch'. In 1745 he entertained 'Company' to celebrate New Year's Eve when they all drank 'pritty hard'.⁴² He played cards (and often lost money) and occasionally hunted with neighbours.⁴³ Sometimes he visited tourist attractions. After church in 1747 he went with friends to visit the ruins of Carigogonell Castle about five miles away. When he was staying with his in-laws near Castlemagner, he visited Lohort Castle, recently greatly improved by Lord Perceval and the object of much local interest.⁴⁴ He played a simple musical instrument, the jew's harp. After his marriage, he gave up playing cards and drinking, and he and Catherine devoted much time to dining with neighbours and receiving them at home.

Peacock makes fairly regular references to attendance at church, but it never figured as large in his life as it did in the life of his Ennis neighbour, Mr Lucas. His own church was at Kildimo, and he also attended the church at Kilceedy with the Hartstonge family.⁴⁵ The services at the church at Kildimo appears to have been somewhat irregular, as he occasionally finds that when he goes to church there is no clergyman, or no sermon, and in 1749 he and Catherine go to church and are not let in.⁴⁶ Peacock's attendance at church was not only for

---

33 17 August 1748. 34 24 May 1743. 35 28 December 1743. 36 22 December 1743. 37 29 May 1744. 38 5 April 1747. 39 2 March 1745. 40 7 April 1743; 23 October 1744; 20 August 1747; 12 January 1748; 21 August 1750. 41 6 May 1744. 42 29 April 1745 10 August 1745; 31 December 1745. 43 27 January 1743; 17 November 1744. 44 15 March 1747; 19 July 1747. 45 15 March 1747. 46 No clergyman, 9 June 1745; no sermon, 4 January 1743; not let in, 26 February 1749.

Introduction

formal worship. He was a member of the vestry, where there were occasional sharp differences of opinion, and had to deal with such matters as the collection of church rates, visitations and the setting and advertising of tithes.[47]

Peacock's diary belies the received view of how Roman catholics were treated, especially those who were servants and peasants. Every year he noted that he gave his servants a day off work to visit a holy well, probably at Patrickswell, and money to spend on drink at a pattern.[48] He entertained the local priest and other neighbours at Christmas.[49] He also made linen towels for the church, and in 1751 the priest, Fr Nicholas Morony, stayed the night with the Peacock family.[50] Peacock's treatment of Fr Morony was in sharp contrast to that of his Rathkeale neighbour, Lord Southwell, who hunted him from Adare. It was also in contrast to his Clare neighbour, Mr Lucas, who while noting St Patrick's day does not seem to give his servants a holiday or money for drink. Perhaps more remarkably, Peacock went to the wedding of the daughter of a Roman catholic neighbour and gave money to the priest and to the poor.[51] Like other masters of the household, he called his servants 'his little family'. He made them a bowl of punch and distributed 2s. 8$^{1}/_{2}$d. amongst them for a Christmas box.[52]

His diary is informative about his dress because at intervals he made lists of clothes and household linen and counted his washing. Peacock was able to fend for himself before his married Catherine Chapman. He made his own sheets.[53] In August 1745 he made some handkerchiefs. He then counted them, and wrote, 'the Lord make me thankfull'.[54] His sister Ireland also copied a shirt for him, which regrettably he found did not fit.[55] As the diary goes on his clothes become more elaborate. He noted that he had waistcoats, handkerchiefs, and gloves, and kept a record of their cost.[56] He had a large number of stockings: in February 1744 he had '11 pair of yarn stockings 7 pair of wosted three of cotton 3 of thread 2 of silk'.[57] After his marriage he noted knee buckles and chequered linen.[58] He carried a watch. For a day and night in August 1745 he suspected that his servant has stolen it, only to find it in his breeches' pocket.[59] When he married Catherine Chapman, his shopping took on a feminine slant when her clothes and necessaries begin to appear in his diary and he bought such items as shoes, ribbons, cambric, buttons, and mohair in Limerick. In 1750 she had a coat quilted.[60] His servants' dress also appears: he had breeches made for two boys and bought them brogues. The tailor visited and made a coat for 'little Jack' and a coat for Matty.[61]

---

47 Vestry disputes, 28 March 1749; visitation, 9 September 1740, 3 October 1745; church rates, 16 March 1741, 29 March 1741; tithes, 4 August 1745. 48 14 March 1741; 17 March 1742; 16 March 1745; 17 March 1746; 17 March 1749; 17 March 1750. 11 July 1741; 15 August 1745. 49 27 December 1743. 50 21 February 1744. 51 1 May 1750; 8 July 1751. 52 25 December 1745. 53 Sheets 11, 13, 15 July 1745. 54 25 January 1744; 22 December 1744; 30 October 1746; 3 August 1745. 55 1 December 1743. 56 5 February 1745; 20 July 1745. 57 25 January 1744. 58 8 December 1747. 59 8, 9 August 1745. 60 20 December 1747; 6 July 1748; 9 April 1750. 61 3 July 1744; 17 January 1745; 17 August 1745; 23 October 1745

Within two months of their marriage Catherine Chapman became pregnant; in October she was unwell, suffering from what might be morning sickness.[62] She does not seem to have any strange desires during pregnancy; six years before, Nicholas had been 'obliged' to give his neighbour Mrs FitzGerald his young sow when she was in labour and longed for sucking pig.[63] In April 1748, she went into labour. Childbirth was a social occasion; neighbours as well as the midwife were summoned and Nicholas asked Alice Hartstonge to be present. The midwife, Mother Handy, was brought from Ballingarry. The room where the birth and the lying-in took place would have been darkened and lit by candles. (Nicholas had made and stored 15 dozen candles a few days before.) He sent to Limerick for rice, lemons, oranges and spermaceti oil to ease and accomplish the birth.[64] These were necessary accompaniments for a successful labour: the spermaceti oil was for the midwife to use, and the oranges and lemons were essential ingredients in a caudle drink for the mother to drink both during and after the birth. The caudle was made from ale or wine, warmed with sugar and spices and was intended to feed the mother without straining her digestion.[65] Catherine's caudle included oranges and lemons, powder and jamaica sugar and rice.[66] Intervention during labour was common in the early eighteenth century. The soft parts of the birth canal were massaged and manipulated by the midwife's hand, greased with spermaceti oil. This was a dangerous practice, which greatly increased the possibility of puerperal infection, leading to the death of the mother. There is no reference to buying spermaceti oil in Catherine's subsequent labours, and it may have been laid in only for the birth of her first child. After a birth, the midwife remained for about ten days to care for the mother and child, and the mother stayed in her room for about a month.[67]

Because of Catherine's confinement to her room she would not have been present at the christening celebrations which were held a week after the birth of Pryce. Nicholas asked his wealthy and influential cousin, Wyndham Quin, to stand as godfather, and the celebrations were correspondingly lavish. Apart from the expected wines, whiskey and punch he killed a sheep and ordered lobsters and asparagus. The cleric who conducted the ceremony was paid, rather late, £10 and the midwife would have received a tip from Wyndham Quin.[68] On the same day, a cradle was bought in Limerick for 1s. 7½d. and fitted up. Women sometimes had a feast during the lying-in period and visitors also drank the caudle; Catherine received some neighbours after the birth of William, her third son, in 1750.[69]

Perhaps surprisingly, at least to us, five weeks after the birth of Pryce both Catherine and Nicholas decided to visit the family at Lisdogan, without the new

---

62 29 October 1747. 63 23 January 1741/2. 64 27–28 April 1747. 65 Hartley, pp 522, 558. 66 Wilson, pp 26–7. 67 Mother Handy returned to Ballingarry on 29 April 1748. 68 4 May 1748; Wilson, p. 26. Nicholas Peacock gave the midwife 5s. 5d. when he stood godfather at the christening of a neighbour's child, 20 August 1747. The clerk was paid £10 on 20 November 1748. 69 5 August 1750.

baby, and they asked their neighbours, Jon Geary and his wife, if they would look after him while they are away.[70] Mrs Geary had had a baby in early January, so would probably have been still breastfeeding and would have been able to wetnurse another baby.[71] With the distance to Co. Cork, and the probability that they rode there, it is not surprising that Catherine was 'very ill' on their return.[72] She may have had post-partum bleeding or engorged breasts due to the sudden cessation of breast feeding. She recovered and they went on to have two more children, George and William, in 1749 and 1750. It was fortunate that they did not have to send for a doctor for all three births, and that Catherine did not suffer from complications afterwards.

Nicholas Peacock himself suffered occasionally from ill-health. He writes of having a 'violent fit' of ague, of toothache and occasional episodes of overindulgence, when he has a hangover.[73] He and Mrs Hartstonge may have suffered from food poisoning when they were both ill after a dinner at Court.[74] He had toothache so badly in 1745 that he was unable to eat, and Catherine Peacock had a bad start to their marriage when she had a tooth drawn by Cyprian Purcell, who broke it and then bled her.[75] When Peacock himself suffered from pleurisy, he sent for medicine from Limerick which cost him 5s. 6d.[76]

The world outside the banks of the Maigue and the walls of Limerick intrudes only briefly, but, because he reads a newspaper, Peacock is aware of the War of Austrian Succession and the defeat of the allied army at the battle of Fontenoy.[77] With the Jacobite rebellion in England, there was a feeling of unease and threat, and locally there were a recurrent fears of the activities of bands of Tories. Peacock sent the Hartstonge plate to Limerick, no doubt to be put in a strong room there. He sat all night on watch when Tories were rumoured to be in Cratloe, Co. Clare.[78]

In December 1745, Peacock made his will, asking 'ye Allmighty God give me grace to mend my life'.[79] Neither the will nor an abstract of it has survived, and we have no record, as yet, of the fate of himself and his wife Catherine after the end of August 1751. But despite the brevity of its style, Peacock's diary is a valuable record of part of his life and of rural Ireland in the mid eighteenth century.

Nicholas Peacock's diary is in the National Library of Ireland, catalogued as an anonymous diary MS 16,091 and part of the Limerick Papers deposit. The diary is written on both sides of 64 unlined pages in a note book 12cm x 38cm (the last 24 pages are 2cm shorter in length). The manuscript is transcribed exactly as written because Peacock's spelling may reflect the pronunciation used at that time. Examples are 'pleat', plate; 'cag', keg; 'futting', footing. For the convenience of the reader, the dates of the entries are shown in bold type, and a space

---

70 5 June 1748. 71 12 January 1748. 72 7 June 1748. 73 16 June 1745. 74 18, 19 and 20 August 1745. 75 10 November 1745; 11 June 1747. 76 20–26 August 1747. 77 14 May 1745. 78 17, 19 July 1745. 79 14 December 1745.

has been inserted where feasible at the end of each month. It is covered with a coarse grey brown paper resembling sugar paper. Inside the back cover is pinned a scrap of a letter with calculations on both sides.[80] The last entry in the diary, for 31 August 1751, comes half way down the last page. It then breaks off.

---

80 25 to bread   0: 6
    26 to snuff   1 1
        to comb   10
    29 to a bit and snaffle   6 6
        to Cate Caffow   5 5
        to Tho Harragan   1:2:9
        Expences   9:9

    6 att   1:2:9   6 16 6
    2 at   3 4   0 6 8
    51 att   1:1   2 15 3

                                1 2 9
                                2 2
                                1: 4:11

[On reverse]
'I am
  most humble St
  Pat Haly

June ye 27 1744   2:8½
Augt ye 13   16:6
att severall times to ye   1:7½
22d of 9br 1744   2:9:0

March ye 18 1744   0 3 0
May ye 2 1745   0 8 0
March ye 26 1746   0 1 1
May ye 3 1746   7 7
ye 24   2 8½
                             3:11:4½

GLOSSARY

This is provided as a rough guide to some of the words used by Peacock, mainly in relation to farming.

all	awl
Allom	alum, a mineral salt, probably used here for fixing dye.
allphabett	alphabet: a piece of furniture for storing correspondence, divided in pigeon holes with the letters of the alphabet on each.
anker	a cask holding $8^{1}/_{2}$ imperial gallons.
axletree	the bar under a carriage on which the wheels are fixed.
balks	rough beams of timber.
ball horse	a white faced horse.
ballsom	balsam, an aromatic ointment made from the plant that grows along river banks to soothe grazes and wounds. Thus, friar's balsam.
bandle cloths	from the Irish measure of an arm's length: *bann*, measure; *lamh*, arm.
bear barley	a variety of barley, used mainly for malting.
bed ticks	a case filled with feathers which makes a mattress or pillow.
binders	those who tied up stooks of wheat with a band of straw.
bleachyard	Before it was suitable for sale, woven linen had to be treated at a bleachyard. There it was first steeped in water, boiled repeatedly in alkaline lye and afterwards steeped in an acid solution. It was again washed. Between these operations it was laid out on the grass to bleach in the open air.
blew sterch	a substance made from flour to stiffen linen.
bleeding	The eighteenth century theory of ill-health due to the body's humours (wet, dry, hot or cold) meant treatment by alleviating these symptoms. Bleeding was a way of reducing heat.

botts	butts: a shooting range.
brimed	a pig in heat.
brimstone	sulphur.
brads	small, thin nails.
brogues	rough, strong shoes made from untanned leather and sewn with thongs of leather.
buckram	a coarse linen stiffened with glue.
buckskin	a fine leather.
bushill	a measure of 4 pecks or 8 gallons
cadis	Irish *cadas*, cotton. Either padding made from cotton wool or floss silk, or worsted yarn.
cadow	caddow, a rough woolen covering.
cag [of salmon]	A keg.
callops	From the Irish *colpa*, a full-grown cow or horse.
caleminco	calemenco, a glossy woollen material woven to show checks, which was imported from Flanders.
cambric	fine linen.
can	a vessel for holding liquid, possibly with a handle.
canvas	a coarse cloth made of hemp or flax.
caperus/coperall/coperus salt	copperas was sulphate of either copper, iron or zinc, used in the case of copper for dyeing, tanning and making ink.
cassanet	possibly cassinette, a fine soft weave of wool
ceeve	sieve.
chine	the backbone of an animal.
chip	the chap, or cured cheek of a pig.
churn	a vessel for making cream and butter.
cleeve	splitting wood.
cobberd	an iron to support a spit in front of the fire.
cock	a hay cock; a conical heap of hay.
collops	a unit based on the number of cattle which would graze a pasture.
comondale boards	deal planks.
cooper	maker of barrels.
corcas	the salt marshes along the banks of a river.
corry coamb	curry comb for grooming horses.
cottening	the process of putting a nap on cloth.
country charges	county cess, taxes levied on the county households to maintain the roads and for other miscellaneous purposes.

Glossary

cowples	a pair of rafters in a roof.
cross ploughing 'gorrowing'	repeated ploughing of fallow land, often to prepare it for wheat the following year. Irish *gorail*
crown	a coin worth 5s.
daging	removing dirt and excrement from a sheep to prevent maggots breeding and eating the flesh.
diaper linen	linen with a diamond weave with dots, lines and leaves in the spaces.
dock	cutting the sinews, bone and tail hairs of a horse (or dog) to prevent disease.
dray/drea	a cart without sides for carrying heavy loads, or a sled.
earnest	money given in hand to secure a bargain, a deposit.
eelfry	young eels.
firkin	a vessel, half a barrel in size, for holding liquids.
flannel	woollen cloth, loosely woven.
flitch	salted and cured side of bacon.
flyer	the part of a spinning wheel that twists the thread as it is wound on to the bobbin.
fossels	possibly fusiles, fine steel for a tinder box.
frieze	a coarse woollen cloth with a nap on one side.
fustian	a coarse cloth made from a mixture of cotton and linen.
futting	footing; placing a number of sods of turf at an angle together, to dry.
gadds	from Irish *gaid*, a rope or band made from twisted fibres from twigs.
gages	quart pots
gale day	The day rent is due twice a year, in May and November.
garde vine	a box with a lock and key for storing wine.
garrott	storage place
gelt	gelded; castrated.
girth	a band of leather securing the saddle to a horse's back.
gowage	a gouge.
greasiors	graziers.
gusmand	gisement was money received for cattle taken in to graze.

hackling	process of stripping and combing out the fibres of flax or hemp.
haggard	stack yard.
harrow	a heavy timber frame with tines used to break up clods of earth.
heffer	heifer: a young cow that has not yet calved.
hemp	*Cannabis indica*, whose coarse fibres were used to make rope or strong material
hipshott	having a dislocated hip.
hoggetts	a boar or sheep in its second year.
hogshead	a large cask for liquids such as beer or wine.
horn branded	marked or branded on the cattles' horns.
horn money	a formerly feudal service, based on the number of horned animals owned by the tenant.
hotspur	an early variety of pea.
Hungary water	rosemary infused in alcohol.
hurlds	hurdles, frames made of wattle, used as temporary fences or gates.
indigo	a deep blue dye obtained from *indigofera* plants
indentures	articles binding an apprenticeship.
jack cloth	a rough linen cloth used for drying crockery.
Jews harp	a musical instrument made from a small iron frame with a vibrating strip of metal down the centre; the frame is held between the teeth and the metal strip plucked by the finger.
kishes	from the Irish *kis*, a basket or hamper used for carrying turf.
lime kiln	an oven for heating and calcining limestone to make lime.
lamp black	a pigment of carbon, made from collecting the soot from lamps.
licorish	liquorice, used for medicinal purposes.
Lisbun wine	white wine from the Estremadura province of Portugal.
little house	possibly an outdoor privy, as in the Welsh *y petty*, the little house
lock	part of gun which fires the charge.
lossid	a kneading tray with a rim, used to make bread. From the Irish *losaid*.
madder	a red dye made from the root of the plant *Rubia tinctorum*.
male	meal, as in corn meal.

Glossary

malt	barley prepared for brewing by steeping in warm water, germinating and then drying in a kiln.
manna oil	a laxative obtained from the bark of the manna ash, *fraxinus ornus*. Imported from the south of Italy.
martingale and cavison	a strap attaching part of a bridle to the girth; a cavesson was a type of noseband. Both used to control difficult horses.
masher	for mashing malt
meads	meadows
metheglin	a spiced drink made from spiced mead, distilled from honey.
moidore	a Portuguese gold coin.
neck trench	a narrow trench.
noging	noggin, a small quantity of liquor, usually a quarter of a pint.
oager	auger, a tool for piercing wood.
pading my mare	pacing his horse.
pantiles	a curved roofing tile in an ogee shape, one curve being greater than the other.
peaper	paper.
peck	a two-gallon measure.
perce	perch
perfected her indentures	completed her apprenticeship.
Permont water	spa water imported from Pyrmont, near Kassel, Germany.
pistoles	a Spanish gold coin.
pitch	a thick dark resinous substance made from boiling down tar and used for sealing between cracks in wood.
pleat	silver plate.
plough iron	the iron blade or share of a plough, which cuts the ground.
pluck	the heart, liver and other internal organs.
pole ax	a butcher's axe with a hammer at one end, used to stun animals.
pottle	two-quart measure.
pounder	an instrument for treating linen, a beetle.
powder and shot	gunpowder and shot for firearms.
powdering tub	a tub in which meat is powdered, pickled or salted.
pumps	light shoes.

quair	quire, 24 sheets of paper.
quern	a hand-worked machine for grinding corn, or a small mill for grinding mustard and malt.
quicks	hawthorn bushes to make a hedge.
rasiers	razors.
reel	bobbin on which thread was wound during spinning.
reep hooks	a curved sharpened blade for harvesting wheat.
reding	ruddle, a red ochre used for marking sheep.
reek	a rick, a heap of hay, wheat or turf, constructed and often roofed with thatch.
resins	resin, used for varnishes.
ronlet	casks, of differing capacity.
rowel	a farrier's treatment of swellings in horses, where a circular piece of material with a hole in the centre is inserted between the flesh and skin to enable pus to discharge.
rug weastcoat	a waistcoat made from rug, a coarse frieze.
russidin	russeting, clothing made of russett, a coarse homespun woollen cloth.
saffron	the dried pistils of the autumn crocus, *crocus sativa*, used for medicinal and culinary purposes. An extremely labour-intensive, but valuable crop.
sallys	Irish, *saileach*, a stick.
sarge	serge, a coarse woolen material.
sayers/soyers	sawyers
schallops	the bent rods used by a thatcher to fasten down the straw. Irish *scolb*.
seare ... punch	a bowl of punch heated with a hot poker
sneaker	a small bowl of punch
pigin	piggin, a wooden half-sized barrel, with one long stave to form a handle.
pott hangers and scheales	scales for weighing produce with a hanging bar above.
puke	a substance to make one vomit.
schane	skein, of thread.
schimer	skimmer.
schrew	screw.
serus	ceruse, white lead, used for painting.
shag briches	breeches made of a cloth with a velvet nap on one side, usually of worsted but sometimes of silk.

Introduction

shilats	shalotts.
shorts	flax.
shrub	drink made from orange juice, sugar and rum.
side saddle	a saddle constructed for women, with both feet on one side of the horse.
slane	a sharp spade for cutting turf. From Irish, *sleán*.
small beer	weak, home-brewed beer.
snuffers	candle snuffer.
spade, Irish	a loy, Irish *láighe*, a one-sided spade, common in Munster and Connaught, which was used to turn the ground.
spermacitty	spermaceti, a fatty material found in the sperm-whale, and used in medicinal preparations.
spigot	a small pin to stop the vent of a barrel.
spools	for winding thread after spinning.
spoke shave	a knife constructed to make wooden spokes, the radials of wheels.
spur rowel	a serrated wheel, attached to the point of a spur.
srline	a sirloin of beef the best cut.
steel hemp	a variety of hempseed.
stock	a man's neckcloth
stone	14 lbs. weight
stone blew	indigo combined with starch or whiting used to whiten linen.
stone horse	a stallion.
strapper	possibly a cow who gave trouble milking.
streaner	strainer.
strike	a measure of one bushel.
sucking pig	a piglet still taking milk from the sow.
sulfer	sulphur. Used to fumigate casks used for wine
suttoote	surtout, a man's greatcoat.
swingle	a wooden instrument for beating and scraping flax or hemp to rid it of coarseness. Another word for scutching.
syth stone	a stone for sharpening scythes.
tallow	rendered animal fat, used for soap and candles.
taster	tester, a canopy over a bedhead, held by posts or hung from the ceiling.

tevanes	from Irish *taobh*. The short beam of the roof, crossing from one rafter to the other side.
tierce	a measure of quantity.
thirteens	the English shilling was worth thirteen pence in Ireland, so three thirteens would have been 3*s*. 3*d*. Irish money.
toe	tow. The fibres of flax or hemp, separated by beating and spun to make string or rope.
torkes	turkeys.
trashers	cuttings from a hedge.
trincher	trencher, a dish.
truckle	a low cart on wheels.
tucking	the processing of cloth from the weaver by stretching it on tenters, wooden frames, to dry it evenly so that it does not shrink
tucking mill	where the process of tucking cloth takes place.
vale	veal.
vardigrace	verdigris, a pigment made by the combination of copper and acid, used in dyeing.
vomitts	an emetic.
ware	weir
warp	lengthwise threads in a loom
wattles	stakes interlaced with branches used to make fences and walls.
wafers	used to seal documents without sealing wax.
weaning	accustoming a young animal to food other than its mother's milk.
wethers	a male sheep, a ram especially a castrated ram.
winnowing	separating wheat from chaff by shaking it in a winnowing sheet.
yeaned	birth of a lamb.

BIBLIOGRAPHY

M manuscript collections R reference works S secondary sources

Adare, Countess of, *Memorials of Adare* (privately published 1865) S
Bailey, N., *An universal etymological English dictionary* (London 1759) R
Barnard, T.C., *The abduction of a Limerick heiress: social and relations in mid-eighteenth century Ireland* (Dublin 1998) S
Barnard, T.C., *A new anatomy of Ireland: protestants in Ireland, 1649–1770* (2003) S
Barry, J.G., *The Cromwellian settlement of the county of Limerick* (Limerick 1900) S
Beggan, Gerard, *In the barony of Pubblebrien: Patrickswell and Crecora* (n.d.) S
Begley, John, *The diocese of Limerick from 1691 to the present* (Dublin 1938) S
Bell, Jonathan, and Mervyn Watson, *Irish farming implements and techniques, 1750–1900* (Edinburgh 1986) S
Bence-Jones, Mark, *A guide to Irish country houses* (revised edition, London 1988) S
Blackstone, Sir William, *Commentaries on the laws of England in four books* (tenth ed., 1807) S
Brady, W. Mazière, *Clerical and parish records of Cork, Cloyne and Ross* (London 1864)
de Breffny, Brian, 'The Peppards of Cappagh, Co. Limerick' *Irish Ancestor* 16:1 (1984), 68–70 S
Burtchaell, G.D., and T.U. Sadleir (eds), *Alumni Dubliniensis* (Dublin 1935)
Burke, *Extinct and dormant baronetcies* (Baltimore, MD, 1985) R
Burke, *Landed gentry of Ireland* (London 1958) R
Burke's peerage (London 1860) R
Carpenter, Andrew, *Verse in English in eighteenth-century Ireland* (Cork 1998) S
The Civil Survey: County of Limerick (Dublin 1938) S
Clare, Wallace, *Guide to copies and abstracts of Irish wills* (Dublin 1930) S
Cronin, Denis A., Jim Gilligan, and Karina Holton, *Irish fairs and markets: studies in local history* (Dublin 2001) S
de Moleyns, *The landowners' and agents' practical guide*, 8th ed. (1899) S
Dickson, David, *Arctic Ireland: the extraordinary story of the Great Frost and forgotten famine of 1740–41* (Belfast 1997) S
Dineley, Thomas 'Extracts from the journal of Thomas Dineley Esquire, giving some account of his visit to Ireland in the reign of Charles II', *Journal of the*

Royal Society of Antiquaries of Ireland, new series, 8 (1864–66), 272–5 (Dublin 1865) S

Dolan, Terence Patrick (ed.), *A dictionary of Hiberno-English* (Dublin 1999) S

Dunraven papers, University of Limerick Special Collections **M**

Dutton, Matthew, *The office and authority of a justice of the peace for Ireland* (1712) S

Dutton, Matthew, *The law of landlords and tenants in Ireland* (Dublin 1726) S

Fennell, Geraldine, *A list of Irish watch and clock makers* (Dublin 1983) S

Ferrar, John *Limerick directory for 1769* (Limerick 1769) S

Garnham, Neal, *The courts, crime and the criminal law in Ireland, 1692–1760* (Dublin 1996) S

General alphabetical index to the townlands and towns, parishes and baronies of Ireland (Dublin 1861) **R**

Gill, Conrad, *The rise of the Irish linen industry* (Oxford 1925) S

Gillman, H.W., 'Index to marriage license bonds, diocese of Cork and Ross, 1623–1750', *Journal of the Cork Historical and Archaeological Society*, 2nd series, 2 (1896), 23 S

——, Index to marriage license bonds, diocese of Cloyne, 1623–1750', *Journal of the Cork Historical and Archaeological Society*, 2nd series, 6 (1900) S

Gleeson, Dermot F., 'The silver mines of Ormond' *Journal of the Royal Society of Antiquaries of Ireland*, 7th series, 7 (1937) 101–15 S

Gribben, Henry, 'The Irish Linen Board' in Marilyn Cohen (ed.), *The Warp of Ireland's Past* (Basingstoke 1997), 73–82

Grove White, J., *Historical and topographical notes, etc. on Buttevant, Castletownroche, Doneraile, Mallow and places in their vicinity*, 4 vols. (Cork 1916) S

Hartley, Dorothy, *Food in England* (London 1954) S

Hayes, Richard, 'The German colony in Co. Limerick', *North Munster Antiquarian Journal* 1 (2) (1937) 42–53 S

Hayes, Richard, 'Some notable Limerick doctors', *North Munster Antiquarian Journal* 1 (1938–9) 115–17 S

Hembry, Phyllis, *The English spa, 1560–1815: a social history* (London 1990) S

Herbert, Robert, 'The antiquities of the corporation of Limerick', *North Munster Antiquarian* International Genealogical Index, Church of Latter-Day Saints, 104, 119 **M**

Journal 4 (3) (1944–5) Spring 1945 S

Journal of the Association for the Preservation of the Memorials of the Dead in Ireland, 7, 3, pt. 2 (1909)–17S

Kelly, James (ed.), *The letters of Lord Chief Baron Edward Willes to the earl of Warwick, 1757–62* (Aberystwyth 1990) S

Kenmare papers, Public Record Office of Northern Ireland **M**

King's Inns Admission Papers, 1607–1867, ed. Edward Keane, P. Beryl Phair, and Thomas U. Sadlier (Dublin 1982) S

Kirkpatrick biographical file, Royal College of Physicians of Ireland **M**
Lamplugh, G.W., S.B. Wilkinson, J.R. Kilroe, A. McHenry, H.J. Seymour, and W.B. Wright, *Memoirs of the geological survey of Ireland: the geology of the country around Limerick* (Dublin 1907) **S**
Lenihan, Maurice, *Limerick: its history and antiquities* (first edition 1866, reprinted 1991) **S**
Leslie, Canon J.B., 'Biographical succession list of the clergy of the Diocese of Limerick', typescript Representative Church Body 61.2.13. **S**
Lewis, Samuel, *A topographical dictionary of Ireland* 2 vols. (Dublin 1837) **S**
Limerick papers, National Library of Ireland **M**
Maunsell, R.G., *History of Maunsell or Mansel* (Cork 1903) **S**
Milward, Rosemary, *A glossary of household, farming and trade terms from probate inventories* (Chesterfield 1993) **R**
Mungret/Crecora parish journal **S**
O'Connor, P.J., *Exploring Limerick's past* (Newcastle West 1987)
Oliver, R.C.B., 'The Hartstonges and Radnorshire: Part I', *The Radnorshire Society Transactions* 1973; Part II, 31–2 **S**
Ordnance Survey letters, Limerick, 2 vols. 1840, Royal Irish Academy 14 D 18 **M**
Phillimore, W., *Index to Irish wills, 1615–1800* **S**
Representative Church Body Library, Dublin **M**
Rutty, John, *A chronological history of the weather and seasons and of the prevailing diseases in Dublin* (London 1770) **S**
Seoighe, Mainchín, *The story of Kilmallock* (Limerick 1987) **S**
The shorter Oxford English dictionary 2 vols. (Oxford 1933) **R**
Taylor, George, and Andrew Skinner, *Maps of the roads of Ireland* (Dublin 1778) **S**
Thirsk, Joan, *Alternative agriculture, a history* (Oxford 2000) **S**
Vicars, A., *Index to the prerogative wills of Ireland, 1536–1810* (Dublin 1897) **S**
Wallis, P.J. and R.F., *Eighteenth-century medics* (Newcastle-upon-Tyne 1988) **S**
Watson, John, *Gentleman and citizen's almanac* (Dublin 1760) **S**
Welpey abstracts of Irish wills (14 vols), typescript in Society of Genealogists, London **M**
White, J.G.: see under Grove White
Wilson, Adrian, *The making of man-midwifery: childbirth in England, 1660–1770* (Cambridge, Mass. 1995) **S**

THE DIARY

1 January 1740–31 August 1751

1740

In ye Name of God Janr. 1st 1739/40 No gift
2 I went home gave ordrs to trash a Monday back to Courte[1]
4 5th at Courte ye 6 home
8th men trashing for Straw
9th I had 10 men a trashing
10th I had the men at Dito. I sould Tho: Bury Esq[2] 7 bushills of wheat at 3s a bushill 4 barrills of barly att 12s a barrill a rack of Oats 7s 8d a barrill and he to trash it and a bushill of Peas for 6s I got some drink and a pinte of Rum from Mr Jon. ffosbery and Mr Bury stay'd to eat some beef and cab: wth me so home by night
11th ye boys putting out dung 12th Hon: Dondon winowing
13th at home reading 14th John Newman a Pileing barly
15th Dan Heas a Pileing 16 Hon Dondon winowing
17th I had 3 horses from Courte 3 from ye tenants 3 from Killcorly[3] tenants 1 from David Nelan and my own 3 horses a carring barly to Courte I got a letter from Hen: Page for Rent[4]
18th I wrote to my Aunt for 10l and got it[5]
19 I sent my Rent to Henry Page and got his Receipt to Courte
20 I orderd a cow to be I killd to home
21st 9 men trashing oats I gave Mr Bury's man 7 bus: wheat four barrils barly and a bushill of peas
22d 5 men a remaking my beans send 4 bus: barly to ye mill
23d 3 men at ye beans went to Courte were I stay'd to ye 30th and got a 4l peice from Mr Bury and 4s from Mcnemarra[6]

Febr ye 1 and 2d a winowing and put up 8 barrill of Oats to Court
3d I gave Tim Lowry 1l: 6s:0d. and owe him still in all ~~10s.6d pd~~

---

1 Court, Co. Limerick, the seat of the Widenham family, was 4 miles north of Adare and 1 mile east of Kildimo. Henry Widenham (d. c.1719) acquired Court when he purchased Crown land in Limerick for £841 in 1703. The house has now gone. See Grove White, vol. 2, pp 166–7. 2 One of the Bury family of Shannongrove, Pallaskenry, Co. Limerick. 3 Kilcorly, a townland south of Kildimo and east of Nicholas Peacock's holding at Caparoe. 4 Henry Page of Askeaton, Co. Limerick. 5 Peacock is probably referring to his great-aunt, Mary Widenham. In her will of 13 June 1736, she refers to Nicholas Peacock as 'her nephew': Welply, Will abstracts RCB 80/B2/1/f.191. 6 There are are two possible McNemaras. John McNemara was a sworn attorney appearing at the Tholsel Court, and a Commissioner for taking affidavits for the courts in London. Mathew Mcnemara was a barrister and freeman of the city. See Ferrar; Voters List for Limerick City NLI MS 16,092.

4th I Darby Longane in full 6:8 and Gave Ned fitzGerald on Tho. Halloran's acct he owed Paul Haly 6s:4d
5th I came home and put up 8 barrills barly and filld 3 to send to Courte
6th 3 men trashing beans
7th 1 man at Dito I got a letter to go to Tough[7] and ye Elogey on ye Petatoes[8]
8th I sent 3 barrills Duch barly to Courte I went to Tough and pd Jon Bork in full 3s:9d pd. Dan: Shea in full and for broges for Jack 5s:5d So to Courte
9th I came home and sent 1s:1d earnest for Cabidge Plants
10th Sent my foale to Courte to be weaned
11th I went to Tough and got 10:6:3 and allows Jon. Bork 17s:11d in full for flanning and for weaveing Canviss for Mrs Widenham then to Courte gave ye money to Mrs Widenham and stayd yt night
12 I got up early and sent home a Horse skin and sent for two pottles of Salte to dress it so home ye little cow calved
12 Hon. Dondon winnowing
14th deliverd Mr. Burys men 7½ barrills of Oats and sent him my acct that he owes me 2l:15s:4d I cleard wth ye bindors and winow women
15 I got 5 barrill Oats claind to put on ye killn[9] went to Courte
16 I walk'd aboute
17th I went to Tough and got 7l:15:5½d So to Courte gave it to Mrs Widenham then sould a cow for Mrs Hartstonge[10] for 2l:10s:0 **18th** I walkd aboute
19th I gave Connr Covard a bill on John fitzGerald for 1 pound 1 shillings worth of linin yarn Mrs Widenham had from him then I got my Oats put on ye kill
20 I began to Plow
21 I came to see it and went back to Courte
22 I sent my Oats to ye mill
26th I took up 5 barrill Oats from Geo: ffosbery[11] and came home and found my Plowing at Capparow[12] finishd
27 I sow'd a barrill and ½ Oats 4 men diging
28 I sow'd 2 barrils and a bus and finishd
29 I had 4 men diging for Parsnips &c

March ye 1st I went to Courte and got a handkercheif from my Aunt
2d I came home

---

7 Tough/Tuogh is 9 miles south west of Limerick on the river Muncairn. Charles II granted 418 acres plantation measure of land and a ferry over the river Maigue between Killcollum and Contaghlagh to Sir Standish Hartstonge. Survey of Tough by William Hunt 17167 NLI Map 16 32; copy of patent granted to Sir Henry Hartstonge, 27 July 1756, Limerick papers, NLI PC 876/3/19/20.   8 *An Elegy on the much lamented death of those excellent patriots and lovers of the country, the family of the potatoes in the kingdom of Ireland, who fell by a general massacre, being confined and starved alive by cold and hunger ... between the 26th day of December, and the 18th day of January in the year 1739* (Dublin 1739/40). Text in Carpenter, p. 248.   9 Because the weather in Ireland was so wet, it was usual to dry cereal crops in a kiln before they were ground at the mill.   10 Alice Widenham, daughter of Henry Widenham (d. 1719) married Price Hartstonge.   11 George Fosbery was high sheriff of Co. Limerick for 1744. *Faulkner's Dublin Journal*, 7–11 February 1743/4.   12 Caparoe, a townland 1½ miles south of Kildimo.

April 1740

3 I sow'd a barrill and 3 pecks beans had 2 men setting Sallys 1 tending ye Plow

4th I sow'd 14 bushills of beans a man tending ye Plow

5 I sow'd 6 bushills of beans and finishd and began to plow ye well Corcas[13] for Oats

6 I sowed 6 bushills Oats **8th** I went to Courte

9th I went with Martr Hartstonge to Church home in ye Eavening and wrote to my Sistr McGan[14] for ½ a stone of sope and Six pewter Spoones[15]

March ye 10th I began to plow for Duch barly

11th att Dito Jon fosberys men trashing beans

12th finishd all my plowing wth ye same horses my own mares Antho: mare hallorans and Shilbred filly

13 I had some turf drawn and deliverd Colonell Taylors[16] 4 bus: beans and Tho: Gibon 1 bus: beans

14th ye men setting Sallys

15 I went to Courte where I stayed **18th** so home and g[ave] Jon ffosbery 3 barrills beans wch he is to return o demand 1d

19th I got 2l:15s:4 from Mr Bury and pd Tim Lowry 11:4½d 10s6 of which I owed him he owes me change 10½d

20th I had a Stone of Iron from my Sist Megan which I am to give her credit for lent ned 10s. pd

22d 4 men diging for English bar I gott a few petatoes I gave Jon. Bryen 1s:1d on acct of his wayges I went to Courte and pd Morroh for bringing my turf and sent for 3 pair of soles a pair of Garters and a pound of Steele wch I paid for **23rd** to church[17]

24th at Courte 3 of Mrs Hartstonge's cows sould

25th to Tough but no good gave Tho Halloran on acct of his wayges 6:8 aboute 12 at night I came home

March ye 26th 1740 I had 13 men trashing beans and set two beds of Petatoes I gave Joan Dally 1:2:9 on acct of her wayges

27 28 and 29th a Sowing duch barly and Sowed in all 3 barrills and a bushill 3 bushills and a peck English barley I gave James Shilbred a bushill of duch Barly to Courte aboute 12 were I stayd to ye 1st of Aprill

Aprill ye 1st I came to Ballylongford[18] and agreed wth Mr Bryan for Portachaca[19] att 12s an acre for 21 years if his lease continues so long and left ye 40s he owes me as earnest

---

13 The salt marshes along the banks of a river which provided exceptionally fertile soil. 14 Mrs McGan, Nicholas Peacock's sister, lived in Limerick. 15 Pewter was an alloy of tin and lead, used by those who could not afford silver or who kept silver for important occasions. 16 Colonel Edward Taylor of Ballynort, Co. Limerick. He married Ann Armitage and they had at least one child, Catherine. Will proved, 1761. Prenuptial deed of Hugh Massy the younger and Catherine Taylor of lands in Limerick, 22 April 1760. Cottesmore Deeds, National Library of Wales; Vicars. 17 Nicholas Peacock generally attended the church at Kildimo, where he seems to have been a member of the select vestry and possibly churchwarden. 18 Ballylongford, a townland 1½ miles north of Adare. 19 A townland in

2d I got Mr Jon fosbery to Rowell my filly I pd. 4s. hearth money[20] and 1:6 country charges

3 I got 2 duch speads sowd my Onyon seed

4th I went to Courte got my Colt gelt and stayd till ye **14th** and bought 2 gallons 3 bowls and a pigin about 10 I came home and found my man and Patt Meny making bounds ye young mare foled

15 My men and Patt Meny trashing wheat

16 I went to Courte and stayd till ye **18th** when I came home my wheat was finishd Patt Meny wth ye men I got ye wheat winowed

20th to Courte **21** I got a chain made for ye bucket so home

21 22 and **23d** I waited for Mr Bryen and went to him he

24th he came to vew ye land I went wth him for a minnit but got none only a promis of one in a fornight before Mr Patrick Peppeard[21] I had 8 men fencing ye medows of Portachaca Sould ½ a peck duch barly for 1:1d I bought a hive of bees from Jon Madigan for 9s gave him 1:1 and a note to Dan Shea for a pair of Shoes

25 and **26th** men fencing ye medow I went to Courte

27th I went to Adear[22] and pd Mrs Quin[23] 20s in full for tythes[24] then back to Courte and stayd to ye **30th** in ye aftr noone

May ye **1st** I took Possession of my new farm and took greasiors

2d 14 men at ye bounds to Court in ye Eavening

3d att ye fare and got 5 cows from Mr fitzGerald pd 5:3 in full for bees a riddle

6½ expences 6 to Courte

5 I accounted wth ye men so home

6: 7: 8th fenceing

9th to Courte where I stayd till **11th**

12 I had 19 horses drawing turf and 7 men and finished back to Courte at night

13 I came home **14** I went to Tough but got no money

15th I went to Court and came back at night

16: 17 I send my Oats to ye mill and 3 barrill more to ye Kill

---

the bend of the river Maigue, just south-east of, and adjoining, Peacock's other holding, Caparoe. The name was often abbreviated by Peacock as 'Porte' or 'Port'. 20 Hearth money was a tax imposed on the number of hearths in each house. 21 Patrick Peppard of Cappagh House, Co. Limerick was the son of Patrick Peppard of Kilmacow, Co. Limerick, who converted from Roman Catholicism in 1739, and Catherine Dowdall. Patrick junior was churchwarden of Ballingary in 1731. He married Faith, daughter of Joseph Standish and Catherine Monckton of Ballynafranky, Co. Limerick. His will was proved in 1751. NLI MS 13,066; de Breffny. 22 Adare is 8 miles south west from Limerick, on the river Maigue. The townlands in Adare parish granted to Henry Widenham were Tuogh (418a.0r.0p.) former proprietor Dermott McMahon; Curry [Curragh?] (196a.0r.0p.) former proprietor Edward Purcell; Rennigarrane (19a.0r.0p.) former proprietor McTullogh Byrne; Killivokane (136a.0r.0p.) former proprietor Edward Purcell. 23 Mary Quin, née Widenham (1682–1776) was the co-heiress, with her sister Alice, of Henry Widenham. In her marriage settlement she brought several townlands originally purchased from the trustees of the Forfeited Estates. She married Valentine Quin (–1744) of Adare. They had eight children: Wyndham, George, Mary (who married Sir William Barker Bt.), Margaret, Alice, Catherine and Anne. 24 Tithes were levied on all landholders, as tenth part of annual produce, to maintain the church. They were levied in income or kind.

July 1740 51

18 I got a young pig from Darby longane and am to allow him 2s: 8d for it in ye beans he had from me

19 I got a young pig from Patt Daniel I spent ye day herding ye men carrying ye oats to ye mill ye rest is not ground

May ye 20th 1740 Jon at ye forge puting ye Plow iron in order to begin to Plow to morrow Tom at ye mill Jack brought his fathers mare and a yearling I bought off Patt Halloran for 8s I markd 18 sheep and 10 lambs I got shorn a Saturday in ye eavening I went to Court where I stayd to **ye 11th of June** in which time I got my land Plowd and cutt my corn wed I bought a file for 1:1 and got a pair of cards from my Sist. McGan wch I must give her credit for and pd. ye Quit rent[25]

June ye 12 13th at ye Plow 14 weeding oats

15 I sent for a ewe my Aunt gave me and a lamb Antho: gave me got a fleshfork and spitt from Tim Bork my 2 mares gott Mr Jacksons horse

16 I began to burn my follow[26] Will Bork turning my Coat

17 Att ye follow and gott a Swarm of bees

18 19 att Dito to Courte in ye eavening **20th** home in ye eavening

21 a fencing

22d to Adear Spent 6½ din'd at Mr Quins[27] from thence home sent 2 turkies to Adear then to Courte where I stayed a sheep shereing and accoumpting wth Will Wall till ye

27th So home half a pound tabaco 5d

28 fencing Caparow

29 at home walking hived a Swarm of bees

30th sent a barrill of wheat to Doctr martin[28] and Sould 3 pecks for 2:9 bought 2 sythes a pad lock 1d needills

July 1 2 3 drawing stones I gave Jon Bryen 3s: 4d

4 5th att ditto

6 a reading and walking I got a Ram from my Brother

7 at ye fallow paid 6½ Countery Charges

8 9 10 at ye fallow

11 at home alone ye Servants at deemo[29]

25 Quit rent was a small sum of money paid by a freeholder or copyholder to the Quit Rent Office in Dublin in lieu of feudal service. 26 Burning off the dead stalks of the previous crop on fallow land. 27 Valentine Quin (1678–1744), son and heir of Thady Quin of Adare, Co. Limerick, married Mary Widenham, co-heiress of Henry Widenham of Court. He purchased the great tithes of Adare and the fee simple of the leases which had belonged to the Ormsby family. From 1721 onward, he bought the whole of the land which had belonged to the Kildares around Adare, and a lease of Carigeen, where the Dysert round tower stands, from the see of Limerick. He had a late conversion to the Church of Ireland in 1739. Valentine Quin's will was proved in 1744. Vicars; *Memorials of Adare*. 28 John Martin practised in Quay-Lane Limerick. He studied medicine in Leyden, where he graduated on 28 October 1733 and returned to Limerick. He married Alice Roche. He died in 1786. Ferrar; Hayes (1938–9), p. 117; typescript copy of Prerogative Will, 29 June 1786. Kirkpatrick archive; Wallis, p. 396. 29 Kildimo (now Old Kildimo) is 4 miles south east of Pallaskenry. The Civil Survey (1654–7) noted that Major Patrick Purcell had a castle in Kildimo, with a bawn, three 'great houses', ten cot-

12 at ye fallow and finishd to Courte in ye eavening
13 to Church back to Courte
14 to Limerick for Mr Hartstonges[30] Pleat pd. for a Sythe Stone and Sand 3 1/2 Sidr 1s E McGan 6 1/2[31] Paul Rahily 10s Patt Halloran for his yearling 8 and Jon Newman 1:1 Jon Donevan 8:1 1/2 peaper and tabaco 5:3 So home at Killdeemo a dram 4d
15 16 17 and 18 John and Tom mowing ye other boys picking weeds 19 all at ye weeds and D welchs boys and Gibons boys I gave Jon Bryn 1:1 on John Newman acct
20th to Courte 21 ned brought me a ~~pair of card and 2 pound of coperall~~ which I must allow my Sistr McGan for
22 I came home Sick aboute 7 o clock ye men at ye hay
23 I continued ill and got a letter from Jon Hartney[32] for his colts a letter from Jon ffosbery aboute Ms Blunt Jon Hannan came wth his daughter I hired her for 5:6 a quartr.
24 I hird Den Bryen for 8s a quartr.
25 I bless God easyer it being a holy day no work[33]
26 ye men at ye trinch 27 reading alone
28 ye men D Nolan and boy at ye hay
29 I bought ye horse James Conway had for 3:14:6 pd Earnest 11:4 1/2 pd Mick Hickey 7:7 in full for last years tyth turf pd Patt: Gready 1:1 earnest for a hondr a turf at 12s a hondrd my men D Nelan and boy at ye hay
30 I went to Killcorly got all consent for Ms Blunt but her own who I did not see D Nelan and boy at ye hay
31 Patt Meny and my own men at ye hay

Augst. ye 1st to Courte and home in ye Eavening P Meny at ye hay
2d I began to Reep english barly and Reep'd only 26 qr P Meny at it then to ye Neck trinch I got a pale from Mr ffosbery and in Return Sent him a line of vale
3d I calld at Tough from thence to Adear Church dined at Killcorly home aboute 5
4th my men and P Meny clainin ye hay yard in ye Eavening Jon Blunt[34] came about Ms I went to Supper to Killcorly and aftr Some talk wth Ms we went to supper wch when ended I came away dissatisfied for 35l wod not mentain her
5 my men and P Meny at Dito I recd 5:5 from D Mullcheer in parte for his cow from Jon Lynch 7s in full for his horse
6 my men and P Meny att Dito
7 I began to Reep oats my men and P Meny

tages, a mill seat, and a salmon and other weirs. *Civil Survey*, vol. 4, p. 346. 30 Price Hartstonge. 31 Edward McGan may have been Peacock's brother-in-law. 32 John Hartny witnessed an indenture between William Bury of Shannongrove and Alice Fosbery of Kilkerily, 23 October 1724. Dunraven papers D/3196/6/26/1. 33 Feast of St John the Baptist and a quarter day. 34 A John Blunt was listed as a freeman of Askeaton c.1701. Transcript of the Court Book of Askeaton, Cottesmore Deeds, National Library of Wales.

September 1740

8 6 men and my own Reeping

9 3 men and my Own Reeping A ffarrill puting new cowples into ye house 2 cows bulld

10 I went to Courte and came back at night

11 2 men and my own a reeping

Augst ye 12th 1740 1 man and my Own at ye hay I J Haloran and I at ye Plow

13 my men Reeping wheat I sent Mr G ffosbery Parte of ye Oats I had from him last Spring 3l:12s:11d

14th I had D Heas 7 women a boy and my own and his binding and fi[nishd] stacking ye Oats and bound some of ye English barley in ye eavening to Courte

15th went to Tough from thence to Adear took my tythes from Bryen McMahon35 gave my note for 2l pd him 4 shill fees pd. for 3 mugs Sidr. 1:7 then back to Tough but did no good

16th to Tough and got 8:0:11 wch I carried to Mrs Widenham

17 I came home 18 my men Plowing and reeping

19 I had 7 men and my own Reeping Oats and duch barly I gave Law Quelly 7:7 in Parte of ye Hundrd of turf I bought from him and his partners

20th 7 men and my own a reeping duch barly and Oats I took my tythes from Paule Rahilly gave him my note for 1l

21 1 man and my own and finishd ye duch barly 3 women binding Oats

22 7 women and a boy binding binding Oats and duch barly

23 4 women binding duch barly and finishd

24 alone reading gave John Bryen 1:1 on acct of his wayges

25 I got Tim Calahan to Settle ye medows I drew in ye English barley to Courte at night

26 I went to Tough and got 6l wch wth hogans bill for wheat and Borks bill for canvis makes 10l odd wch I gave a receipt for so to Courte gave up my acct aboute 10 at night I was oblidged to go for Mrs Quin my Aunt taken so ill on wch acct I was oblidged to Stay at Courte to ye **1st of 7br** Paid on acct of ye horse 1:1:11 so home

7br ye 2d I drue Home Caparow Oats Jack went home Sick I had Tho Gibons boys and will Newmans boy tending ye horses

2d I began to Stubble follow T Gibons boys at ye hay

3: 4th Plowing my own and T Gibons boys at ye hay

5 I finished ye fallow ye mending ye ends of some of ye ridges and cutt some bullrushes

6 Stacking Oats

7 to Ahacolare36 spent 1s:7d and got 2 sheep from James finucan so home

8th I got an Atteach37 for Made had his distress taken so home and went to Courte

---

35 Possibly Henry M'Mahon, a commissioner in Limerick for taking affadavits. Ferrar. 36 Probably Aghakillare, Co. Limerick. 37 The levying of a writ to seize goods in order to redeem a debt.

9 I went to ye visitation[38] pd 3:4 sent 2:1 so to Courte
10 I came home earley and had a man and woman pulling beans and stacking barly
11 I had 29 men women and boys Pulling beans and stacking barley I bless god finishd tho late
12 att hay and breaking ye fallow
13 att Dito and Jon. Dondon 14 att home reading
15 I recd 1:9:3 from Patt Hogan for medows and 7s from Rich: Dillane and John Bryen pd on acct of ye horse 1:14:8½ pd for a Sive 6d ye boys cutting reed[39]
16 I got half a barrill barly from Mick mead ye boys at ye reed
17 I began to Sowe barly Sowed Meads ½ barrill and got a barrill from Mr fitzGerald I gave Dennis 1:1 on acct of his wayges 6d for Salte
18th I Sowd 7 bushills of barly and agreed for 2 bushills of wheat wth will boyle for 1s:6 pd 1:1 earnest
19 I Sow'd I had from fitzGerald 4 bushills barly
20th breaking land for wheat and trashing wheat so to Courte where I got wch I owe for pd
21 to Tough got 11:4½ from Shilbred wch I gave Will Boyle on acct of ye wheat back to Courte home in ye Eavening
22d I began to draw barley I had 2 men D welchs horse and boy T Gibons 2 boys 2 horses and boys from Courte
23 I began to draw but ye wettness of ye day hinderd me
24 I had ye above horses and men and will Newmans boy finishd ye barly and Oats and began ye beans
25 I had ye above men boys and horses finishd ye beans and began to draw hay
26 had ye above horses 6 men T Gibons 2 boys welchs 2 boys and finishd all my harvist so far as to enclose ye hay yard I got 1:8:9 from Den McMahon for his medows
27 I gave John Bryen 1:2:9 then to Courte were I stayd till ye 29th pd Mrs Widenham 5:3 I borrowd of her to pay ye visitation fees gave ned 2:2 to buy tabaco and Oyle
30th I sent for ye Oyle and tabaco my men wth Mr Bryen

8br ye 1st I began to plow for wheat I got some money for medows
ye 2d I finishd ye Corcas wheat sowd 2 barrills 6 bus barley and 3 bushills of wheat I got 4 payr of stockings pd 1:7½ for knitting them
3 I pd Gerge fosbery 1:15:5 in full for his Oats will Boyle 1:0 in full for wheat to Joan Daly on acct of her wayges 5:5 Sow'd some parsnip Seed pd to two of ye binders 1:3 so to Courte where I stayd seeing ye turf drawn till ye 11th and bought ye tyth of Kilmoran[40] turf pd 1:1 earnest and gave a token to get 5:5 more

---

[38] William Burscough, bishop of Limerick from 1725 to 1755, would have held an annual visitation which the diocesan clergy and churchwardens would have attended. Leslie, 'Limerick diocese'. [39] There was a reed bed on the banks of the Maigue, just north of Kilcorly in 1851. 6" Ordnance Survey map, County Limerick volume, sheet 11 (Dublin 1866).

November 1740

from John Nash pd D Halloran on acct of ye petatoes 1:1 to ye binders 1:3 to brandy 1:1 tabaco 5d so home and found ye hay and corn made up but ye beans
12 alone reading
13 I began wth ye turf bought a stack from M Hickey for 1s:1d from Gready 7s.
14 1 man and my own at ye turf 15 and 16 att Dito
17 att Dito Tho Gibons 2 boys I bought 40 kishes from L Quelly 40 from Gready and 20 from barrott I measured 120 Kishes to Ned Daniel I had 3 horses from Tough D Heas and D Welches horses and my own drawing ye turf and 4 men
18 att Dito 1 man Gibons 2 boys D welchs horse and my own
19 to Courte and back att night
20 I took ye Sheaf and tyths of D Nelans gardin and got from John Nash 13:11 from Mr John fosbery 3:0:0 I gave Tho Halloran 13:10 on acct of his wayges then to Courte
21 I went to Limerick pd J Bonfield[41] 1l:9s:10d in full and in full for a handkercheif and yard of linin 3:6 to N:E 1:7½ to E Mcgan 6½ to Coambs 7 paid for my wig 7:7 to me 2:6 then back to Courte
22 I came home 23 I sent oats to ye Kill 24 got it home ground
25 I pd Ned Daniel 4:6 in full for drawing ye turf pd ye binder 5:3 2 men and my own reeking ye beans 26 att home
27 I began to sow wheat in ye gardings pd W tobin 4s.
28 29 att Dito and finishd 2 men each day
30 I went to Courte
31 I had J Ryan Curriheen[41a] men and my own a diging ye Petatoes I bought from P Halloran I carried 7 peck to Courte sent ye same quantity home

9br ye 1st 1740 I went to Adear to see Mastr Quin[42] and gave 4d for a dram and remove so back to Courte
2d I got 7:7 from J Bryen and gave him for his Son 5:5d
3d I orderd a cow to be killd
4 I came home got my wheat winow'd John and 2 horses wth anthony ye boys diging petatoes
5th I got 17s:6d from D Welch wch wth 8:6 for two sheep I gott from him makes 1:6:3 on acct of greaseing heffers
6 I gott 1:3 from James Shilbred wch wth 11: wch he gave me makes up ye half years greasing his cows
7th ye boys finish'd diging ye Petatoes 1 man wth them John wth tim Bork I pd 2:2 for a pottle of Parsnip Seed
8 I sowed wheat in ye petato ground
9 att home reading and pd Mary Hannan in full 5:6 I gott 9: 2 from Tho: Harragan in full for ye greasing

40 Kilmoreen/Kilmoran is 1½ miles south-west of Kildimo.   41 John Bonfield, merchant of Limerick. Will proved in 1757. Vicars.   41a Curraheen, Co. Limerick, near Kildimo.   42 Probably Wyndham Quin, Valentine Quin's heir. Wyndham Quin (1717–89) married Frances Dawson of Dawson's Grove, Co. Monaghan, in 1748.

10 1 man and my own soweing Parsnip Seed
11 to Courte and got 10:10 from Mrs Hartstonge she owed me
12 I made 3 bushills wheat be sowed
13 I went to Limerick pd ye Quitt rent pd 8:1½ in parte of a gone to knife 4d to my expences 9 to Permont watr for Mrs Hartstonges 1:7½ ~~wch she owes me~~ So to Courte
14 I gave up my acct and my Aunt gave me to pay ye horses for drawing ye turf 10:6
15 I came home and found my fallow turn'd up I gave ye men a shilling to drink so back to Courte where I stayed to ye 30th
30 I came to Church from thence hom in ye way bought a cow from Connr Quin for 2:14:0 and brought her home I got 6 from Jon for ye gras of Cenedys cow

xbr ye 1st I got some oats putt in and had ye gate of caparow mended
2 3d I got ye oats trashd 4th got it winow'd
5 I went wth 4 barrill to ye Kill and stayd till ye 8 to see it dryed then sent it to mill and came home dung wett and in dred of an an overflow 9 got it home
10 killd my cow and sould ye hide for 9:5 but did not get ye money I Sowd more Parsnip Seed
11th I wrote to Mrs widenham for ye Price of ye wheat and she sent it to me 2:7:6 ye men puting out dung
12 diging for cabidge but could not set them for ye frost I gott a lossid from Anthony
xbr ye 13th 1740 I got a pig stye made and had ye rest of ye oats put in
14th I wrote to Colonell Taylor aboute my rent and w ye Answer he sent me a Cornits Commission[43]
15th I went to Courte bought some flax gave 3:3 and sent it home and a heffer I got from tim Bork for Corn
16 I pd 3:6 for a hay knife 10d for tabaco pd Connr Quin on acct of ye cow 2:8:0 17 I came home
18 I went to Tough then to A Ryans and got 4s:10½d Church rates so to Hally park[44] but got no money pd M Feigh[45] 3s I owed him and pd for a dram at Will Covards
19 I got 5:16:2 from will Taylor he ~~still owes me~~ 3:4 I walkd all day for Church rates but got none spent 7d
20th a continued Snow 21 at home reading
22 wrote to Jon fitzGerald but got no answer got 3s greasing
23 I finish ye tongsing flax and trashing ye stak of oats ye flax weighd 56 pound

---

[43] Peacock's commission would have been in the militia. where the cornet carried the colours. In the 1740s there were rumours of the intended invasion of England by France, and this would have led to militia arrays; in November 1745 Peacock went to Lord Southwell's militia array. [44] Holly Park. A house belonging to the Taylor family in the parish of Kilcornan. It was sold to the de Vere family. Lewis. [45] Michael Feigh gentleman of Aghakillare. Will proved, 1766.

1741

24th I went to Courte where I stayd to ye **3d of January 1740/1** in wch time I got 6s from Mead 3s and 2 hondrd of gadds in full from farill 6s in full from Mrs Hartstonge I sent some sticks home to make a plow soll and some beans from Antho Widow when I came home I got 5s from Edmond Naghtin on acct of greasing then made up my Rent

4th of Jan 1740/41 I went to Ballynorte[1] pd my Rent and aftr breakfast home

4th tho Sunday I sould will bork a bus of beans 4:6 a bus of Oats 2:8$^{1/2}$ Recd. 5:5 of it got 8s James Russill owd me and got my Frize home cottend 5 I got 2 shoes on ye bay mare

6 I bought 9 bandills of lining to make a sheet to sowe my Corn I pd 5:4$^{1/2}$ ~~and owe~~ 10$^{1/2}$

7th had my own P Daniels D Conways and Anthonys wifes horses a drawing petatoes I went to see Franklin Clampitt[2] who was cutt a thirdsday in a quarrill wth John ffitzGerald

8th I began to plow for beans and gave 3$^{1/2}$ barrills of oats for my petatoes

9 plowing for beans so finishd ye beans and began to plow for Oats sowed 3 pecks

11 went to Courte in ye morning and home at night

12 13 a soweing Oats

14 I was sent for to Courte to drive and Sease Jon. fitzGeralds stock got a letter of atturny[3] so to Tough and seased 52 cow 16 horses and 120 sheep wch stock Ned fitzGerald and James Bryen gave their note for their being forthcoming on demand So back to Courte

15 I came home calld att Ballylongford found franklin extream ill sent an acct of it to Courte pd 3 9$^{1/2}$ for hackling my flax I got 5:5 from Martin Quain on acct of greasing Tom Gott 9s 10d from E Naghtin wch he recd on acct of his wayges aftr nightfall I was sent for to Courte I went Mr Clampitt died

16 I was sent to Tough about ye Stock wth men to drive them away I left them on Edm Tho fitzGerald[4] and James Bryens note. So back to Courte and gave ye note to Mrs Widenham I spent 2s on ye men

19th I was sent for by Mr Bryen and gave him my note for 11l payable ye 20th of Febr so home

---

1 Ballynorte is 2 miles north-east of Askeaton. 2 Franklin Clampitt of Killknockan, Limerick. Will proved, 1740. Phillimore, vol. 3. 3 A letter of attorney is a legal document where a person appoints another to act on his behalf. 4 Thomas and David Fitzgerald leased Gortdrohid, called Glessans, for 25 years from the Kenmare Estate in 1751. Kenmare Estate papers PRONI D/4151/B/1/1/ f 52.

18 I went wth ye funerall beyand Adear and spent 5d
19 and 20 Sowing oats and finishd ye Corcas oats
21 I went to Courte
22 I went to Tough and by much ado we got 4:5:8$^{1}/_{2}$ from ye tenants so to Courte gave it to Mrs Widenham I gone
23 I got a bushill half a peck and a strike home pd 10s for them 10 for a pound tabaco
24th I came home ye men trashing Oats they gave Matt Halloran 2 bus beans Sould $^{1}/_{2}$ a peck for 1:1$^{1}/_{2}$ to will bork a peck I sould a peck for 2:2 I got my Pig killd
25th reading alone
26 Matt Hallora brought a cow to swap for Corn I gave $^{1}/_{2}$ a peck of Oats for toe ye men trashing oats
Janr ye 27th 1740/41 4 men and my own a trashing and seting beans sould 1 bushill gave Antho Widow a bushill Oats pd Peter doil 10$^{1}/_{2}$ in full for his bandle cloath gave an act of their wayges to T Halloran 5 bandills Joan Daly 5 bandills Mary Hannan 4 bandills of flanning
28 trashing beans I gave out 3 bushills
29 I had 4 men and my own trashing barly gave out a barrill and a peck I got 2 bed ticks for 2 bushills of beans
30th I had 6 men and my own a trashing barly I bought a cow from Harry Supple for 6 bus Oats 2 bus beans and 5:5 in money I bought 3 Chests for 11:6 earnest 1:1 I gave out 5 bushill a peck and a half of beans six bushills Oats
31st 4 men trashing barly and setting beans and parsnip

Febr ye 1st 1740/41 at home alone
2d I had 2 lambs yeaned pd Connr Quin in full for his Cow 6 2 pd Ned Bryen in full for his flax 4$^{1}/_{2}$ for three chests 11:6 I was sent for to Courte aboute Tough affairs
3d I went to Courte and stay'd till 4 then my aunt gave me 2:3:4 to pay for ye Oats she had from Mr ffosbery[5] so away calld at Clorane[6] gave ye mony to Mrs Russill[7] to give her brother so home and found my barly piled and 9 barrills winnowd 5 men att work I gave Jon Bryen $^{1}/_{2}$ a peck beans
4th I sent Jon Bryen to Limerick to try what he cold get for ye barly aboute 8 he came home wth an acct that I can 22s a barrill he brought some Iron two shovels Salte &c 4 men

---

5 William Fosbery of Clorane, Co. Limerick, was the eldest son of Francis Fosbery of Kilcooly, Co. Limerick, who was said to have emigrated to Ireland in 1690 and died in 1717. His third son William married Jane, daughter of Frank Evans, and they had three sons. One member of the Fosbery family married Alice Hartstonge, a niece of Henry Hartstonge (d. 1719) and is mentioned in his will. Burke (1958). 6 Clorane, a townland 1$^{1}/_{2}$ miles north of Adare, south of Kildimo. The Bury family owned land here. Survey of Kilgobbin by William Hunt, 1767, NLI Map 16 H 32. 7 Mrs Russell was Elizabeth Fosbery, daughter of Francis Fosbery of Clorane. She married Philip Russell, the younger son of Nathaniel, who married Mary Harris daughter of Sir Edward Harris, MP for Clonakilty and chief justice of Munster. Nathaniel was killed at the siege of Limerick in 1650. Philip became high sheriff of Limerick in 1744 and died 13 March 1762.

February 1741

5th 8 men att work I gave Joan 1:1 to Jon: Tom: and Den each 6$^{1/2}$

6 I had 8 horses drawing barley to Courte ad deliverd 12 barrills & Daniels horse at hire

7th 4 men trashing barly 8 at home alone

9 3 men and P Halloran trashing barly I gave Tom: Hall a bushill Oats

10 a wett day no men but my own Pileing barly John at ye forge a making a plow Iron

11 4 men trashing Oats I deliverd 8 barrills barly at Courte gave Tim Bork a barrill I got my Irish plow sull made by Den Madigan but did not pay him

12 my Frize cloaths makeing 6 men at work

13 I wrote to Limerick for more trimings Will Flahavan here 4 men at work

14 5 men diging for plants and Parsnips I sent 2 barrills Oats to Ballynorte

15 I went to Courte in ye morning back at night Tom gott 3:3 from Carly

16 I began to cutt my fallow but ye wettness of ye day made us stop I bought a pott for barley I bought 10 bandels of bandle cloath from Jon. Newman gave Mary Hannan 8 of them

17 I sent 4 barrills Oats to ye Kill got a drake from Court I gott 15 4$^{1/2}$ for 9 bushills from Jon Bleach

18 att ye Plow I gott in chloaths done gave ye taylor 2:8$^{1/2}$ and ~~owe~~ him 2:8$^{1/2}$ more I went to Tough but did no good

19 I went to Courte and got 11l from my Aunt in part of barly

20 I sent 15l to Simon Kent[8] and he sent me ye bill and a receipt

21 I bought a horse from Patt Daniel fo 9 bushills beans

22 I came home in ye Eavening ye boys sould 3 pecks beans Tom gott 3:3 in parte for his heffer Jon. Bryen 1:3 on acct of his wayges ye little cow calved

23d ye men trashing I sent 4 barrill oats to Colonell Taylor 9 bushills to John bleach

24th I bought a cow from David Ruddle for 2:9:0 for wh Mrs Widenham will credit him and I her sent Colonel Taylor 4 barrill Oats in full of ye half Schore I promised I gave Tho: Lynch 6 bushills of barly in parte of ye bar I had from Mr fitzGerald last 7br I gave ye Taylor 2:8$^{1/2}$ in full to Tho: Halloran 6:2$^{1/2}$ in parte of his heffer

25 I Danl and my own men trashing oats

26 Jon Newman and my own at Dito and finishd all ye Oats I gave Newman 5d I owed his brother Pattrick

27 my men trashing beans Will Bork died

28 putting in English barly and making a cabin for ye lambs gave Darby Longane a bushill beans pd Tom in full for his heffer 5:6$^{1/2}$ to Joan on acct of her wayges 6:0 to two chissills 1:10 recd for 2 bushills beans 9 shillings

---

[8] Simon Kent was a commissioner in Limerick for taking affidavits for the courts in London. Watson.

March ye 1st 1740/41 alone reading pd 1:9½ for making ye sull to Dennis on acct of his wayges a peck beans

March ye 2nd 1740/41 I began to plow at Caparow I pd Jon. Bryen on acct. of his wayges 2:2d I gave Matt Halloran 4 bushills of barly a bushill of beans wch more than pays for his cow by 5:8 wch he owes me

3d ye plow and harrow going I gave T Halln a peck barly

4th ye plow and harrow going I gave D Ruddle a bus beans pd 4:6

5 ye plow and harrow to Patt Daniel 4 bus oats 2 bushills beans pd by Smiths work

6 ye plow and harrow going I gave Tho Gibon on Jon Bryen acct 2 bushills ½ a peck of barly wch comes to 7s:10½

7th I had ye plow and harrow going and finishd E Daniels mare 3 days N Lowrys mare 4 days I gave Danel a bushill and half a peck of English barly I gave ye men a shilling to drink

8th att home P Halloran brought me a little pott for 2:2 I gave him 2 8½ to pay for it I sent to Courte to see them I gave Nick Lowry a bushill beans

9th I began to sett my petatoes I gave Brig Bork a peck beans Jon Bryen wth Patt Daniel

10 a Setting petatoes 1 man Jon wth P Daniel

11 a Sowing pease a peck of wch I got from Geo: ffosbery ye other from Jon Haly 2 hird horses and my own plowing I gave Jane Rourke on acct of her pot 1s

12th 2 men and my own and finishd ye petatoes began to fence ye garding

13 4 men 14 2 men and my own att Ditto

15 to Courte in ye morning back att night

16 to ye fare to collect Church rates[9] and spent more than I gott I pd Jon ffosbery 2:8½ on acct of ye gon

17 to Tough for Church rates collected 6s so back no pott having drank it yesterday[10] I bought a pound tabaco for 5½d

18 I sowd my English barley 1 man att work

19 1 man att work I went to Courte in ye eavening

20 I went to Limerick recd 20l from Ned Merowny[11] then I paid willm Richardsons[12] bill gave E McGan 6½ to tobacco 10 to my expences So to

---

9 Church rates were levied on householders within the parish, principally to support the poor, but also to repair the fabric of the church. Churchwardens had a duty under ecclesiastical jurisdiction to keep the building in repair. The parish also levied money to keep up roads within its bounds. 10 St Patrick's day. Every year, Nicholas Peacock gave his servants time off to go to Patrick's Well and money to spend on drink. St Patrick's Well/Patrickswell, is 2 miles northwest of Adare on the road to Limerick. The well stood about 400 yards to the west of the graveyard of St Patrick's church, in ruins by 1840; 200 yards east of the old church was St Patrick's Seat, six small stones laid on the ground almost in a circle. Ordnance Survey letters, f.544. 11 Edmund Morony was Valentine Quin's lawyer in the 1730s. He was the son of Piers and Catherine Morony. She married Thady Quin, Valentine's father. Morony to Quin, 24 May 1736; 19 November 1736. Dunraven papers D 3196/A/4/1, /2; *King's Inns Admissions*. 12 William Richardson, grocer, china-, glass- and earthen-ware merchant,

April 1741

Courte gave my aunt ye receipt and ye remainder of ye mony so home 4 men at work

21 I pd Rourke[13] 4s and ow him 3d 3 men att work and Geran[14] surveying

22d I went to dick fitzGeralds to look att a fatt cow from thence to Courte gave an acct of her and her price so home acompanyed wth Geran

23 he finishd ye Survey and finds ye land contains ye Spur included 26:2:20 att 12s an acre[15] I went for ye cow and paid 3:5:0 full for her carried her to Courte 2 men at work

24 I came home and found Jon ffosbery dead ye break and harrow going Den and Tom sick I hird Joan Coagh for 20s a year gave her 6½ earnest

25 I being so unwell could not go to ye funerall Mary Hann left me I gave Jon Bryen 8d in full of her wayges to Salte 5d

26 I began to sow duch barly but Johns illness hinderd me I only sowd 13 riges 1 man & Daniels horse at work

27 I gave Jon 1:1 to buy wine his illness hinders ye plow & E Daniels horse harrowin 1 man trashing to bridgt Bork ½ a peck beans

28 I finishd ye beans 1 man at them I gave B Bork ½ a pek D barly

29 to Adear to Church p 1s:8d for wine I owd Jon Mcquier and 5 to ye P I came home and dined and went to Courte where I stayd to ye 31st then I walked home from thence to ye vestory gave up ye Book tho not ye acct[16] I dined att Mr Quins then to Tough but got no money I sent 3½ barrill Oats to ye Killn

Aprill ye 1st 1741 I pd Darby Harragan in full for a chest 8s:10d pd Mary Coagh in full for picking duch barly 1:1½ 2 men and my own trashing beans

2d I sent ye Oats to ye Mill 2 men at work I sowd Carrotts onions shilotts in ye bed opposite ye window I gave Tho: gibons a busl beans in full for his boys hire last harvist

3 I went to Tough but no good went to Courte breakfaste there to home 1 man att work to Brig: Heas a peck of Engl Barly on acct of a quern

4th I had 2 men fencing and diging and Sowd Carrotts onions and leeks in ye plott att ye end of ye house I gave Tim Bork 2 bushills of beans I had 17 dow candills made

5th I gave P Daniel 9 bushills beans in full for ye ball horse Eliner Hines came to me

Aprill ye 6th 1741 att ye duch barley & Daniels horse

7th att Dito 1 man 1 horse went to Counsillr Mension[17] gott 30l so to Courte where I stayed till ye 9th in wch time ye men sowd ye hemp seed ye [illegible]

Main-Street, Limerick. Ferrer.   13 John Rourke was probably steward to Richard Thornhill of Rockfield, Co. Cork.   14 John and Maurice Gerin/Guerin were surveyors. John surveyed land at Coolegown, Co. Limerick in 1733. NLI Map 18 H 3.   15 Land was measured in acres and rods or poles. A pole was 5½ yards long, and an acre was 40 poles long by 4 poles broad.   16 As a collector of tithes for Kildimo parish, Peacock had to keep an account book which he rendered up to the parish vestry.   17 Probably Matthew Moynham, an attorney of St Peter's Cell, Limerick. Ferrar; *King's Inns Admission papers.*

Oats and a little english barly I bought a ram from Brig: Heas for 4s Jon gave her a bus of english barly ye men put out dung on ye petatoes

10th I gott my lambs Den: tom and D Nelans boy setting beans Jon att ye forge and brought home 3 new Irish spades

11 ye men making ye walls of ye hay yard I gave B Bork a peck of Duch barly to ned fitzGeralld 4 bushills duch barly at 5:5 a bushill

12 att home alone

13 two men and my own trenching petatoes

14th 1 man and finishd all my Seeds but ye flax Tho: sick I got my filly docked and bled my two horses

15 1 man and my own fencing caparow tom earthing ye plants

16 I finish at caparoe and let all my men and maids go to Jon Newmans wifes funerall I gave a peck barly to ye [illegible] bro man

17 and **18th** a thatching

19 I went to Courte and stayd to ye **22d** in wch time I markd 12 yearlings for Mrs W and so for Mrs H I borrwd 1:1 of her and gave it for thread and tabaco when I came home I gave Tho: gibon on Jon and toms acct a bushill barly my Sist. Mcgan carried a barrill of barly I got my 2 horses removed

23d John woke me to let me know he gott 2 cows in ye wheat I orderd them to pound[18] Den ye plow tackling and 2 horses wth Nick Lowry B Bork brought a pound of tread

24th 1 man & Daniels horse and my own a plowing ye Sally bed for flax seed

25 I sowd ye Sally bed 1 hird man

26 I went to Adear house to get ye warrant to collect ye church rates but Mr Quin refused so to Church thence home gave a peck barly for N Lowry

27th I went to Clorane got ye warrant signed by Mr ffosbery and his parte of ye church reates so home ye plow att work for ye flax but ye ground so hard we could not sowe it Ed Daniels horse at work I gave B Hease a peck of barly wch more than pay for a pott a quern and ram I bougt of her

28 Tom Den and I went to collect ye church reates and allmoste finishd my shear I came to Adear Hongry and dry spent 2:8d so home all most D

29 I sent y two yr olds 3 yearlings to Caparow ye man a drawing water to ye west park I gave Jon bryen 1s I owed him wch he pd for ye Ram

30th I sent John to Rathkeal[19] market but he did no good ye boys at ye flax ground I got 2 new spooles wch cost 8d or 6d

May ye 1st 1741 recd 1:5:5 for greasing pd Jon Bryen on acct of his wayges 5s:8d

2d I gott my sheep wash'd and drew ye water to ye west park Sould 2 bushills duch barly for 10:10 gave Joan Daly 2:8½ to Dennis 2:8½

---

18 Landholders were able to fine those who allowed their animals to trespass on their land. If payment was not forthcoming, the animals were put in the pound. 19 Rathkeale, a market town 14 miles south west of Limerick on the river Deel. It was formerly a stronghold of the earls of Desmond. Lewis.

May 1741

3d I went to Courte where I stayd to ye 6th made up all ye acct gave Mary Sammon 1:1 pd 1:1½ for tabaco allowd John Newman on Den's acct 1:1 recd 2:4½ from Newman got 2 new duch speades made so home

7 pd Mrs Quin 40s in full for tythes pd Mrs Fosbery 1l:3s:0d on acct of Church reates due to her husband

8th I gott my 2 foles and sow gelt pd 4:4 for them and 1:1 for shearing my sheep I had ye remainder of ye flax seed sowed ye men were parte of ye day at ye Bounds

9 I weand my lams and sent them to Caparow ye men went to pound wth Killcorly sheep I got 6:7 from Martin Quain in full for greasing

10th att home reading and walking

11 I had ye two year old brought home I bled and branded them and put them in ye Corcas and ye cows in ye little medow

12th ye men att ye trench I gott 3:3 from E dondon for trespas of sheep

13 my little mare foled I went to Courte from thence to Tough did no good back to Courte

14th to Tough got 6:7:6 wch I gave Mrs Widenhem I paid for a pound tabaco 10d

15 pd for a girth 1:1 sent Jon ffosber by Ned 5:5 in full for a gon and all accts

16 I came hom in ye morning ye boys at ye trench I gott from mrs ffosbery for my Aunt 6:11½

17th I wrote to Jon fitzGerald and sent his answer to Mrs Widenham

May ye 18th 1741 my big mare foled I went to Courte and agreed wth Mr Standis Gread[20] for his stone horse on my likeing him to a nogin of oyle 3d 19 I came home

20th I went to see ye horse but do not like him so home and spent 3:1

21st att home and a barrill of barly to mill

22d I recd a letter from Mr Hartstonge to go to tough I went but did no good I went to Courte stay'd till aft Dinner so home to N Lowry a peck duch barly

23d at home and lost ye swarm I took a Monday

24 to Courte

25 got ye sheep washd so home

26 to Courte to meet jon. fitzGerald he only sent 3:5:8½ aft dinner I came home gave B Halloran a bushill and half a peck of duch barly 27th at home

28th to Courte to get ye sheep shorn but cold not it being a holy day[21] I gave 1:1 for hives

29 I got ye sheep shorn and pd ye men

30 I sorted ye wolle

---

20 Standish Grady of Elton (?–1784) married Catherine Quin, Valentine Quin's daughter. He served on the grand jury for the Limerick summer assizes in 1744. He was an executor of Henry Hartstonge's will. Kenmare estate papers PRONI D/4151/B/1/1/f.24; Grand Jury Lists for Tipperary and Limerick 1776–1836 NLI MS 7331 f.10.   21 Feast of Corpus Christi.

June ye 1st 1741 I came home for a clain shirt back to Church thence to Courte pd Mrs Hartstong 1:1 I ow'd

2: 3d wth ye men at ye bog

4 to Limerick paid ye Quitt rent gave 2:8½ earnest for a vestory book 1:1 to my god Son and 1:1 expences

5 to Killgabon[22] aboute bullucks but did no good so back to Courte and aftr dinner I came home I lent Mrs Widenham 4:2 to pay ye tinker

6th I went to Ballynorte but did not ye Colonell so home in my way back I bought 2pr of buckles and a pair of studs for 1:1

7th I sent Jon Bryen to Limerick wth ye flaning and thread he got 2pr of Cards and some oyle from my Sistr he bought 8 hondrd of Plants 2 ounces of indigo 2 pound of Coperus Salte and weafers wch cost 5:5

8th I got a letter from colonell Taylor to Ryan

9 I went wth ye letter to Ryan and he shewed me where I should cutt turff I agreed wth James Shilbred to cutt me a hondrd and half of turf of 18:6 so home and gave Shilbred a peck barly and a peck beans as earnest **10th** I gott ye Plants sett

11 prepareing to draw my turf

12 I had 11 horses and 3 men att it

13 I had 8 horses and 2 men ye bay horse was taken wth a fever but being bled is pritty well I kill a Ram to put his skin on ye horse I made and end about four I had 19 horses viz: my own 42 days D Nelans 2 days T Calahan 2 days M Hallorans 2 days sulivans Burks and Donevans each one day

14 I went to Courte where I stayd to ye **27th** in wch I gave John Bryen 6½ to a pound tabaco 10 to timber for my barn 1:5:6 to ye men 1:1 to me for a dram 4 I went to Tough and took up 40 weathers for 20l four cows for 7:13:3 and 40s for my own use all wch I gave in acct to Mrs Widenham so home and found my men drawing stones Jon Bryen att Limerick wth butter wch he sould for 1:0:3 and paid 5s 3d for dyeing and pressing my flaning 1:1 for Salte 4d costom &c

28[23] att home alone and pd Joan Coagh 4:5½ in full of her wayges and sent her away and hired Mary Haragan in her pleace

29 I sent Jon. Bryen to ye fare to buy truckles he got none so home and Patt Boyle 1:1 on acct of his wayges

30th I began my barn and had Jon Heas and Tim Calahan at it att 5d a perch and theyr diner I putt undr ye southeast corner stone wch was ye first stone laid 6½ Den and Patt tending them ye rest and John Daniel wth Mr Bryen Reeping Reap

July ye 1st 1741 my men wth Mr Bryen I gave ye Masons each a bushill of barly

2d I reepd ten sheaves of barly

---

22 Kilgabon/Kilgobbin is 2 miles north east of Adare, where the Quins had a dower house. 23 Quarter day. Feast of St John the Baptist.

July 1741

3 I sent Jon to ye fare and he bought me 2 truckles for 11:1d tom and Patt wth tim Calahan

4th I had 10 men 5 women and Reepd my field of barly I acoumpted wth Dan Shea and owe him 2s and ye price of a new pair of bootes pd 11¹/₂d Countery Charges

5th I went to Killgabon pd Mr fitzGerald 20s due to Jon Brown of ye church rate so to Church dined at Mr Quins so home

6th my men Reeping ye women binding

7th ye men at Dito and cutt all that was fitt and headed ye Stookes Tim Calahan rough hewing ye roof of ye barn he gave me a stick

8 I went to Courte pd 10d for tabaco 3d for sand

9 I made up ye acct between me and Mrs Widenham and Recd 3:15:2 from her in full of all acct between us so home in ye eavening

10 I took my tythes for 7 pound and procters[24] fees I pd 1l:1s:2¹/₂d for Iron

11 I gott my lambs shorn and send Mick Hickey 7:6 in full for last years tyths turf gave ye boys and ye mades 1:7¹/₂ to drink at ye Pattern[25]

12th pd Paul Rahily 1l:0:0 in full for last years tythes 1:5:2 for him and Vall fitzGerald in full for procters fees for this year att home alone

13 my men and J Daniel trashing Reap

14 I had 3 women 1 boy and my own stacking barly

15 I had ye last of my barly reepd

16 2 of my men wth Mr Bryen ye walls of my barn made

17 I sent my truckles to Courte to be shod

18 2 of ye men wth will Boyle ye other 2 att ye forge I gave Jon. Bryen 2:8¹/₂ on acct of his wayges to Patt Daniel to buy Collers 1:1

19 att Courte Mrs Widenham gave Tim Lowry 10:10 wch wth 4:10¹/₂ she gave him before makes 15:8¹/₂ remains due to him of last years hire 7:3¹/₂

20 I went to Tough but got no money back to Courte aboute 12 I came home and found my men mowing

21 got my truckles home and 2 barrill of lime from Courte

22 ye men mowing and staking ye last of ye bear barly

23 I got James Flahive to bleed me I began to reep wheat

24th 2 men and my own and 1 woman reeping wheat and finishd ye gardings tho unwell I rid to meet my sister but she stayd att Courte

25 my sistr sent to see me I sent my sheep and lambs to Caparow in all 46 I sent for more dye stuff wch cost 2:8 I rid to meet my Sistr but missd her so hom

26 being unwell all night I lay abed till 12 and continued unwell

27 still uneasey 4 and my own reeping corcas wheat and finished I had 2 women binding 2 pulling flax

28 ye men att ye hay

---

24 A tithe proctor was responsible to the clergy for assessing crops and collecting tithes.   25 A pattern was a meeting at a holy place or holy well to pray and take part in worship.

29 in ye morning reeping duch barly at ye hay ye rest
30 I had 7 men and my own and reepd all ye English barly I had 5 women and my own binding and pulling flax
31 I began to reep Oats none but my own men and women 3 women more at ye flax Mr Bryen sent for ye weathers he bought from me last spring

August ye 1st 1741 I had 5 women and my own turning flax and pulling hemp I sowd some cabidge seed
2 to Courte in ye morning home at night
3 I had 21 men 10 women and 5 boys my own included and we reepd bound and stook Caparow ye well corcas and parte of ye other
4 I had 4 men 3 women 1 boy and my own at ye oats
5 I had 3 men 2 women 1 boy and my own at ye Oats
6 2 men 2 women 1 boy and finishd ye Oats
7 to Tough no good thence to Courte where I Stayd to ye **9th** so home my men claining ye hay yard
10 3 men 2 women and my own heading Stookes binding flax and pulling pease
11 I had 1 woman my own men and women stacking Oats at Caparow and wheat at home
August ye 12th 1741 ye men wth Harding I sent half a barrill of barly to mill
13 I sent Jon Bryen to Limerick aboute my diapour not wove my men and horses and 2 boys drawing wheat I pd 2 of ye binders 1s Jons expence 6d
14 1 man 3 boys and my own drawing barly
15 a holy day at home alone[26] 16 to church home at 3
17 1 man 3 boys and my own drawing barly and finishd
18 I had 2 men 3 boys 4 women and my own puling beans
19 I had 1 man 4 boys 4 women and my own finishd ye beans and began to Stack oats
20 I finishd ye Stacking ye Oats and began to Stubble follow ye Corcas
21 ye plow going I stackd ye duch barly
22 to Courte where I stayd to ye **27th** wch day I gott 8 ginies from Mr fitzGerald wch I gave to Mrs Widenham when I came home I found ye medows finished ye follow turnd up and ye flax pulld at wch they had 4 women
28 I had 2 boys my own men horses and boys drawing Caparow Oats drew it all home and a stack of ye other
29 1 men 3 boys and my own drawing oats
30 at home alone lent my truckles to Courte
31 3 boys 1 woman and my own puling hemp and makeing hay

7br ye 1st 1741 I had 2 men and my own at ye hay 4 boys and my own burning Stubble 1 woman and my own pulling of hemp

26 Feast of the Blessed Virgin Mary; Peacock would have been alone because his servants would have gone to attend mass.

September 1741

2d ye man trashing wheat and finishd pulling hemp

3 I had 4 boys my own men horses and boy and drew ye English barly and a reek of Oats

4 ye above men boys and horses finishd ye Oats and began ye beans I got 12:5 from Tho: Harrogan ~~wch~~ I gave to Tom Halloran 11:4$^{1}/_{2}$ 7s:9d of wch is ye rest of his wayges he left me to be married

5th I had ye above men boys and horses and 2 men from Mr ffosbery finishd ye beans and most of ye duch barly

6 at home alone and missed a sheep

7 I had 4 boys my own men and horses finishd my Corn mad up my pease sent 4 bus barly to Mill

8 I went to Courte **9th** at Courte and got 1:10:0 that Ned fitzGeralld owed me for barly

10 sent to town for my diapour and got 40 yards ye weaving is 7$^{1}/_{2}$ a yard I sent 1:2:0 to pay for it I agreed wth Conner Harragan att 10:6 a quartr or 40s if he stays ye year

11 att Courte pd 6d for garters

12 att Courte gave Jon Bryen 5:5 to pay ye binders I pd 1:1 for tabaco I went to Tough but no good

13 att Courte

14 I went to Tough and drove ye whole land and geathered 62 cow 30 young Cattle and 138 sheep all wch were released on his promise of paying ye money or giving up ye distress in 8 days[27]

15 I came home and found ye men att ye Corcas land where ye tyde overflowd this morning 3 boys burning Stubble

16 ye plow att work 2 boys burning Stubble Rob Peacock lay wth me but did not pay him ye hearth money[28] **17** he went away and carried ye 1s 2d part of Gusmand wth him I cut up my diaper into 8 table chloaths and 5 towels

18 I finishd ye 2d plowing and trashd some wheat gave Mr Bryen a bushill I got 1:2:0 from Gerge ffosber wth directions to Mr Bryen to allow me 1:17:3 being in full for 3 acres of medow he had from me

19 I sent Jon wth ye giney to Mr Bryen I gave his man 5 bushills of wheat ye harrow att work I had a sheep killd **20** alone reading

---

[27] Peacock has been attempting to collect rent from John fitzGerald at Tough for some time: see entries for 20 July, 7 August and 12 September 1740. His remedy was to impound the stock of the tenant and if not satisfied within a specified time, to take legal action including the sale of the beasts in order to recoup some of the debt. [28] Hearth money collectors both assessed and collected this tax, with the aid of parish constables. Collectors had 'walks' allocated to them, based on several baronies. Each year, a collector would send to Dublin a note based on each parish of the total numbers of hearths and houses for the previous year, and the numbers of new hearths liable to be charged, exempt houses and arrears. The salary of a collector was £40 a year and as state employees they had to be protestants. D. Dickson, C. Ó Gráda, and S. Daultrey, 'Hearth tax, household size and Irish population change 1672–1821', *Proceedings of the Royal Irish Academy*, 82 C 6, pp 132, 136–7.

7br ye 21st 1741 ye men trashing wheat I gave Mr bryens man 2 I got my old frizecoat turnd into a weastcoat and briches for David I had 60 bandills of blanketing from ye weavers at 3/4 a bandill come to 3:9 wch I am to allow him for in medows

22d I began to sow wheat and sowd 3½ bus E Daniels mare harrowing

24 I sowd a bushill and finishd and sowd 2 bus of barly

25 sowd 4 bushills of barly

26 sowd 1 bushill of barly sowd in all 7 bushill of barley and 7 bushills and a peck of wheat to Courte

27 I gott 4 crowns from Mrs Widenham wch I gave ye mowers home in ye eavening

28 cutting reed accted wth will Hickey recd 1:5½ in full

29 I went to Courte and Mr Hartstonge accted wth Mr Jon. fitzGerald and allowd him 40s I gott ye 15 of June last and a bushill of wheat and 2 barrill of barly he gave in this year so home and Jon gave me 40s he got from W Covard and Den Bryen

30th recd 1s:5 from E Daniel he ows me still 1s:7d I got 18:1 from Den Mcmahon I paid Mr Bryen 3:10:8

8br ye 1st 1741 I gott 6:10½ from Mcmahon I had 3 men 3 boys and my own men and horses drawing hay I pd Jon burns widow 1:1 in full

2d I had 1 man 3 boys att ye hay and finishd I sett them all all to trash ye flax in ye eavening to Courte were I stayed to ye 5 so home and borrwd 16:3 from E hines I gave Patt Boyle 10:10 Dennis 8½ Mary 3:9½ I finishd ye making my 8 tablecloaths and 5 towells

6 7 and 8 drawing turf at tough I had 27 horses 17 men my own included 8 hired boys to tend ye horses I pd all ye horses pd Shilbred for ye turf and pd for ye milk gave my men 1:1 in Rum so home

9th making a Reek of wheat Mary went away ye horse rider took my Mare in hand I borrowd 1 2 9 from D bs

10 I finishd ye wheat and made up ye gaps

11 att home gott 12:6½ from Tho Haragan gott a message to go to Ballylongford in ye morning

12th I waited on Mr Bryen and pd 1:8:10 in full for May rent gott a Minicett[29] signed and sealed aftr dinner drank shear of 3 bottles of wine gott 1:7 from Ned Daniel in full so home

13 I went to Courte where I stayed to ye **17th** in wch time I clear'd off ye mowers and Tim Lowry to May last I gott 15:2 for 3 bushill of wheat pd 4d to Gerge ffosbery in full for pease for tucking ye blanketts 2:8½ for irons 1:7½ to tabaco 1:1 I gave Jon 11:11d to by board for ye barn doar Jon gave a wether to Mr Bryen for 10:10

---

29 A minute.

17th when I came home I found E Hines ill of a fever wch ocationd my staying away to ye **20 of 9br 1741** in wch time I gott my barn Roofed beam fill new doar and window shutt and half ye barn thachd my frize spun and ye women pd all ye binders boys paid tim Calahan gott a barrill of Oats Mr Bryen got 2 bus a peck and half of wheat a barril of barly I sould Mrs Widenham 10 barrills 5 of wch is deliverd Jon sould David Welch 2 barrills of English barly and a barrill of Oats for 2:19:0 pd by ye giney I borrwd ye 9 of 8br a moidr and french crown he gave John bryen I sould Doctr Martin a barrill of wheat and 2 bushills of pease att 5:5 a busill wch come to 2:14:2 of wch I red 1:2:9 wth wch I pd Mick Cosgary ye giney I owed him wth ye money I got from Welch I pd Mr Jon Bonfield in full 1:4:0 pd for 3 Handkercheifs 2:6 Quitt 7:8½ a barrill 3:3 earnest for a horse I bought from Brother Mcgan for 4l:2:8½ expences 8d tabaco 1:1 to Jon on acct of his wayges 5:5 Den 1:7½ Patt Boyle 1:1 to ye horse ridr 11s:0d more to him 1s:1d to Patt Boyle 1:1

xbr ye 1st 1741 I began to fallow for duch barly I gott a letter from Will Flahavan aboute an affair of Courteship I gave ye horse ridr 1:1 and sent him a hunting wth my mare I took Phissick
2nd ye plow going Ned and Conr trashing to Patt Slatter 1:1 on acct of his hire I gott a letter from Ballynorte for my Rent
3 I sent 3 barrills of barly to Courte and wrote for mony but gott none I sould a cock of hay to John Mcquier and gott 2:8½ earnest I gave it and 6½ in parte for a pott hangers and pair of scheals I bought for 5:5 ~~due 2s:2d~~ I sould ye hay for thre pounds
4th ye Plow att work I put in ye Pease and trash'd them I sett in my frize to be warped and made them pull 120 bandills in ye peice and gave a stone of fleece woll and a stone of lambs woll to be made into flaning
5th I gott my pease winowd and have aboute 10 bushills I sent my frize to ye weavers finishd ye turnings of my fallow gott a killn made to dry my hemp and got my mare removed
6 att home Reading
7th I finishd ye winowing in barly got ye garden beans trashed began to fallow ye Orchard Sould a busl of barly for 2:6d 2s:2d of I gave John Bryen to pay for ye Schales and pott hangers in full Jon Heas made a 28 a 4 a 7 a 4 pound weight 3 women and my own at spinning ye flanning
8th 9 10 11th att Courte
12th I went to Adear to See Mrs Quin from thence to lough and gott 10l:6s:0 so to Courte gave it to my Aunt and She paid me 7:14:6 on acct of my barly She owes me 15s:6 Still Mastr Hartstonge ill
13 I came to Church and from Killdeemo green sent John Bryen to Will Taylor for ye money he owes me and gave him ye 7:14:6 I gott yesterday to carry to Ballynorte at night he came home and brought me a receipt for 9:2:9 and 4 ginies parte of Will Taylors money he owes still 1s3d

14th I got up before day and gott my cow Killd and sent Jon wth ye hide and gave him 2 ginies to give my Sistr on acct of ye horse and a giney to buy things in town I had Petter Miller and an cottier a dressing my hemp Jon did not come home till aftr we went to bed he Sould my cow to Jon Mcquier for 3:10:0 and he recd 1:0:0½

15th as Soone as I was up Jon gave me Mcquirs money and an acct that he pd 1:9 for salte 5s for a lock for ye cheste 1:4 for a padlock for 29½lb of Irish Iron 5:2½ Costom and turnpike 2 his expences 6½ he brought me a receipt from my Sistr For ye 2 ginies he Sould ye hide for 5:3 then I gave Elinor Hines that I borrowd of her in 8br last 17s:10½d I gave Patt Hure 2:8½ in full Jon Mcquier sent for his cow and I gave her to his man ye hemp going on Jon began to Plow Portachaca corcas for Oats at night I gott my Cow salted and ballance my years acct

16th ye hemp and Plow going I got my frize home from ye weavers I had 116 bandills att a ½ a bandill ye weaveing comes to 4s:8d

17th ye Plow going on at night they finishd ye hemp it weighs from ye Swingle 1c:0qr:14lb ye breaking swingling and drying costs 7:3: Mr Bryen return'd seven bushills of ye barly he had from me last month I am to charge but a bushill

18th I sent 2 barrill of barley to Courte in full of ye half Schore and ½ a barrill for my own use I pd 5:5 for dressing ye hemp and sett ye Pallitine[30] to hackle att 1:1 a Stone I pd 1:2 in full for Spining my flanning it weighs 3lb4 as it is

xbr ye 19th 1741 ye hemp a hackling ye men trashing I got some of my Frize tucked

the 20th to Courte they gave Tim Calahan ½ a barrill of oats

21 I orderd a cow to be killd and came home aftr diner Patt Daniel had a barrill of Oats ye men at making dunghills and winnowing Oats I gave Ned 2s:8½d to buy me some oyle lam black tabaco nails and peaper I accted wth Dennis and gave him 3s:3d in full for a year and quarters wayges and over pd him 1½d

22d I Recd 2:16:0 in full for 4 barrill oats I gave Antho ffarrills widow 2 bushills Oats in full for ye beans I had from her xbr last I gave Brigett Halloran a bush Oats and sent Mick Cosgary a bushill wch he has not pd for after night fall I got a quart of Oyle 10d a pound tabaco half a quair of peaper and a barril of lam black I lent David Welch 4s:4d ye Plow going ye hackler at work

23d I gave John Bryen on acct of his wayges 6:0½ to Jean Daly on acct of her

---

30 In 1709, 821 Palatine families (3,073 people) came to Dublin and were sent to settle on estates all over Ireland. Each man, woman and child received 8 acres of land at leases of 3 lives and 5s. an acre and they received 40s. a year to buy stock and farm implements. The first plantation was a failure, and by 1712 more than half of them had left the country. But 1,200 remained and their settlement was concentrated on the Southwell estate around Rathkeale, Courtmatrix, Killiheen and Ballingrane, with small groups in Adare, Pallaskenry, Askeaton, Kilfinane, Ballyorgan and Castleisland. Their rent was paid by the Irish government; they were given a musket (a 'Queen Anne') and 40s. a year for seven years. Sir Thomas Southwell gave them materials to build houses. They cultivated flax, hemp, oats and potatoes. Hayes (1937).

December 1741

wayges 2:8½ to Patt Boyle in full of his half years wayges 6s ye Plow going and ye hackle

24th I acct ed wth Jon Heas and setteld Tim Calahans acct I had ye Plow tackled but ye frost hinderd them then they fell to trashing I gave Peter Miller 4:10½ on acct of ye hackling so to Courte where I stayd to ye **28th** and expended on Tho: Jackson[31] and Nell glin &c: 4:10½ when I came hom Jon gave me 14s he got for a barrill of Oats he gave B Halloran a bushl R Welch a peck Peter miller a bushill Oats and a pound of ye hemp to spin

29th I got up early walkd to Ballynorte where I was allowd 9l for ye Oats I deliverd last Spring and 9:2:9 I pd ye 13th and I pd 1l:17:3d in full for my years Rent wch became due ye 1st of 9br last and got a receipt after breakfast I came to Ballyengland[32] and pd 7:0½ for a half hide of Leather so home and pd M feigh 6d for two drams a terrible hard frost ye hackler att work

30th I sent ye same of my barleys to Ballynorte a sucking Pig and 4 turkeys ye boys pounding ye hackler att work I sent 60 bandills of flaning to ye weaver I gave Briget Halloran 6lb of toe to spin and 8lb of toe to Mary Holehan and ½ peck Oats

31st I sent to Courte to borrow ye Reel and gott it I accted wth Dan: Shea and owe him on all accts but 1s7d I got Tim Calahan to put new ends in my bed I pd Patt Slattery 1:1 in full of all accts I sent half a bush oats to ye mill and got it home ye hackel and pounder at work so ended ye year ye Allmighty God give me greace to mend my life

---

31 Thomas Jackson, chandler, of Parade, Limerick. Ferrar. 32 Ballyengland, 1½ miles north-east of Askeaton. The castle was the seat of the Hewson family and was renamed Castle Hewson. The 2-storey house has an old tower-house at one end.

1742

Janr. ye 1st 1741/2 God gave us ye blessing of a thaw wch tho: wellcome was like to overflow for ye banks run over in severall pleaces but God be praised I have never a gap broake I had my Petatoes try'd and Severall sufferd by ye frost ye Richball died I removed my bed and had my Room clain'd out

2d I had tim Calahan a mending ye white chest but did not finish ye men trashing Oats no hackler I gave John Bryen 16:9½ and on accting wth him for two years wayges ending ye 19th of 9br last he owes me 3s:1½d

3d I sent by John to Rathkeal for a hondrd of 6d nails I gave brigett Halloran 1s:1d to pay for ye flyers of her whele I spent ye day a reading Coursey[1] sent for his sheep and got them

4th ye hackler at work ye men trashing Tim Calahan att ye Chestes I gave out a pound of hemp to each of my maids a pound to Millors wife a pount to Sarah Hodge a pound to be spun to two dowsin yarn

5th I gave Watt Tobins daughter 30lb of hemp Brig Hall one pound I gave Brigt Welch a bushill Oats in pledge of wch she left me a Ring that she says is gold ye men trashing aboutt 11 att night ye hackler finishd I have out of my 9 Stone of rough hemp 9lb fitt for 4 dowsin yarn 4lb 4 fitt for 2 dowsin yarn 4lb 4 of tow and 2½lb of ye pullings of ye fine hemp and 5lb of fine towe ye Hackling come to 7s:4d wch I pd in money and oats att 11d a Stone Tim Calahan att ye Chests

Janr ye 6th 1741/2 I got ye 4:4 I lent david welch 22d of last month I Recd a letter from Will Flahavan aboute ye barly I acct wth Peter Millor for ye Hackling and he owes me 4:6½ besides a bushill of Oats his wife had I gott my fryse from ye Cottners 51 bandels and pd a ½ a bandill for Cottening it wch come to 2s:1½d I gott 1:1:0 from Connr Harragan for a barrill and half Oats I measurrd 16 bandills of ye frise to John ye 10 I give him and charge ye six to his acct

7th I had Jon Newman and my own men trashing oats Harry Evans making Stirrop leathers and bridells he cut out 4 pair of Rains 4 pair of Stirrop leathers a gyrth and I have enough to make another gyrth and Saddle straps I gave Joan Lowry 3d on acct of her wayges Mr Bryen sent for a fatt sheep and I orderd it for him

8th ye man trashing Oats Harry Evans went away I gave him 9 bandills of frize

---

1 Coursey Ireland was married to Nicholas Peacock's sister Anne. He entered TCD 9 July 1724, aged 18, the son of Courcy Ireland, gentleman of Kilkenny. He died in February 1743. *Alumni Dubliniensis*.

9th I went to Ballylongford to meet Jon Mcquier but his not coming hinderd me from clering 9br rent after breakfast I came home Sent 5 barrills Oats to mill gott a letter from Ballynorte for ye barly wch I answerd then I walkd to Courte 10 to Church then back to Courte Jon sent me 1:14:1$^{1}/_{2}$

11th to Limerick acctd wth ye widow Corbitt gave my Sistr 5l:2s:0d on acct of ye horse 11:4$^{1}/_{2}$ to Hen Evans and laid out 1:4:9 on lock powdr shot Cambrick &c so home I owe my Sistr She lent me 1:4$^{1}/_{2}$

12th to Tough and gott 6:13:0d pence from mr fitzGerald I went to see my corn ground and gave Hon 1:1 so home and gave Mrs widenham my acct wth ye money I recd in ye way to Courte I bought Old Robins gun for 15

13th att Courte and gott 5l from Mrs Widenham and pd Weights[2] Procter in full for her tythes

14th I came home and sent a new spead home and pd 3d for a dram and found my corn att home ground two men and 2 horses a drawing wheat att Ballylongford Brigett Halloran brought in her thread and I gave her a pount of hemp

15th I had Harry Evans att work att ye bridells I got my male sifted ye men trashing B Halloran brought hom ye toe thread and I gave her 4lb of toe

16th Harry finishd 4 bridells 2 girths 4 pair of stirrops and 12 saddle straps I got him to make up 3 collors of ye old rains and stirup leathers Sarah Hodge brought home her pd of threat I gave her a pound of hemp and a peck of Oats Tim Calahan at work ye men trashing

17th I spent ye day a reading wth Harry Evans I gott a letter from Ballynorte to get my barly ready as soon as possable ye duch att 2s1 a barrill and ye English at 1l:3s:0 a barrill

18th ye men finishd trashing ye Reek of Oatts aboute 12 Henry Evans went away I mead them fall about winowing ye Oats Tim Calahan att ye Quern I pd 4d$^{1}/_{2}$ for a pd of butter and sent it and sope to ye mill to gett my frize tucked Will Wall came to help me to trash

19th I had my Oats winowd and a little duch barly trash it was a heading ye English barly Tim Calahan finishd ye Quern I accted wth him and owe him 2s:4d towards half a barrill of Oats I gave him

20th I had John Newman and Will Wall and my Own men a trashing E Barly I got 66 bandills of flanning from ye weaver wch I am to allow him a $^{1}/_{2}$ a bandill wch comes to 2s:9d

21 I finishd ye trashing ye E Barly wth Willm Wall and my own men B Halloran brought her pound of thred I gave her a pound of hemp Walter Tobin daugter brought her 3 pound of thred I gave 3 pound of hemp and a bushill of oats

22d I gott my pig Killd ye best part of ye barly winowd and began to sow Oats

---

2 Archdeacon Richard Wight, second son of Edward Wight, alderman of Limerick. He entered Trinity College, Dublin in 1700 and took BA in 1704 and MA in 1714. Prebendary of Ballycahane and Killeedy, rector of Rathronan 1736–45, archdeacon of Limerick 1740–51 and prebendary of St Munchins, 1745–62. *Alumni Dubliniensis*; Canon Leslie, 'Diocese of Limerick'.

February 1742

in Portachaca Corcas Willm Wall left me, Tim Calahan made a rale for ye taster of my bed and putt 4 locks to ye 4 chests

Jan ye 23d 1741/2 I had my E Barley Piled and 2d winowed and have 7 barrills I got ye Duch barly put in and sowd aboute 5 bushills Oats I got an Iris linin wheel from Maddin and 6 pigins wch I owe for I had a man come from Tough for a Sucking Pig that Mrs fitz was in labour and longed for it I was obliged to give my young sow I gave Brigett Halloran a bushill Oats on rect of spining I gott my pig Salted ye 2 fliches and hams weighd 1lb:1qr:13lb ye chine weighd 2lb8 and ye head 20 in all

24 alone reading Patt Halloran brought a schane of 3 dows of thread ye hemp wod not spin finer my sow was brimed to a pottle Salt 1½d

25th my own men trashing D Barly ye harrow att work

26th ye men trashing ye harrow att work and finishd all that was plowd I sowd aboute 9 barrills I had a letter from Colonell Taylor for Oats I gave Darby Harragan half a barrill Oats on Connrs acct

27th I distrubitted ye flaning amoungs ye Servants some I bistowd ye rest I charged to their accts ye men trashd all ye barly that was wth in and winnowd some of it Tob: Cox called to see me Brigett Heas brought in her tow thred She had spining and wants 3 pounds of ye weight

28th in ye morning I gott ye remaindr of ye barly winowd and Piled but on acct of ye days turning to a violent Storm of both wind and rain had much ado to save it from being wett I gave David Wellch 2 barrills of English barly and a barrill of Oats in full for ye money he gave me last Spring Brigett Halloran brought her pound of thread I gave her a pound of hemp

29th I gott my barly winowd and riddelld ye men att it I began to make up my Cortins I gave maddin ye Coop half a barrill of Oats and allowd him in part 2s:2d for a wheel and 1:6 for ye six piggins

30th I sent Jon to Limerick wth half a barrill of oats and sent a turkey to my sistr and another to Doctr Martin aboute nightfall he came home and gave me an acct how he laid out ye 6s he got for ye Oats to ½ a Stone sope 2:9d to Salte 1s:9d to buckram 6½ to moulds 1d and to thred 2d turnpike custom and measuring 2d he gave me chance 2½d my sistr sent me a bottle of Shrub ye men puting on dung I was employd aboute my Cortains Harry Evans dined wth me and tould me my mare had a very greate Cough Jon brought a crock wch cost 3d.

31 alone reading pd 1½d for caperus

Febry ye 1st 1741/2 I began to trash my flax seed and bought a knife and fork for 6½ pd 9d for 2 knives for ye men 2½ for needills about 12 Colonell Taylor sent for barly and I deliverd William Celeger[3] 7 barrill of duch barly and one of English barly and sent them a way then I gave Rob: ffaril half a barrill Oats on

---

[3] Possibly William St Leger, captain of dragoons and son of Sir John St Leger of Doneraile and Louisa Pennefather. He died in 1748. Burke (1860) Doneraile.

acct of ye gon I bought from him ye 12th of last month then I gave Peter Millors wife 3 peck Oats for wch she gave me 2s:0d I gave mary Holehan a peck Oats for wch she is to give me work or money She brought her toe thred home

ye **2d** I deliverd Willm Celegar 4½ barrills duch barly and 4½ barrills of English barly I deliverd in all 11½ barrills of duch barly att 2s1 a barrill 12l:1s:6d and 5 and ½ barrills of English att 2s3 a barrill 6l:6s:6d he owe me in all this day 18l:8:0d about 12 I went to Courte

3d att Courte

4th Recd a letter from Colonell Taylor for Oats after diner I came home and brought my mare and found B Hall: and ye mades att ye flax seed ye men diging for Cabidge plants

5th ye men trashing bear barly B Hallorn. And ye maids att ye flax seed Watt Tobins daughter brought 3 pounds of thread I gave her 3 pound of hemp I wrote to George ffosbery for a certificate for ye flax seed and got it wch being a new thing I to know how to wright one on Ocation[:]

> County Limerick
> Jon Bryen came this day before me and made Oath that flax seed by him intended to be deliverd att ye Trustees Store in Limerick[4] a mounting by Estimate to twelve bushills is all of ye groath of this present year and no parte there of Purchasd from any Other person Sworn before me this 5th day of febr. 1741 forty one

febr ye 6th 1741/2 I sent Jon and Connr wth ye flax seed to Town and gave him 2s:8½d I got some Oats in and gave B Halloran a bushill oats att night Jon came home and tould me he was oblidged to borrow 11:4½d to buy bags for wch he pd 6d kepd 6s to measureing 1d for a printed receipt turnpike 1½d tabaco 10d his expences 6d ye Receipts I insert wch are as follows

> No 83
> Recd from Jon Bryan three sacks containing three barrills of sound, dry flax seed for ye use by order of ye Right Hon and Hon ye Linin Board as wittness my hand in Limerick this 6th day of febr 1741 Rob Turner

> No 83
> Recd ye above 3 barrills of flax seed and 3 sacks into ye stores in Limerick for ye use of ye Right Hon and Hon ye trustees of ye linnen manifacture as wittness my hand this 6th day of febr 1741
> Tho Vincent

> No 83
> Jon Bryen came before me this day and made Oath on ye holy Evangelists that ye flax seed deliverd at ye Right Hon and Hon ye

---

4 The Linen Board made grants for spinning wheels and reels, and gentry distributed spinning wheels to poor women. Gribbon, pp 74–5.

February 1742

trustees stores in Limerick amounting to 3 barrills is ye very same seed he recd from Jon Bryen ye person mentiond in ye affidavid by him now deliverd wth sd Seed and that since he Recd it there hath been no other seed mix'd there wth nor any of ye same imbezld Sworn before me in Limerick this 6th day of febr. 1741 Thos. Vincent wch receipts is all I gott

7th alone reading
8th Dea Hallinan came to make Davids Cloaths I had Jon Nash and his Brotr trashing Oats and 2 of my own men Jon Newman and ye other 2 puting out dung I deliverd three barrills Oats to Colonel Taylors man for wch I gott ye Colonells Orders I gave Nell and Joan to each a pd of hemp
9th ye taylor att work I had 2 men trashing Oats I gave Briget wellch a peck Oats on Jon Bryen acct
10th ye taylor att work I had 2 men and my own diging for plants I got my Oats winow'd and more put in I sett 548 plants I gave watt Tobins daughtr a bus Oats B Halloran brought home her tread I gave her a pound of hemp
11th I had 2 men and my own 2 trashing ye other two finishing ye Plowing and Sowing Portachaca Corcas I got a letter from Colonell Taylor to go to Courte to buy him a fatt lamb I went and Mrs Widenham bistowd him one and sent it by his man I stayd att Courte and wrote to Mr Bryen for more of Ballylongford land att 11s an Acre I sent him ye Storekeepers receipt for my flax seed wth direction to receive ye Price of it ye letter missd
12 I came home had 2 men and 2 of my own trashing ye other 2 plowing ye boy and I finishd ye planting of 12½ hondrd of plants I set out some parsnips for Seed and sowd some radish lettice cabidge and Carrott seed and sett some Shalotts sould half a peck Oats for 6d to Mary fitzGerald and gave her 2 pound of hemp to Spin
13 ye plow going and ye trashing wth ye men I had yesterday I had a letter from Colonell Taylor for Oats and deliverd his men 7 barrill wch wth 3 I deliverd a Monday makes 10 barrills att 13s a barrill wch comes to 6l:10s:0d wch wth 18l:8s:0d for barley make up 24l:18s:0d wch sum Colonell Taylor owes me insted of my oweing him Rent ye Lord make me thankfull and forgive me my sins and enable me to pay my debts wch I bless god are but few Tobins daughtr brought her thread and I gave her 3 pd of hemp Hary Evans went away in ye morning ye Taylor finishd his work I Owe him for making Hary Evans a coat for David a coat 2 weascoats 2 pr of briches a coat and jerkin for ye girl a little vest for Den
febr ye 14th 1741–2 I had a lamb yeaned I satt at home all ye servants att mass I sent Mr Hartstonge 7 bandills frize
15th ye plow going ye other 2 men fencing ye boy B Hall and ye maids winowing Petter Miller brought me a little turnop seed Parsley Seed a pint of ye glory of England and 40 grains of Partridge pease

16th ye plow going I harnissd my 3 yr old colt to ye harness ye forenoone and in the plow ye aftrnoone I finished ye winowing Oats and put in some beans I sould two bushills oats and recd 3:6 I gave a bushill for 8 yards of canviss and 4 bushills Oats to Mr Bryens man and got my lining washd

17 ye Plow going ye other men trashing beans I made a sheet to So ye seed of some of ye canvis I bought yesterday ye Snow and frost hinderd ye harrow I made 2 collors ye maids dryd half a barrill Oats

18 ye Plow going I finishd ye trashing what beans I had in and got them winowd I coverd ye teaster of my bed ye little cow calved ye Snows continueing in Showers hinderd ye Seed Soweing

19th I put in some Oats ye men ye Snow thawd

20th ye men att ye Corcas Bank I went to Courte where I gott a letter from mr Mellsop and an acct that no good could be done in Dublin for Harry Evans

21 I came home aftr diner and got 3:7$^{1/2}$ from Thos. Halloran he owed me I pd. For ye Churn I had last harvist a Shilling and 8d for ye ferkin I gave John Bryen 4s: 4d to pay ye Taylor in full 3 lambs yeand

22 I had my 2 yr old colt harnissd and wth him ye bay colt and ye young mare sett ye Plow and two harrows doing and sowd abute 10 bus Oats and 4 busls of beans I got in my linning yt was washd he 6th ye frost continues one of ye harrows I borrowd from Mrs ffosbery 1 man trashing

23d ye Plow and ye two harrows att work I sow'd aboute five bus beans and 15 busls Oats I had a cow calved and a lamb yeand 1 man trashing

24th I bought a truckle from James Maddin for 5:3 and bespoke an other for ye same and rails yt will contain 2 statute kishes each att $^{1/2}$ a peice I gave him 6 bushills of ~~barley~~ Oats I had a lamb yeand I got ye towe thread weighd it weighs 38 pound

25 a wett morning ye men trashing ye Plow going in ye aftr noone

26 ye Plow and ye two Harrows going I finishd ye beans and began to plow ye well Corcas I finishd ye harroing ye big corcas wth oatts and sow'd aboute 4 barrills and aboute 15 bushills beans Ned diging in ye Sally bed I gott my canvis warped and left 3 pound and half ye thred for filling in answerd 48 bandillls and all moste 5 hondrd I sent it to ye weavers I gave Mr Bryens Man 4 bushills Oats

27 ye Plow and 2 harrow going and finishd ye Sowing my Oats all to ye head ridges Ned diging in ye Sally bed

28th I went to Courte home att night

March ye 1st 1741/2 I finishd my Oats and Sowd 3 pecks beans by ye well I had ye fields fenced I sould Pattr Daniel my big mare and her fole for 5l he is to pay half ye money ye 25th of July ye other half ye 1st 9br Watt Tobins Daughter brought her thread I gave her 3lb of hemp and a bushill Oats 3 lambs yeand

ye 2d a second making my Oats and Some of ye wheat I gave Jon Connily a bristol barrill of Oats on Mr Bryens acct B Halloran brought her tread I gave

March 1742

her a pound of Hemp Mary fitzGeralld brought her 2 pound of thread wch is so ill Spun yt I nither gave her more or farther employd her I recd 3s:10d in Parte from N Lowry he owes me a shilling

3d I finishd ye wheat 3 of ye men diging in ye Sally bed ye other helping to winow corn I got 5 barrills claind to Send to ye Killn[5] and half a barrill dryd 3 lambs yeand I have now yeand 15 lambs I pd Jon Nash in full for his and brothers work 5s

4th I sent ye men wth ye Oats to ye Killn but it was taken up before I had a cow calved and 2 lambs yean'd I sett ye men to dig for Parsnips aboute 12 to Courte where I stayd to ye 4th and got my 2 Plow irons mended I got a quart of Pease att Ballygeasy[6] to tabaco 5d

March ye 7th 1741/2 I came home and recond in all 19 lambs yeand this year

ye 8th I had 2 of ye men diging for peas ye other two att ye mill I sow'd 8 rows of ye glory of Eng I gott from Den: Cormock and 8 1/2 of ye Seed I Saved 60 grains of ye Patrige pease and a row and peice of ye blew pease I had a few of my Petatoes dug I got an invitation to go to a Christning to Tough

9th I had 5 men and my own diging Petatoes I got a receipt from Paul Rahily for 1:10:0 he gott from John Mcquier I gave Conner Harrogan a bushill of beans on acct of his wayges

10th 4 men and my Own diging Petatoes

11 my own men finishd ye Potatoes I gave B Hall a pound of hemp she brought her tread I sowd Kidney beans and carrotts in ye Sally bed

12th I went to Courte and came home by 11 I went to Mr Bryen aboute more land but did not agree ye men all ye Plow I gott 6 cloath stocks from Madm Hartstong

13th I went to Adear and dun'd Jon. Mcquier who promi to get me money against ye 18 came back aboute 12 and Tobins daughter brought her tread I gave her 3 pound of hemp and a bushill of beans ye Plow going

14th att home alone

15th I sent Jon to Town and he brought me for ye flax Seed 4l 1s:0d I gott 47 bandills of canvis from ye weavers wch cost for weaving 1/2 a bandill and comes to 2s 11 1/2 we ballan'd accts ye men diging for beans I turnd out my mare

16th I sent 14 weathers to ye fare and only sould 3 I gave Tho Collins a bushill beans

17th I went to Ballylongford and settled accts wth Mr Bryen and gave him 4 1/2 ginies in full for last 9br. Rent I sent a barrill and half of barly to Darby Longan to be maulted and a peck of wheat to be ground I gave ye Servts 2s:8 1/2 and sent them to drink their Patriks Pott Patt: Halloran brought some of ye thread I gave him ye remaindr of ye hemp

---

5 From a later reference (16 December 1745) Peacock is here referring to a purpose-built kiln. This was brick- or stone-built, with a firing chamber under a drying-room which had a platform of pierced tiles and a funnel above. A fire was lit, and the hot air directed upwards through the tiles and out through the funnel. The grain was spread over the warm tiles and was regularly turned. Brunskill, *Traditional farm buildings*, p. 94. 6 Probably Ballyashea, 2 miles west of Kildimo.

ye 18 I went to Adear to Jon Mcquier but did no good ye Plow going 2 men diging
ye 19th I gave a bushill of wheat and ½ for 30 bandels of bandle cloath I finishd ye Crosscuting and got some beans set
20th to Courte John Den and ye Plow and horses wth David Welch ye other 2 men a setting beans
21st att Courte gave Ned 1s:1d to buy me tabaco and sleeve bottons I sent home my 4 bushils mault and sould my Barly to Mrs Widenham for 1s6 a barrill in ye Eavening I came home and got aboute 3 pintes of Pease yt I sent for to Mr Jackson I got 5:5 from David Welch on acct of ye hire of my men and horses
22d Jon Den ye Plow and horses wth David Welch ye other men a diging for pease fencing and 2d diging Petatoes I sow'd ye Pease I got from Mr Jackson in 11 Rows I gave Dennis 2:8½ and Ned 2:8½ on acct of their wayges
23d a plowing in ye Petato garding
24 a plowing ye narrow garding and sowd 6 ridge of Nakeed oats[7] about aboute 2 bushill of pease I got 6:3d for wheat I gave Jon Bryen 2:2 Connr 6½ Den 6½ and Ned 6½ and 2½ bandills of frize
25 ye men at ye fare I had a pair of broge for David wch cost 1s:2d a pair of soles 3d
26th puting out dung
27 I had 9 men and my own a diging for petatoes I pd. 3 of them
28th att home
29th I agreed for trashing my barly att 7d a barrill my own men Puting out dung and trashing wheat
30th ye men diging I gave 2 bushills of wheat to 3 of ye men for 5s:5d a bushill and got their note I gave them a bushill beans on acct of their Hire
March ye 31st 1742 ye hired men trashing my own and James Wellens diging and trashing wheat ye Millo heffer calved

Aprill ye 1st ye hired men trashing my own diging I got 3 barrils of barly winowd and sent to Courte I gave robin Mc Tomas a note to Mr J Johnston for half a barrill Oats wch if he gets is in fall for his gon I employ'd me yesterday and to day eating petatoes
ye 2d I finishd diging ye north garding and finishd ye reek of barley and Sett ye Harrow going Mr Bryen sent for a lamb wch I sent him I gott 4:6 for wheat
ye 3d I break going I had my duch barly trashd ye bear barly Piled and some winow'd 4th att Courte home at night
5th I sent 9 barills of barly to Courte and had in ye reek 14 barrills my Sistr McGan came wth her son Harry for Baddam bond[8] wch I gave her and gott a receipt for she brought me a bottle of Shrub and I gave her a peice of beefe and another of bacon in return Patt Halloran and John and ye maids when they fin-

---

7 Naked oats: a French term for the rhealba oat, which has an atrophied husk which is smaller than other than other varieties. The husk is easily detached after the crop has been harvested, thus preserving the kernel.   8 Possibly part of an inheritance from Brettridge Badham, a relation of the Hartstonges

ishd ye winnowing a cutting Petatoes I sent a horse wth my Sister and She sent me a hondrd of Oysters and sent home my frise I owe her for ye pressing

6 I had P Hall Jon Hall Brigett Hall my own 2 maids a man and boy Setting Petatoes ye other 3 men and I sowing duch barley and sow'd 4 bushills

7th I had ye above men and women at ye above work and finishd ye garding I pd ye men yt trashd my barly and got 2:8^{1}/$_{2}$ for wheat I gave Patt Halloran a bushill of barly and 1s:1d in cash

8th I finishd my duch barley I got from Mr Widenham 11:2:1

9 I went to Courte lent Tho Jackson 2:8^{1}/$_{2}$

10 I went to Limerick and gave my Sistr 11:4^{1}/$_{2}$ yt Jon Borrowed ye 6th febr 2:8^{1}/$_{2}$ for pressing and dyeing 7s for a Pype and 9s in full for ye horse 1:1 to Kelly I pd Jon ffosbery 5:5 for mending ye gon bought of Jon Bonfield a coat two fustian weastcoats and a peice of linning for Shirts pd him 3:17:6 and owe him 1l:3s:10^{1}/$_{2}$ pd for a hatt 6s for a wig 7s for a Prossis[9] 5d Snuff 2 ye mares Serjant 6 my expences 1:6 Paid Tho White on James Pursills acct 4:6

11 and **12th** att Courte

ye **13** I came home and found my beans trashd I served Jon. Mcquier wth a prossis I have 7^{1}/$_{2}$ barrills

14th I had P and T Hallorran diging for Petatoes 5 men trashing Oats I gave 4 bushills beans for notes one to ye cooper and one to one of ye trashers

ye **15** I sent Jon to Limerick wth a barrill of wheat to Doctr Martin he gott 11:4^{1}/$_{2}$ from him he pd for hops 1:1 peaper 10 tabaco 10 costom turnpike and expences 4^{1}/$_{2}$d he brought my chloaths but did not pay for making them he brought ye Pype and my Sistr sent me a table and 2 bottls of Shrub ye above men diging and ye above trashing I gave

16th I had ye above men trashing and ye above men setting Petatoes

17th ye above men trashing I finishd my Petatoes and measurd ye Oats and they trashd 18 barr

18 att home

19 I sent 11 hogetts to ye fare pd. Mrs Quin 4:13:4 in full for tythes went to Mcquiers and wth fisher spent 2 bottles of sidr so home and Jon sould ye wethers for five and 2 each then I went to Courte and found my Aunt very ill and Mr and Mrs Barker wth her[10]

20th I came home in ye Eavening ~~Darby~~ Connr Harragan got ye barrill Oats I ow'd him Patt Halloran got a barrill Oats and a bushill beans on acct of Spining

21st I finish ye trashing my Oats and pd. Ye men for trashing them and ye beans my own men trashing wheat and sould a bushill sould David Nelan and Jon Cark 10 bushils of Petatoes sould ye trashers 2 bushills beans 3 pecks Oats and agreed wth them for cutting and Saveing a hondrd and half of turff and pd them 9:3 in parte of 18s gave Brigett Halloran 1:1 for ye ^{1}/$_{2}$ peck of flax Seed

---

9 The means by which a defendant is compelled to appear in court. 10 William Barker (1704–70) married Mary Quin (1708–76) who was the second daughter of Valentine Quin and Mary Widenham, and Nicholas Peacock's cousin. The Barkers lived in Kilcooley, Co. Tipperary.

22d Aprill 1742 I went to Adear to try if I could do any good wth Jon Mcquier and got a note from Wid: Quin[11] wch I accepted of so came home sick ye men fencing and diging a bed for carrott wch I could not sett I gave out 3 peck beans
23d Still so unwell yt I could not geld my lambs but got Tim Lowry to geld them and ye pig ye men fencing Caparow I sould 4 bushills petatoes for 4:4 I gave Martin Sawn a busl. Oats on Elenor Hines acct
24th I sent 5 barrills Oats to yet Kill got some wheat claind Ned and Connr began to dig for flax seed I sould a bushill Oats for 2s gave out 3 bushills beans
25th alone unwell
26th ye men att ye mill I gave Tim Calahan on Jon Bryen acct 2:6 to Ned agoing to Limerick 6$^{1}/_{2}$ to Stone blew $^{1}/_{2}$ for Phissick for Jon 4d lent Jon Newman 1s:1d Robin Pea lay wth me sent 5 barrill more of Oats to ye Killn
27th Rob: went on his Collection ye men sowd ye flax seed I gave out a bushill beans on Jon Browns letter and recd 2s for a bushill Oats Rob came at night
28th Robin took his leave Joan and 2 of ye men at ye mill I recd 3s for wheat and pd ye hearth mony I sow'd ye last of my Carrotts and Kidney beans and a few pease I agreed to clain out ye hay yard for 1:6
29th I gellt my 2 calves and 2 yearlings ye horse rider doct my 3 colts I gave him 1:1 and I gave ye men for clearing ye hay yard 1:6 they owe me 1 $^{1}/_{2}$ I got my bean halm burned and sent 21 bandill of linning I intend for briches to ye weavers I gave Mat Halloran 2 bushills of beans on acct of work or greasing cattle for me I gave Patt Wellch on Connors. Word to pay me 6s:6d ye first of August
30th to Courte and accted wth ye men

May ye 1st 1742 att Courte and to a list of ye tenant Cattle pd 1d for a pair of buckles lent Mrs Harts 6$^{1}/_{2}$ pd 7d for steele
2 I gave home in ye eavening
3d I went to ye fare and sould 20 lambs for 2l two hogetts for 10s:10d pd for a truckel 6s:6d for hoopes 3 dowsin of Ceeler hoops 4:6 2 dowsin of Can hoops 6d to ye men to drink 2:8$^{1}/_{2}$ to a little can 3d my own expences 5s:5d to home
4th send a barrell of beans to ye mill I gave 53 schenes of thread to be wash'd I gott my sheep washd and sent by Math: Halloran to greas 3 two yr old bulluck two yr old heffers 2 bullsshin yrlings and 2 heffer yearlings I gave Patt Cleary tim Art and Darby Doolin 2 bushills of beans on acct of cutting turff
5th I had a taylor making briches Dennis att ye mill Ned and a man trenching Petatoes Conner att Limerick for Iron for ye truckles aboute 6 he came home and pd 1:4:1 $^{1}/_{2}$ for Iron Salte 1s:9d to 2 bandils for Jon 5d to nails 3 expences and turnpik 1:3 $^{1}/_{2}$
6 I gott ye 1:1 I lent Jon Newman wch I gave Conner on acct of his wayges I had 5 men and my own trenching Petatoes I got my truckle and hoops home I got my colt gellt and pd 1:1 for it ye taylor at work I got one briches made

---

11 Possibly George Quin, who married Caroline (or Mary) Cavendish, and may have pre-deceased him. Burke, 'Dunraven and Waterpark'.

May 1742

7th ye Taylor finishd ye other briches I pd him 1s:4d I had ye Cooper and his son at work I had 3 men and my own at ye Petatoes I gave Math: Halloran half a barrill Oats and a peck to Peter Miller I pd 2:8½ for Staves I lent Patt Daniel ye ball'd horse and sent ye barly by Simon

8th I had ye cooper and his son att work and mended all ye vessells but ye powdring tub and male tub wch were full he made 2 new pecks I gave him a bushill Oats and sould a peck oats 3 men and my own at ye petatoes and finishd them

May ye 9th 1742 I went to Courte in ye morning and back in ye eavening lent ~~Sally 5s:5d~~ pd

10th I sent my truckles to be shod connr and David at ye forge ye other 2 men att home sheep shearing

11th I bought 2 ceeves a barrill a Stone and Measher 2 Stellins a Stoole for ye ceeve and a pott from Pattr Day for 15s:2d pd of it 12 shillings I gave Willm boyle on Joan Dalys acct a bushill Oats a bushill beans Conr and David att ye forge ye other 2 claining out ye barn and puting in wheat

12th David and Conr at ye forge and finished ye truckles and got ye 4 horses shod Ned and Den trashing wheat Tim Calahan mending ye Closett I gave him 4 bushills Oats of acct of building an house at ye same rate I pd last year I gott 4 bushills of Mault ground I gave Brig: Halloran a bushill Oats a peck beans

13th I had Tim Calahan making gages and ye Cooper mending ye vessells I went to Courte where I stayd to ye **15th** and gott 2:8½ I lent Tho: Jackson ye 9th of march I lent Tho Hanrahan 6½ I pd Patt day on acct of ye vessells 2:8½ when I came home I found my drink brew'd ye truckles and harniss ready to draw ye turff Tim Calahan att them Conr went to Limerick wth wheat he sould ye 4 bushill for 18s:2 and he pd for indig 1:5½ tar 9 tallo 1:1½ vomitts 5 turnpike costom measureing 7d

16th to Courte and pd Patt Day in full

17th to Limerick and pd ye Quitt rent pd 1:6 for a bras cock 1 shilling for a lock 6 for hinges 7s shilling for making my Chloaths expences 2s so home

17th I gott Mrs Widenhams sheep shorn

18th I gave in ye acct of ye sheep pd Jon. Donevans sistr. In law 6s:8d in full for ye canvis I got last summer aftr. Diner I came home and found my men att ye turff and Tim Calahan every day a making it up and Patt welch and will Newman two days each a Shovelling ye Petatoe trenches

19th ye men att ye turf and Tim Calahan making it up **20th** att Dito

21th I was hindrd from drawing by ye wettness of ye morning I got my flax bog'd

22d I finishd my turf early and have a reek 10 feet broad 10 feet hight and 32 feet long

23d att home

24th I went to Courte where I stayd to ye **29th** in wch time my wheat was trashd I had Patt Welch 6 days will Newman 5 days I got 15 10 for ~~woole~~ butter for woll 1:1:10 For wheat 18s I pd Joan Lowry 5:5 to Den 1:1 to Betty Grime on acct of making my shirts 2:0 to buy brandy and lemon for John 1:9

30th I went to Courte where I stay'd **5th of June** in ye eavening in wch time John pd Terence Bryen for ye use of Mr Den O Bryen 3:8:3 Patt Welch was 6 days at work I got a drea made att Courte ye horse rider took my bay colt in hand

June ye 6th to Adear Chirch dined att Mr Quins and came home I spent 6d
7th I gave Ned fitzGerald 10s 10d on acct of his wayges and loyterd aboute
8th to a vestery to adear Raisd 17l:10s on ye Parish so home
9th to Courte where I stayd to ye **21st** in wch time ye men sould as much wheat as came to 3:10:1 and a ferkin of butter wch was sould for 1:3:1 wch I set down on acct when I came home I found my lime killn burnt and emptyed and Jon Heas breaking Stones to fill it again Jon pd terence for Mr Bryens use 2:5:6 hapd in parte for timber 5:5:0 I went 2:9 for it still
22d I went to Ennis Couse[12] to see my brotr and his wife he gave me a pair of gloves aboute 4 I left it and spent 10d I got home aboute 8
23d I went to Courte and stay'd till ye **26th** in ye eavening then came home unwell wth a cold I gave 2s:11d for Iron 2:6 for ½ a stone of Sope and 3d for an ounce of saltes gave Luke Caromady 2:7 wch he is to work
27th of June 1742 att home unwell gave Ned on acct of his wayges to pay ye taylor 1:7½ Jon Recd in two payment from Tim Lowry 5s:5s 2s:9d of wch is to give in full to ye woodman ye Remaindr he accted for
28th ye men Settling ye foundation of ye new house
29th tho a holy day[13] Praparing to draw turf God willing in ye morning I gave Martin Sawn a bushill Oats on Nells acct to Tim Bork a bushill beans to Patt Wellch on acct of work a peck Oats a peck beans Cleary brought me and acct of ye turff as timothy Mallony measured it and is 3:3:19 kishes
30th I had 4 hired men 12 hired horses a drawing out of ye bog my own men and horses drawing home I dined att Tough

July ye 1st 1742 I had my own 4 horses and Patt Hanrans horse drawing home 9 hired horses drawing out of ye bog 4 hired men Math Hallorans mare was taken wth a fever I came home aboute 12 in ye Eavening I sent ye 2 ball horses to Law Daniel to have a shoe put on each Will Leamy gave me my tythes of ye turff
2d I had ye above men and horses att Dito work and finishd drawing out of ye bog
3d I had my own 4 rails Patt Hanranhans 2 rails and Tho Harragans rail and 9 horses wth baskitts drawing home but did not finish
4th to Courte and back in ye eavening
5th I setting my own 4 horses to draw turff and finis I spent in all 25 horses wth rails 9 horses wth baskts and drawing out of ye bog 30 horses and 22 men my own included I sent 5 barrills of lime to Courte
6th I begin to Reep my barly ye wettness of ye day stopt as none but my own men
7th my own men and willm newman reeping ye 2 maids binding I gave Pattr Hure 1s:1d

---

12 Enniscouse is south west of Rathkeale.   13 Feast of St Peter and St Paul.

July 1742

8th ye above men reeping and women binding I went to Tough gott 7l:17s and od pence from Jon fitzGerald from thence to Courte and gave ye money to Mrs Widenham and got 4 10 from her to pay for ye remaindr of ye Sleats

9th I gott up att 3 went round to Cunraty and counted 10 thousand and half and pd for them and got home att night fall

10th I came home aboute 12 and wrote to Tough but got no answer ye barly stooked

11th I gave my men and women lave to go to Deemo[14] I satt att home reading I gave Jon a note on Jon everitt for 3s worth of drink and money to pay 2 of ye horses I had drawing ye turff I gave Dennis on acct of his wayges 6$^{1}/_{2}$

12th I sent Jon to adear and he got 11:4$^{1}/_{2}$ in full from Tim Lowry he pd fran: Naghtin 2 for his horses hire at ye turff when he came home I made up 1:2:9 and sent it by Jon to terence Bryant I sent a bushill of wheat to Jon Everitt and two to Tim Lowry

13th I made ned begin to trash ye flax I gott my lambs shorn

14th I had Ned att ye flax 4 women and Nell a tongesing it Jon Heas laid ye foundation of ye stable

15th Ned and Connr and 5 women att ye flax Jon Heas pointing ye barn Tim Calahan making door cases I accted wth him and do owe him 4:9

16th Ned and Conr and 2 women and Joan att ye flax Tim Calahan att ye house Jon Heas finishd pointing ye barn aboute 10 I gave him a note to Jon Mead to gett his hammer mended aboute 2 I Recd an express from Courte and went there ye business was to garde ye Pleat to town from tories tis said yt are in Cratto wood[15]

July ye 17th 1742 I went wth ye Pleat to Limerick left it and a letter wth Mrs Perrys made so home in ye Eavening

18th att Courte

19th I went to Tough but got no mony from thence to Adear and took my tythes for 6l:13s:0d and Procters fees I sent 20d so home and dined then Jon Bryen gave me ye acct of ye money he recd for ye butter he gave Terence Bryen 1l:2:9d to Salte 1s:8d to pay for Joan's flaning 1s:2d to Costom and turnpike 2d to oyle 4d then I went to Courte

20th I went to Shannon grove to see if Mr Jon Bury[16] had any Sidr to sell he tould me had so back to Courte sent for a sample of ye sidr but it was not like in ye aftr noon I went to Tough and back to Killdeemo where I found Jon. fitzGeralld in Company wth archdacon Weight who met ye Parissoners to sett

---

14 It was a Sunday. 15 Crataloe, Co. Limerick. There was a proclamation on 19 June 1742 that a number of named people were Tories, Robbers and Rapparees out in Arms. Two, John and Denis Nugent, were in Co. Limerick. They were to be ordered to surrender or be apprehended on pain of death. *Pue's Occurrences*, 15–19 June 1742; *Twenty-third report of the Deputy Keeper of the Public Records in Ireland* (1891), p. 66. 16 Shannongrove is 1$^{1}/_{2}$ miles north of Pallaskenry, in the parish of Ardcanny. The two-storey house was built around 1709 by John Bury senior, and completed by his son, William. The architect was probably a member of the Rothery family. John Bury of Shannongrove, Pallaskenry was the son of William Bury and Jane Moore, daughter of the first baron Tullamore. He was born in 1725 and married Catherine Sadleir in 1761. He was a freeman of the city of Limerick in 1748. Bence Jones; Herbert, p. 104; Burke (1976).

Tythes he invited in and I staid till Mr fitzGeralld wrote to Mrs Widenham then to Courte but got no money only a Promis he would send some in ye morning and offerd to let Mrs widenham have a hogshed of sidr

21st I loiterd aboute John ffitzGerald sent 1:18:9

22d I came home early and found my barn Plaisterd Tim Calahan and Ned att ye barn ye flax finishd and ye barly a stacking

23th 24th 24th 26th att Courte

27th I came home and had Jon Nash and brotr Tho Congan and my own 3 men reeping wheat John Daniel tending ye masons in ye eavening back to Courte

28th I came home had ned: connr Phill fitzgerald and John Daniel reeping Dennis tending ye Masons I gave Jon Williams 1:1 wch more than pd for ye coambing two balls of wosted in ye eavening to Courte

29th having satt all night on ye watch I left Courte att 3 and came home I had Connr ned M Halloran L Caramedy P fitzGerald and 2 men from Mr Harding aboute 12 I finishd reeping and Stooking ye white wheat and sett ye men to remove ye Stones and aftr to ye hay I gott a letter from Mr Jon ffitzGerald and 8:0:2 to give to Mrs Widenham I weighed 34:$^{1}/_{2}$ pound of thread to send to ye weavers so to Courte gave ye money to Mrs Widenham

30th att Courte and got 1:2:9 from Patt Daniel

31st I sett my trackles to draw a Stack of turff I bought J Nash from Courte and pd him 10s:10d in full for it I had J Nash and brotr Jon Daniel an another man reeping duch barly and cocking ye hay Den and Ned mowing L Caramady tending ye masons and Connr in Limerick aboute 5 to Courte

August ye 1st att Courte

2d I came home and had my own men J Nele L Caramady Jon Bork Phill fitzGerald James and Ned Casgary reeping duch barly and finishd it Connr gave Terence Bryan in Limerick 10s he bought a Stone and Sand I gave Jon Heas a bushill Oats

3 att Courte ye men reeping ye red wheat and at ye hay

4th I came home and went to Killbreedy[17] and Clonans to demand money and recd 4:6 from Vallon in full 5:8 in full from nele and 2:2 in full from fisher I spent 1:3 and pd 4:8$^{1}/_{2}$ in full for church rates then home and fund my Stable walls finishd to Courte

5 I came home and wrote ye coppies of Prossess and servd them and gott 3:3 from Pattr Hanranhan I pd Terence Bryen on Mr Bryens acct 12:9 wch wth ye 10 shillings Conr gave him a Saturday makes 1:2:9

6 I came home and sett John James and 3 women to stack my wheat Sent David for my 2 yr old Connr and Jack drawing Sand Ned and Den a mowing I gott 3:9 from Tho Collins so to Courte where I stayd to ye **26th** in wch time I Recd from Garratt Pursill 9s from James ffitzGerald 9s Jon Quin 6:6 James Maddin 1:1 from Quirk 10s from Patt Daniel 1:2:9 recd earnest from Ned Hikey 14:6 from

---

17 Kilbreedy townland is half a mile north-east of Caparoe.

September 1742

Will Lynes 6½ I gott a list of ye hired men and women interd money Paid ye servts and Spent ye day drinking wth Jon fitzGerald

27 to Courte where I stayd to ye **3d of 7br** then came home and gott 14s:0½ from Jon he pd for 1cwt 25½ pound of Iron 8:7½ wch wth ye 14s I Recd makes 1l:2s:8d ye Price of 1cwt 6qr 9lb of butter he gave me 6:4½ he got from T Lowry in full for wheat I pd 10d for tabaco I lent Madm H 3s 10½d

7br ye 4 att home ye men stacking Oats

5th to Church from thence to Courte home at night

6 I began to draw my hay

7th to Rathkeal to buy Sheep for Mrs Wid: but not geting money could do no good but Spent 4:4 then I went directly to Courte where I stayd to ye 10th in ye Eavening and deliverd ye wolle and recd Ye Pryce of it wch I gave to Mrs Widenham I gave Rourk on acct of mowing 15:0d½ pd 10d for tabaco and 8d for corks when I came home I found my hay finishd ye duch barly and wheat at home and moste of ye bear barley

11th I finishd ye barley and began ye Oats I gave Joan Daly 5s:5d in ye Eavening to Courte where I pd off all ye hired men and women and staid at Courte to ye 13th then home and sett ye men to making a ditch aboute ye haggard I agreed wth Connr for an other year att 4s2 ye year and gave him 2:2 besides I sent him to ballynorte for Quicks I acct wth Jon Heas and gave him 6:8

13 I came home

14 I went to ye fair and bought 9 sticks for 19s6d and five hondrd of wattles for 2:4 and pd for both wattles and timber I spent 1:5 pd Jon Nash in full so home

15th I went to Adear to meet Mr Badham[18] and spent 6d and lent Coursy 2:8½ so home

16 I mett Mr Badam at Killfinney[19] and came wth him to Adear and offer'd him 7s:9d an acre for Killmoreen So to Courte I spent 2:2

17 I waited for him att Courte he came in ye Eavening

18 I gave him a proposall in writeing as I offer'd he went a way and I came home and found 5 men and 4 horses of Mr Hardings wth wch and my own men and horses I finishd ye drawing home my harvist

19th att home and gave Cate Carty 12s:3d wch I charged to ye boys acct I gave nell Hines 13s wch more than pays her years wayges

20th I Recd. Money for medows and pd John Bryen 1:7:1 to Conner in full 7:10½ to Dennis 10s:10d to Joan Lowry 5:5 I acctd wth Callahan and gave

---

18 Brettridge Badham of Rockfield, Co. Cork, was the son of Alderman Thomas Badham of Cork, and Jane, daughter of Roger Brettridge of Castlecope. He was born in about 1682, and married Elizabeth, eldest daughter of Henry Boyle MP of Castlemartyr in 1707. His second wife was the Hon. Sophia King, daughter of the third Lord Kingston. They had two sons (who died young) and one daughter, Sophia who married Richard Thornhill of Rockfield, Co. Cork. Brettridge Badham was MP for Charleville 1713–14 and for Rathcormick 1743–44. In the 1720s he managed the affairs of Lord Orrery. C.M. Tenison, 'Cork MPs', *Journal of the Cork Historical and Archaeological Society*, 1 (second series) (1895), p. 41; will, 30 July 1741, proved in 1744. IGI; NLI GO 139; NLI MS 4177.   19 Killfinney is 4 miles southwest of Adare on the road from Croom to Ballingarry.

him 6:6 and settled all accts to this wch he spent mending ye truckel and making Quins ye 4 horses Den and 3 Boys wth Mr Harding ye rest making up beans peas and ye red wheat yt grew in ye garding

21 ye horses wth Mr Harding ye men cutting thatch I went to Mr Bryen and pd him in full for ye May Rent and got a recept so hom

22d to Hally Park wood and drew out my timber to Mick feighs and spent on me and boys 4s:4d so home

23 and 24 ye men drawing ye timber home I gave them 1s:1d to drink Connr Luke Caromady and Patt Welch each day cutting thatch

25 drawing home thatch and trashing ye broaken Corn John Heas filling ye Killn he has 3 days breaking Stones and 4 days labourers hire

26 att home I went to Courte in ye eavening to Tough from thence home Spent 6d

27th att home ye men thatching ye corn and got a sheep killd

28th to Courte and got my geldeing and Mare shod

29 att Courte and saw 5 barrills of Coles measurd

30th I horn branded all Mr S Hartstonges stock and gott a calf for my trouble I lent Mr S Hartstonge 5s:5d so home and found ye Plow going

Friday 8br ye 1st ye Plow going ye men making bounds

2d ye [?] going ye morning ye men att ye bounds aboute 2 a violen storm rose wch true down ye tops of all my reeks and had much adow to save them I bless God I did not receive much damnage[20]

3d att home reading I gave Jon Heas 1:1

4th ye men Settling ye reeks and thatching them

8br ye 5th 1742 I was sent for before day from Courte I went to Courte and found yt Mrs Widenham departed this life between one and 2 a clock in ye morning aboute 12 I went to Limerick and bespoke a coffin and all other Preparasions for her bural so back to Courte

6th I wrote Lists and sent them away then I went to Town and Sent out ye herse and all other things belonging to it and Sett a sute of morning a making and bought a pair of wosted Stockings for 4s:6d laid out of my own money 8:3 so back to Courte

7th aboute 10 I dresd me in morning and helpd to give out ye scharves and keept an acct of ye Linnin and gloves aboute one she was taken out and carryed to her long home I went back to Courte hartily wett

8th Roch[21] and I compared accts and agreed then ye glover and I agreed in accts then ye bill for ye herse was given in all wch acct I gave Mrs Hartstonge and aftr dinner came home

9th I began to sow wheat and sow'd 3 peck in ye Rush Corcas and finishd it ye men at ye trinches Mr and Mrs Hartstonge owes me 17s:6½ 9br ye 2d she pd me

---

20 October 1742 was a windy and wet month; the 2nd, 10th and 11th were stormy. Rutty. 21 Philip Roche, woollen draper, Main Street. Ferrar.

10th at home reading
11th ye Plow att work and ye men att ye trenches
12th I finishd ye long corcas of wheat and went to Courte where I stayd for ye most parte to ye 20 then home and got ye bulluck killd I walked to see ye Corcas banks yt ye men were imployd in since I went from home
30th I sent Jon to Limerick for hops and Iron then I went to Courte where I stayd to ye **14th of 9br** during wch time I pd my Quittrent and gave for a snuffers sizers thimble allmanack and buttons 1:1 to peaper 6 to tabaco 1:1 to morning buckles 1:4d I lost 1:1 att Cards
15th of 9br being unwell and in dred of ye C I took Phissick John gave me an acct of ye money he gott for ye wheat and Parte of ye Pryce of ye butter he left in my sisters hands 10s:9d towards paying for a saddel I got from Henry Evans and Jon brought home ye Saddle and left him 1s:1d to buy stirrops
16 I gave Patt Hure 11:4½ and 2:8½ for Schallops and gads to Jon Bryen to give Joan Lowry in full of her years wayges ending ye 13th 3:9½
17th ye men covering ye Petatoes
18th ye Plow turning ye fallow ye men trashing
19th 20 and 21st att Courte and lost 1:4 att cards
22d I left Courte earley calld att Clorane from then I went to Tough thence home and sett my flax a hackling att 10d a Stone
23 I sent 5 barrills oats to ye kill ye men at ye bank
24th I sent Jon to ye fare for spreadtrees but he got none ye men at ye bank
25th ye men att ye bank
26 att ye mill and bank frank away
27 ye meal came home I paid John Brodeen six shillings for hackling 3 stone tear one Stone 13 pound shorts and 3s:4d of toe
28th Courte from thence to Church then back to Courte where I stayd ye night I lent Tho Hanrahan 2:8½
9br ye 29th 1742 I came home and Sett John Donevan Tim Callahan James Maddin and [?] to hew ye timber for ye roofe of ye Stable
30th all dito ye same men all but young Maddin

xbr ye 1st I had Jon Donevan and Tim Callahan putting on ye roofe and they allmost finishd I wrote to Severall people for money and got but 5:5 from Darby Longane ye weaver brought me home 42 yards of cloath wch wth 45 yards I had before makes 85 yards att 3d a yard comes to 1l:5s:8d wch I paid him I borrowed 9s9d from Hen Evans Fbr 16 pd to make up his money I gave Cate Carty 5lb of flax to spin att 7 hondrd cloath
2d tim Callahan hewing beams for ye stable ye men claining it out
3d Tim Callahan att Dito and mending ye harrow I sent ye men to ye wood for wattles but could not gett them I put in a reek of Oats
4th I sett Tim Callahan to build a prop for ye house and Brigett Halloran to warp ye flannin I have 48 yards bandills ye men cutting wattles

5 I walked aboute ye Cattle and Cornfields I sent for a quartr of tabaco and a peny worth of Needills to adear and pd for them

6th Tim Callahan finishd ye Prop ye men trashing Oats I made a new bag

7th I began to plow for Oats att Portaca Tim Calla a settleing beams in ye new house ye men cutting wattles I got a Shew on my bay gelding att E Daniel

8th I sent Connr wth wheat to Limerick but he could not Sell I gave him 6½ and he brought me a pound of Sugar ye Plow going ye men wattling ye stable Tim Callahan a fixing a manjer I sent my gelding by Patt Hure to Shannnongrove to be sold he did not come back I gave out 28 Scheans of fine thread to be washd and bouild

9th ye Plow and harrow going ye men thatching ye Stables Tim Callahan at ye manger ye men trashing I ~~borrowed 1:1~~ pd to pay John Brodeen for hackling toe B Halloran made me a pair of Mittins

11 I had ye Plow and harrow going and finishd ye Plow ye men wattling ye rest of ye new house aboute 2 to Courte where I stayd to ye 16 in wch time I gave 3s for 2 bottles of rum and a pound of tabaco lost 9d att Cards and ~~lent~~ Mr Hartstonge ~~4:8~~ so home ye men a trashing I gott my shirts washed I gott my mare shode

17 ye men trashing Tim Callahan mending ye truckles I killd a cow

18th I rid my young gelding to Killfinney but got no mony I spent 2s:1 att adear I sent Jon and Connr to Limerick wth wheat and Oats and sent ye old pewter wth him he brought 2 new dishes 2 pleats and 3 spoones[22] he gott 10s:8 for wheat 7s10d for Oats he brought 3 deal boards wch cost 6d he pd 2d for foles expences costom &c 10d

19th att home unwell I gott 5:11 for my cow hide

20th ye men trashing sent Jon to ye forge to gett hinges for ye stable doar tim callahan making it

21st I sent Mr Hartstonge 18:5 ~~3:3p of wch I borrowed~~ of Jon Bryen James Madin att work I gave him a bushill of wheat tim Callahan finishd ye doar ye men puting in ye peas and winnowing

xbr ye 22d 1742 ye men trashing pease James Madd making a cowle Tim Calahan mending Chairs he has 14 days including this I gave Tim Nele half a barrill Oats on Callahn acct

23d I got some of my Pease winowd and ye gardin beans trashd I got 42 bandill of flanning from ye weavers and allowd him 10 wch ballancd accts betwen us Cate Carty brought home ye 5lb of thread I gave her 5lb of flax

24 I sett 3 of ye men to Clain out ye stable and 2 to dig petatoes I gave Patt welch a barrill and 3 bushill oats in full for his hire to this day

25th att home and gott a gridiron from Pat Daniel and he pd me 7s:7d

26th I went to Courte where I stayd to ye 3d of Janr 1742/3 in wch time I pd 1:1 for a corry coamb 6½ for a lock 3 for buttons lost Cards 2s when I came home I found my seed beans trashed and a peck of beans sow'd and some parsnip seed

---

22 Old pewter was exchanged for new when it was worn or damaged – pewter was a very soft alloy – or when the seller wished to exchange it against a more expensive or fashionable item.

1743

4th I had a warm dispute wth nick Lowry aboute a lamb of mine he took as his own yesterday and oblidged him to bring it back ye men trashing barly

5 ye men trashing barley

6th I wrote to Courte to see how they were ye bearer brought me an earthen dish and a peice of a fruiet Cake I washd my bottles and by much ado made up 2 dowsin wch I filld wth ale accted wth Dan Shea and owe him on ballance 7s:10d

7th I went to Clonshere[1] to excuse my not paying my Rent calld at Tough from thence I went to Adear and calld at ye house and Spoke to Widenham then came to Adear and pd John Griffy 1:1 I owed him and in company wth Sale and Barns I spent 1:7½ so home pritty well in fort ye men finishd ye reek they had in I went to bed early

8th ye men diging for beans and Parsnips I made a new line Harry Evans went home

9th to Courte from thence to Church then aftr prayers and no Sermond I went back to Courte

10th I got ye 2:8½ I lent Tho. Hanrahan 28th of 9br and 4:8 I lent Madm Hartstonge wch wth 2:5 I got from Patt Daniel and 4d of my own I pd Dan O Bryen 10:10 in full for a Cobberd I bought from him aftr dinr I came home and found my barly ~~trash~~ winowd and ye peas gardin Plow'd

11th I had ye Plow going till 11 then sent John to bay bush[2] to try to sell my barly but could not ye men puting out dung

12th I sent Jon and Ned to Limerick wth a barrill of barley and 2 bushills of peas ye barly he sould for 8s and brought home ye Peas ye men puting out dung and diging for cabidge

13 ye men finish ye ground for ye Cabbidge and began a new dich down ye lane I wrote to Mastr for money but got none I gave Tim Callahan a peck of wheat and a peck of Oats to Mary Hohehan for spinning toe I gave her 18lb of toe wch she is to spin att one penny p pound I bought 6 hurdles for ~~3 wch I owe~~

14th God blessd me wth an increas of pigs for ye Sow farrowd seven pigs ye men att ye dich I had all ye barley I have trash'd winowd and have 6 barrills 5 of wch I got filld into bags to send wth Gods permission to Limerick in ye morning ye frost hinderd me from Setting ye rest of my Shalotts one of ye little pigs died

---

1 Cloonshere/Clonshire is 1½ miles west of Adare, on the road from Limerick to Rathkeale. 2 Bay Bush is 1½ miles south of Adare, in the parish of Drehidtarsna.

15th I sent my barly to town and spent ye day aboute home I got my tallo renderd at night fall ye men came home and brought me 40s for my 5 barrills of barley they pd markett costom 8d gateage 4d and turnpike 2½ coperas ½ and my Sistr sent me a little parsnip seed

Janr 16 I went to Courte and in my way I gott a fall and hurte my fott where I stayd to ye **23d** and spent 3s:3d when I came home I found my barly trashd and some plants sett

24th I paid Tim Callahan in full 4s:3d I sett ye men to making ye dich Patt Welch is paid for his hurdles

25 and 26 Mrs Greadys man came for ye chainge of a frize wch I got him

27th ye men finishd ye dich and fencd it I got my barley winowd and riddeled James Maddin brought me six new chairs att 8d each I gave him 4 bushills Oats David Bruen and Mr Williams dined wth me on their way to Charlevill[3] when I have drink I find People come to drink it

28th ye men a trowing down ye dich of ye Sally bed I gave B Halloran a bushill oats and 9lb of toe Patt Hallorans wife brought home a pair of Stockings I gave her a ball of coam'd woll to make more

29 I got up before day and gave Jon 3s:9½ and sent him and ye rest of ye men to Limerick wth barley I gave James Maddin a peck of wheat and Jon 2 bushills Oats I gott 34 dowsin of Candills made att night Jon came home but did not sell ye barly he pd 2:1½ for Iron he took up in towl 2½d turnpike 4d gateage

30th I wrote to Mastr Quin for money he sent me word if I wod go to him in ye eavening he wod as well as he could satisfie me wch I did he gave me a bill on one Nihill and a mug of Sidr so home

31st I sent to Jon FitzGerald for money aboute nine one of ye heffers calved no money from Tough but 1:8 from Longane I went to Courte yt family were all att Shanongrove

Febr ye 1st 1742/3 att Courte

2d I went to Limerick demanded ye bill of Nihill wch he refused paying I borrowd 7l:14:4 from my brother and dined wth him from them to Crottelo to buy timber but did not agree then back to town and delivered 7 barrills of barly to Mrs Johnston to be maulted then I came to my Sisters and pd my reconing and 10d I owed her for tabaco I spent in all in town 6s:7 then back to Courte and did not go to bed till 2 a clock

3 I gave John Bryen 10l:1s:3d to carry to Clonsher on acct of my Rent I got 1l:2s:9d from Pattr and he gave me 18:8 he got for peas

4th I had 6½ barrills of barley brought to Courte and gott a receipt for ye 10:1:3

5 att Courte

---

3 Charleville, Co Cork. A market town 29 miles north west of Cork which was founded by Roger, first earl of Orrery and lord president of Munster in 1661. Lewis.

February 1743

6 I came home and gott 2:4½ from Jon he got for pease and a bus: oats another of ye heffers calvd

7th I borrowd a cross cutt saw from Tho Drury and sett James Maddin to work wth itt ye men trashing

8th Maddin att work wth ye Sawe wch I sent hom in ye eavening ye men trashing I gave ned 1:7½ on acct of his wayges I sent home ye pound of thread I borrowd from Mrs ffosbery

9th ye men trashing Cate Carty brought 5lb of thread I gave her 5lb of flax

10th Maddin a boreing and cutting wattles I got my Oats winowd in ye eavening I was sent for to Courte

11 att Courte Maddin cutting wattles ye men clearing a way ye Rootes and tops of ye sallys

12 I came home Maddin slitting hoopes I gave him 4 bushills Oats I gave Jon Sulivan on Ned fitzGeralds acct a barrill Oats and sent 3 barrills of Oats to mills

13 att home alone reading I gave Jon Bryen one of ye 3 thirteens I borrowd of him to lend Madm Hartstonge he wanted it to pay for his wig I got ye money I sent Jon fitzGerald

febr ye 14th 1742 I sent Jon to ye mill to get ye Oats ground I went to Ballynorte got a receipt for 9l rent as I came home I met some company wch made me spend 1:3 so home

ye 15th I had James Maddin making hoops and Dan mayd att ye botts but he did no good but weast powdr I gave him discharge and he went to be married I gave him 2:8½ to himself and wife some drink I lent or gave Conr 1:1 and lave to go wth him I got my Male home ye geese began to lay some chicking hatchd ye men diging for beans ye boys sewed them

16th Maddin att ye hoops ye men diging for beans and making a lamb house I sent Jon to Limerick wth peas some of wch he sould for 3:6½ I sent Hen Evans 14s:4 in full for ye Pewter and all accts I sent by Jon for a pound of gon powdr he gave 1s:3d he pd 1d turnpike ½d gateage 1d towl Colonel Taylor sent for a lam wch I sent him B Halloran brought home her thred I gave her a pd of flax I got ye male sifted

17th Maddin finishd ye hoops he made 687 ye remaindr of ye day he spent a blowing up ye blocks I sent ye Plow going ye men at ye orchard dich

18th Maddin and 2 men a slitting ye blocks I gave him half a peck of peas ye Plow and harrow att work I sent my brotr three turkies and some pigs I had 2 bushills of beans

19th Maddin att work a burning blocks and cleeve: ye Plow going but was obliged to stop ye harrow one of ye horses being lame I sent a pig to Ballynort and got 2000 quicks

20th I sent a pig to Courte and they sent a horse for me I got there by 12 nothing new I got my buckskin home

21st I came home and had ye Plow and harro going to about 12 then wettness of ye day oblidged me to stop I sent 5:5d by John to Terence bryen in acct of Rent ~~2:2d of wch I borrowed~~ ye men at ye dich

22d I tackled ye plow and harrow but only Plow'd one ridge when ye wettness of ye day stopt me ye men att ye dich I gave Patt Boyle a peck of Gardin beans for 3 days work

23d ye wettness of ye day hinder'd ye plow I had ye men ye dich I got a letter from Norry Ireland[4] to desire help to remove his mothers goods I sent ye horses but they did not send anything Patt Daniel brought me a shoe for my mare and sett it on B Halloran brought her toe thread I gave her half a stone of toe

24th being a holy day[5] ye men did no work but put in some oats Mary Holehan sent in ye toe thred but she wod take no more toe I gave Darby longane a sucking pig he is to give me another when his Sow pigs

25 ye Plow and Harrow going ye men diging for beans ye forenoone trashing ye aftr God bless me wth to Cows calved I have now 6 calved Maddin making Staves

26 ye plow and Harrow going ye men diging for beans madin making staves his son slitting twigs I sould 4 bus: Oats for 3s I had two other cows calved

27th I gott a letter from Sistr Ireland for help to move her things a letter from Harry Evans aboute his eff he imagins in my hands a letter from Mr Hartstonge to go to Courte wch I obey'd when I came to Courte I got 6 Doctr Martin owed me for peas

28th Mr Hartstonge gave me 4 ginies I went to Crottalo and bought 40 sticks for him wch stood mee three ginies every way I came to Limerick dung wett I spent 2:6 and 8 before I came to town I gave H Evans 17:½ to snuff 1:1 to Jues harps 3 So to Courte

March ye 1st 1742 I came home and stay'd an hour I went to Coursey Irelands funerall aboute 4 back Courte and found ye house full of quallyty

2 and 3 I waited to have ye timber come home but it not coming I came home ye 4th while I was att Courte I lost 1:1 att Cards and pd 4d to Leary ye pedler wch a peice of gold I gave him wanted when I came home I gave Mr Bryens man 2 barrills Oats John gave Ned a bus gardin beans Mary Holehan a peck Oats and J Maddin a bus Oats

5th of March 1742 ye men at ye dich Maddin a making laths for ye break besides wch he has 3 days this week I gave him 2 bushills oats a peck and ½ of barly wch I charged to his acct aboute 12 Mr Jackson came to me to direct I would go to Courte aboute business when we eat a couple of egs I went to Courte did ye business stayd there to dinner so home

6 7 and 8 att inniscouse came hom in ye eavening found my own men and Patt welch at ye bounds ~~Maddin yesterday and today~~

9 ye men att ye bounds and trashing maddin making barrills I went to mellane[6] aboute business from my Brother on thence to Courte where I stay'd that night

---

[4] Possibly a son of Anne Ireland, Nicholas Peacock's sister. [5] St Matthias. [6] Melan/Mellon is in the parish of Ardcanny, 2½ miles north east of Pallaskenry. Seat of the Westropp family.

March 1743

10 I gott up att 6 and came directly home Maddin att ye barrills ye men trashing I bought 13 bandills of Canvis from Sistr Ireland and pd 2s:2d for it I bought 7 bandills more from Jon Daniel and pd for it to a peny all wch I made into three bags I have now 4 new and 8 old bags

11th ye men trashing my Brotr sent for 2 barrills Oats wch I sent him I sent a barrill to Mr Bryen I got a letter from Mr Hartstonge to go to Courte Ned fitzGerald and a man from P welch a trashing on acct of ye beans

12th I went to Courte from thence to Limerick wth Mr Hartstonge and got 4 barrills of flax seed for Mr Har bought aagers chissills and gowage for him so to Court

13th att Courte I sent 2 cows to my brother

14th I went to Ballinemonna wood and bought 200 of wattles for Mr Hartstonge for 10s so to Courte I sent for my mault to Town gave Jon 10d

15th my horses came to Courte for ye mault and brought my cloath wth wch I went to pallas[7] to gett it and a piece of Madm Hartstong's bleached I gott a receipt for them I pd 10d for a pound tabaco spent 7$^{1}/_{2}$d so back to Courte

16th I orderd my hoggetts to ye fare they sould 8 of them for 2:5:6 wch Jon brought me and a letter from my brother yt my sistr molly[8] was very ill I orderd him to send to see her in ye morning

17 aboute 8 ye messinger came to me wth an acct yt my sister was easeyer

18th I lent Madm Hartstonge 1:2d$^{1}/_{2}$ /pd in ye eavening Jon brought me word yt my sistr died in ye morning

19th I went to my sisters funerall aboute 3 she was buried so I came directly home I gave my Brother 1:2:9 pd Mr Johnston in full for making ye barrills of mault 1$^{1}/_{2}$ expences: 6$^{1}/_{2}$

20th att home reading and walking

21st I had 3 men and my own diging for parsnips peas and cabbige ye plow a sowing peas I sett 2 rows of ye partrige pease 5 rows of blew pease and of ye glory of England all wch I intend for seed besides wch I sett a plott of peas for eating Brigett Halloran gott 4 bushills of mault she gave 1s:1d earnest

22 ye men att ye dich I finished ye peas and gave Jon 6$^{1}/_{2}$ and orders to carry ye Cattle to ye fare a thirdsday I went to Courte in ye aftr noone

23 att Courte and pd for a pound of hops 1s:2d and noging of oyle 3

24th I went to Limerick spent 2s:2d so home to Courte

25th att Courte 26th I came home in ye morning John gave me an acct he sould 2 yearlings for 1l:18s:0 he gave my Brother 1:10:0 wch wth 1:2:9 I gave him ye 19th two barrills of Oats I sent him ye 11th and 2 cows I sent him ye 13th make up 7:14:4 I borrowed off him ye second day of Febr last, the other 8s Jon gave

---

7 Pallaskenry is a market town 12 miles west of Limerick in the parish of Kilcornan. It was a centre for linen in the eighteenth century. Philip Guier, who had settled in Ireland in 1709, kept a bleach yard at Shannongrove for many years before his death in 1778. Lewis.   8 This may be Molly Quin, granddaughter of Henry Widenham, who left her £500 'for her advancement in the world'. Will, Dunraven papers D.3196/K/1/2.

me wth ye price of a bushill of Oats Jon gave P Boyle 3 bushill Maddin att ye rate of 14 a barrill

March ye 27th 1743 I was sent for to Courte I came home in ye eavening I send Mcgan by Patt Daniel to be sold I gave Jon 4s:2d for fere he would want change
28th Tho: Bork came to make cloaths for Jack Rohan ye men trashing duch barley I have 10 pair of yarn stockings and and old pair 4 pair of wosted I had 7$^{1}/_{2}$ of linnig thread wound and order'd it to be warped
29th I went to Courte I lost att cards 4$^{1}/_{2}$ I stay'd there till ye **31st** in ye eavening when I came home Jon gave me an acct of ye money I left him and of 1:4:4$^{1}/_{2}$ he got for a 2 year old bulluck he gave Joan 20$^{1}/_{2}$ I borrowd of her he gave Frank Burn 3s:9$^{1}/_{2}$d on acct of his wayges and 6$^{1}/_{2}$ in earnest for gads he gave Mr Bryens men while I was away a barrill and half of Oats all ye toe thread came hom but Tobins

Aprill ye 1st 1743 I pd 3d for Salte for a ferkin I pd John Bryen on acct of his wayges 5:5 I pd him 2:2 in full of ye money borrowd of him to lend Madm Hartstonge and 2d I owed for milk I sent him wth a barrill of Oats to Mr Bryen I had ye butter made and nere filld ye ferkin I sent 54 bandills of bandill cloath to ye weaver it weighed 8$^{1}/_{2}$lb
2d I sent a barrill of Oats to Mr Bryen my sister Ireland gave me a pair of wolling gloves I sett better than 300 of plants ye men remaking a stack of Oats I gott 4 bushills of ye Limerick mault brewed I gott a pair of shoes from Shea and David another
3d I went to Church from thence to Courte and back in ye eavening I gave Mrs Hartstonge 11:3 to pay for ye weaveing of my lining I gave Mr Bealan 5:5 for wch he is to give his 10:10 if I can prove ye Coocoo to sing on or aftr ye 15th of July 1743
4th I went to Enniscouse and I scharce gott there when my Brother was Blessd wth a son I stay'd there till 3 so directly home to a dram and turn 2 I sent 9 chickings to Ballynorte
5th I sent 3 three year old bullucks 1 two year old bulluck to Courte to greas I sent 1 two year old heffer and 2 yearling heffers to greas wth Phill fitzGerald on Courte lands I gott a lamb killd Tobins daughter brought her toe thread I gave her 5lb of flax to spin my sistr Ireland came to me
6th ye men and I spent ye day remaking wheat a vast Quantity of mice above 2 thirds of ye wheat lost my Brother wrote to me to go to his sons Christning my sistr gave me a little woden bole dish
7th I went to ye Christning and had ye favour of being Godfather wch cost me 10s:10d wch I borrow'd of my brother aboute 4 I came directly home I gave Luis moor 1:1 to turnpike 1 ye men att ye wheat
8th I gett my calves and lambs I sent John and a horse home wth Joan I accted wth her and owe her 4:6:7$^{1}/_{2}$ of wch I gave her 1s:2d Patt Hure took my gelding in hand to bitt him ye men trashing I had a cow calved a red heffer calf

April 1743

9th I gott a saw oager and reep hook from my sistr ye men diging Petatoes I sett a rowe of peas yt has Short hottspur in ye Northend long hottspur in ye south end and dwarf peas in ye middle I gave a bushill of Duch barly and a peck of gardin beans to Thos. Farranan I had gosslings hatched

10th tho Sunday I sent a man and horse wth my sistr Ireland to Rathkeal and another wth some of her things I went to Courte and Stay'd there my Brother son died yesterday

11th aboute 12 I came home and found Tho:farranan Patt Halloran Patt welch Patt Boyle ned fitzGerald and my own men a trashing Oats I gave Patt Boyle 3 bushs of Maulte

12th I had tho faranan E fitzGerald P Halloran and P welch trashing I gave T farranan $1/2$ a peck of peas Brigt. Halloran 3 bushills maulte

Aprill ye 13th 1743 I and my one men second making Oats I gave Mr Bryens man half a barrill of Oats I agreed wth Margt Quinlevan for half a year for 10s gave her 1d earnest

14th my Brother sent for 2 barrills Oats wch I sent him my own men and Den: Carroll a trashing and removing Straw B: Halloran and another woman winowing I gott 3:9$1/2$ from Hen Supple in full

15th I had Martin fitzGerald Patt Boyle Den Carroll P welch and my own men and 2 women att ye Petatoes ye Plow going I was sent for from Courte in ye eavening

16th att Courte I had 2 men from ffitzGeralld and Denis Carroll and Connr and frank and ye 2 women att ye Petatoes ye Plow going Jon gave Henry Supple a Barrill of Oats to Den Carroll a bushill B: Halloran a peck I lost 2:8$1/2$ att Cards

17 I came home in ye morning and Spent ye day mostly reading I gave Phill fitzGerald a barrill of petatoes on acct greasing

18th I finishd ye cross cutting my fallow and had P Boyle Den: Carroll Jon Nele and my Own 2 men and 2 women att ye Petatoes and finish ye diging I gave Jon Nele 4 bushills Oats P Boyle 4 bus for his father a peck oats to Mary morphy and half a barrill duch barley to Mr Bryen

19th I had 4 men from Mr Harding my own 2 Patt Boyle D Carroll Jon Nele and Mart. FitzGerald Second diging Petatoes 2 women picking them up I gave Mr Bryen four bus of duch barly James Maddin 4 bus Oats and I sould 4 bus of Oats for 3s Maddin carried 2 dowsin of hoops and staves to make ferkins and pecks

20th I had 3 men from Mr Harding P Boyle D Carroll M fitzGeralld Jon Neve D Harragan and my own men att ye Petatoes and all most finish ye Second diging I finishd ye Breaking ye fallow P Boyle had 3 pecks Oats I gave Mr Bryen 4 bushills duch barly and a peck and half of my sisters beans I gave Darby Harragan 3 pecks Oats

21 I began to soe duch barley but ye wettness of ye day was a great hinderance my own men and Darby Harragan a trashing Oats I gave Brigt. Halloran 5lb of flax to spin

22d my Mare foled ye men a trashing Oats and duch barley I gott 3 peck home from Madins yt he made of 63 of my own boards and 6 of my own hoopes and a small ferking I gave Margett a pound of flax to spin I gave Dennis Carroll a busl of Oats

23d ye Plow and Harrow going ye men trashing and winowing I gave James Griffy half a barrill of Petatoes wch he is to give 4s for and another wch he is to give 2s:6d for I gave B Halloran a busl Oats

24th I spent ye day reading and walking I recd $10^{1/2}$ from Martin Sawn for a peck of maulte I pd $2^{1/2}$d for Salte and tabaco

25th ye Plow and harrow going D Harragan and my Own men fenceing I gave coghlen Grady a barrill of Oats on his note I gave Mary Morphy on her brothers acct 2 bushill maulte I gave P welch a peck of Oats in full I sould Mick Ryan a peck Oats for 5d I gave mary Holehan a bushill Oats in full for spining toe I gave James griffey a peck of gardin beans

26 tho a very wett morning I finishd my Plowing ye men diging for Parsnips and jobing I gave D Carroll a peck Oats in full and sould him a bushill for 10d I sent my gelding to be removed and Patt Daniel sent for a barrill of oats wch I gave him I gott home a new churn from Maddin

Aprill ye 27th 1743 ye men owing gardin beans I sent Jon wth a ferkin of butter to town wch weid 2qr2lb he sould it for 14s he pd for sope 2:5 iron 2:3 salte 1:6 tabaco 10 a main and file and nails 6 cost of turnpike and expences $6^{1/2}$ I pd hearth mony 4s to ye 2 women picking ye Petatoes in full $11^{1/2}$

28th I went to Courte ye **29th** and **30** my Plow harrow man and horses sowing flax for Mr Hartstonge

May ye 1st 1743 I went to Church back to Courte

2d I came home in ye eavening Jon gave me an acct of corn he gave out and 2s:1d in cash

3d I wrote to my Brother and sent him my gelding by Patt Hure and gave him 1:1 then I went to Courte from thence alonge wth yt family to Shannongrove so back in ye eavening I gave will Porter 1:1 and lost $6^{1/2}$

4th I accted wth ye men ye **5th** wth ye Cowboy so home and gave P Daniel 2 bush: of Maulte Jon and den att ye mill ye other two wth Mr Harding Jon gave me an acct of money he recd and of Corne he gave out wch I interd I had ye male of 5 barrills of Oats come home

6th I had a man from Martin ffitzGerald Dan: heass son and my own trinching Petatoes I gott hom two new peck from Maddin wch have 42 Staves in them

7th my 3 men wth Mr Harding ye other and J Nele P Boyle D Heass son and ye 2 Nashes att ye petatoes I gott 15s:2d from Patt Daniel 8s:5d I pd Mr Wilm ffosbery in full of Ned fitzGeralds wayges I gave James Maddin 4 bus Oats and a peck of peas

8th I wrote to Colonell Taylor for lave to cutt turff and gott it and spent ye day reading

May 1743

9th I had Tho: farranan Darby Harrogan Jon Nele Patt Boyle ye two Nashes and my Own men at Petatoes and finishd them aboute 2 then I made ye men dig ground for Plants I pd ye two Nashes in full for this day and last Saturdays work I pd 3d for 2 pound of pitch

10th I gott my Sheep washd my own men and Darby Harragan a trashing duch barly I gave John 10 and orderd him to go to Limerick in ye morning wth butter barly and wheat then I went to Courte

11th I gott a bill of 70:6:10½ from Mr Hartstonge and went to Limerick recd ye contence of ye bill paid his and my own Quittrent did some other things spent on me and ye men and snuff 2:6½ so back to Courte D Harragan and J Nash wth my own men a trashing

12 I gave Mr Hartstonge an acct of his money and wittnessed his lease of ye dear Park then home and gave me an acct yt he sould ye butter for12:6 and ye wheat for 6:10 pd my Sistr on ye mens acct 1:1 he pd her on Sam Evanss acct 4:4 he pd for turnpike costom and wayage 7d½

13th I gott my sheep and lambs shorn pd 10d for shereing them D Harragan and my own men trashing I gave Tim Bork 4 bushill of barley he pd me in parte 2:8½

14th I had Jon Nele and D Harragan Den and frank trashing Jon and Conr in Limerick they sould 2½ barrills of barly for 1:6:0½ two stone 10lbs of woll for 1:1:7½ pd for Iron 1:4 for garters 4 costom turnpike and waying 9d I pd frank burn on acct of his wayges 10

15th I went to Courte from thence to Church back to Courte and stayd that night

May ye 16th 1743 I had D Harragan and my own men fencing Caparow I spent ye day wth them and finishd I gave Jon Bryen on acct of his wayges 10s **17** my own men trashing wheat I agreed wth Martin fitzGerald to cutt and save 200 kishes of turff for 24 shillings gave him 1s:1d earnest

18 I had D Harragan Jon Nele Conr Den and David trashing wheat I sent Jon to ~~Pallas~~ Askeaton markett[9] wth barly he sould 9 pecks for 5d a peck wch comes to 3:11 Patt hure took my bay gelding in hand

19th I sent Conner wth barly to pallas and Dennis wth barley to Rathkeal I went to Eniscouse and stay'd there till 4 then home from thence to Courte I spent 16d

20 21 22 att Courte

23 I went to Rathkeal to ye Cant and godd 2 chears a table and an allphabett for Mr Hartstonge all wch cost 1:17:10 I spent 1:8 so to Courte

---

9 Askeaton is 16 miles south west of Limerick, on the river Deel. The castle and the monastery, by the 1740s in ruins, were built by the Fitzgeralds, the earls of Desmond. Goods were shipped from the quays here to travel to Limerick and beyond. Joshua Wight, writing in 1752, described the abbey as 'the finest and strongest work I ever saw, the coign, the Doorcases, Windows, arches and pillars both above and under ground together the great vaults spaceous large and high stone pillars and Collums the best of marbel so clear and transparent ... the finest piece of architecture in Ireland'. Wight, 19 October 1752. Lewis.

24th back to ye Cant and carried a horse for ye things my brother paid for them before I came I sent them away and bought some prints for 5s:6 I gave my brother 5 half crowns wch wth 3:3 I gave earnest comes to 16 9½ ~~I owe him 1l:1s:1½d~~ I gave ye men 5 so home

25th I came home had my gelding shod Conner sould ye barly he had att askeaton today for 3:4 in my absence Jon gave Dennis on acct of his wayges five shillings to pay wch I gott 3:2½ for barly 1:1 from Tho Collins and 9d from Brigett Halloran I pd Frank Burn in full for 3 quarters wayges 3:7 I had Dar Harragan and my own men trashing wheat Dennis gott 2:7 from Patt Boyle on acct of his wayges

26 I had D Harragan and my own men trashing wheat I sent Den to Rathkeal wth barly he sould 14 pecks for 5:8 I lent Patt Daniel a duch spead I spent ye day makin a winow Sheet of 40 bandills of Canviss

27 I had D Harragan and my Own men att ye wheat I had Pett taken in hand to be rid

28 I had D Harragan and my own men att ye wheat and finishd I gott 2 shoes on Pett

29 att home

30th ye men earthing ye cabige and puttin oats I got 2 shoes on pett

31 I had Jon Daniel W Holehan M Nash J nele P Boyle and fitzGerald and my own men trashing Oats I went to Killbreedee and orderd Tobin to Summon Cleary and partner I spent 6½ so home and sent my Sistr Ireland a firkin of butter wch weighgd 2:26 pound neat wch comes to 18s:3½ but got no money

June ye 1st I went to Adear aboute selling a horse and spent 6s I gott a letter from my brother to creditt my Sistr Ireland for 18:3½ I sent him by Mauris 2:9 wch wth ye 3:3 earnest 13:6½ I gave him makes up ye money he paid for me ye 24th att ye cant I gott 5:5 from Tim Bork on acct of barly he has had in all 12 bushills he pd me 8:1½ I sent Conr to Askeaton he sould a peck of barly for 3s:0d I had J Nele Will Holehan P Boyle and my own men trashing Oat

2d I sent Den wth Carley to Rathkeal and went to Courte home in ye eavening Den sould ye barley for 4:½ pd a penny turnpike

3d of June 1743 I had Jon Nele P Boyle Jn Nash Jon Daniel will Holehan and my own men att ye Oats I went sent for to Courte

4th att Courte Jon Nele P boyle Mic Nash Jon Daniel W Holehan all ye Oats P Halleran gott 4 bus Oats in acct of ye table &c. I had from him Whitsun tewsday

5th att Courte pd 6½ for snuff lent Madm 2½ I came home in ye eavening

6th P Boyle Jon Nele and my own men att ye Oats I gott 2:3½ from B Halloran on acct of Maulte

7th I had P Boyle Jon Nele and my own men att ye Oats I gave Mick Caromady a barrill of Mault for wch he gave me his cash note payable on demand I gett 18s:1 from Patt Hure on acct of ye horse

June 1743

8th I had my own men and women winowing I gott 2:8½ from Thos. Farranan in full and 2:8 from Ned Vahan in full aboute 6 to Courte

9th att Courte they sould in Rathkeal 10 peck of barly for 4:2d a peck Oats for 4s

10th att Courte from thence to ye blachyard and gave ye men 1:1 to sidr 4 straberys 2 pd Jon Bonfield one pound 3:10 in full of all accts

11th att Courte they sould att askeaton 12 pecks Oats for 3:6 I went to ye dearpark and back to courte

12 I came hom in ye eavening

13th Jon gave me on acct of ye above Corn and ye money wth 3:3 he gott from Mary Morphey for mault and 4d in parte from Tho Collins I gott 6 bushells of mault brewd I gott 3:5 from B Halloran

14th I gott my brothers horse shod ye men weeding I gott a hondrd ferking from Maddin

15th I sent my Mare wth ye bushills of Oats to askeaton wch was sould for 3d a peck when ye mare came home I had her harnissd to ye truckle she would hardly draw ye men and 2 horses puting out straw

16 I sent ye mare wth Oats to Pallas wch was sold for 3:2½ I sent Oats to Rathkeal wch came home I measured out ye medows ye men making dunghills

17th I wrote to Widw Quin[10] for money I sent in word he wod gett my note from his Mother and send it to me I had a letter from Colonell Taylor for money wch I sent by Jon to my brother for mony he sent me 3 att 1:18:8 each wch comes to 5:16:3 I gave James Maddin a barrill of Oats ye men and horses claining of ye yard

18th I sent Jon to Limerick wth wheat and Conner to askeaton wth oats I went to Ballynorte and pd my Rent Breakfasted went to asketon from thence to Robin Hunts[11] dined there calld at Ballyengland so home Jon sould his wheat for 5½ a stone Conner sold his Oats for 3½ a peck

19th I went to Courte **20** there **21st** home I bought a quair of peaper a pound of tabaco Jon thatching ye men weeding I pd Tho Bork 1:3 for making Jon Rohans cloaths

22th I gave Jon 9d and Sent him and Conr to Limerick wth butter and wheat and Dennis to askeaton wth Oats wch he sould for 3:8½ I went to Courte

23 att Courte was envited to Shannongrove

24 ye family of Courte and I dined there I gave 1:6

25 att Courte **26** to Church back to Courte dined there so home and pd 3d for a dram

27th Jon gave me an acct yt he sould 42 Stone of wheat for 19:3 and 1c:1qr:2lb of butter for 1:8:6 he pd 1:6 for Salte costom turnpike &c 11d I had Ned Naghtin and my own men thatching I gave John Nele a busl of Oats

---

10 Possibly George Quin, son of Valentine Quin and Mary Widenham. Born 22 February 1718/19, he married Caroline (or Mary) Cavendish. They had one daughter, Mary who married Thomas Taylor. Burke. 11 Robert Hunt leased the glebe at Askeaton from the Kenmare Estate for 999 years from May 1753. PRONI Kenmare Estate papers D/4151/B/1/1 f.51.

June ye 28th 1743 I went to Courte and lent Mr Hartstonge 1:2:9

ye **29th** att Courte and sent home money to pay Countery Charges we all dined att ye dear park

ye **30th** att Courte lent Ms Nancy 1:1[12]

July ye 1st 1743 We dined att Coroheen[13] wth Mr Burys family I gave 1:7½ for Sider I lent williams 5:5

2 att Courte lost att Cards 6½

3 att Courte gave ye Cook and house maid 2:8½

4 I came home in ye morning Jon gave me an acct yt he gott for wheat 6:6 for 2 bushills Oats 1:2 from Mick Ryan 1:1 earnest for medows and 1d from J newman I gott my calf killd

5 I was sent for to Courte went from thence wth Mr Perry[14] to Limerick pd Jon Melsop 1:10:3 for a heffer to land [?] Nick Magan 1:1 my expences 1:1 so back to Courte

6 7 8 and 9 att Courte

10 I came home in ye eavening

11th taken ill wth a pluricy **12th** ye Doctr came and wrote for things to Limerick wch cost 6:6

ye **13** and **14th** ill **15** I bless god much better **16** eased

17th I just got up and orderd 5 barrills of Oats to Mill pd my Sistr Ireland 1:9 in full for beans and pecks

18th God enabled me to get up some time

19th I bless God mostly up Jon gave me an acct yt since my Absence and illness he red for 14s peck of wheat 15:2 for Oats in Limerick 2:2 for Oats from Nash and Nele 1:9 for a pig 1:8½ for Mault from B Halloran 2:9 from Mick Caromody 14s for a barrill of Mault of wch he laid out 1:9:4 for boards Costom and turnpike Severall times 10d Sythes 5s

20th I sent Jon to Limerick wth a barrill Oats wch he sould for 4s he bought 5 board for 6 a stone Iron 2:4½ five hondrd of Nails for 3:3 a pair of broge for Jack for 1:3½ soles for David 4d Costom &c. 4d Paul Rahily and son came to vew my tythes I pd them ye money I owed Mr Buckner[15] for last yers tythes and got my note

---

12 Ann Hartstonge. 13 Coroheen/Curraheen is 1 mile east of Pallaskenry. 14 The Revd William Cecil Pery, who was born in Limerick in 1721, second son of the Revd Stackpole Pery. Educated in Limerick, he entered Trinity College Dublin in 1738 and took BA in 1742. He became vicar of St John's Limerick. He became dean of Derry and chaplain to his brother, Speaker Edmund Sexton Pery. He married Jane Walcot of Croagh in 1755, and their son Edmund became 1st earl of Limerick. He married, as his second wife, Mary, sister of Sir Henry Hartstonge. Because of the failure of the male line in the Hartstonge family, their land passed through this marriage to the earls of Limerick. Limerick papers, NLI PC 875/2/15/56. 15 The Revd William Bucknor was the son of Richard Bucknor, a gentleman of Waterford. Educated in Limerick, he entered Trinity College, Dublin in 1720 and took BA in 1724. He was rector of Croom, Adare and Dunamon from 1739 to 1783. Leslie, 'Diocese of Limerick'.

July 1743

21st Jon and Den att ye mill I gave out 11 Skeans of fine thread and 21 of course
22d to Courte wher I mostly stayd till ye **10th of 9br** and haveing seen Mrs Hartstonge and family settle in town I came home dung wett during my stay Courte I rec and an acct of ye Pryce of ye medows wth parte of ye money I gave ye Servants parte of their wayges pd Mr Bryen 7:15:8 wch wth 14:3:10 in Corn make 21:19:6 ye foll of a years rent end May 1743 I pd 3:3 for blew sarge I sould my 4 bulluck for 11:10:5 and made it up 12l wch I gave Colonell taylor on acct of 9br Rent I gave 6s:6d for a hatt 7s:7d for a wig 16d for gloves 14s:9d for bleaching my cloath to Pursill[16] for flowring my Room and making a dresser 18s to woden ware 10s:10d
9br ye 11 1743 went to Corrohan from thence to Courte where I stayd yt night settling ye affairs of ye house
12th I sent home ye Apples Mrs Hartstonge gave me and came home at night
13th att home alone reading att night fall a Man came wth a list to go to Mrs taylors funerall and stayd all night
14th I spent a Settling out my room and Cutting out my linning I have 5 pair of New sheets and Six new Shirts
15th I sould Jon Nele a bed of Petatoes for 9s worth of woll[?] and I went to Mrs Taylors funerall gott a Scharf and hatt bands so home and spent a shilling
16th I went to see my brother came back att night turnpike 1d I had 7 men sowing wheat sould Jon Nele some petatoes 9s
17th I had 5 men sowing wheat and diging petatoes
18th I had 7 men diging Petatoes I made one of my sheets
19th I went to Coraheen to agree for making ye derepark wall and Burn said he would make it for 4s a perch six fott high and find all ~~conveniences~~ stone or elce yt he wod finish ye wall and Mr Hartstonge find all things for 3:6 a perce from thence I went to Courte and gave ye quittrent drivers 2:8½ for not driving me and Courte then veiwd ye sellers then away and got home by nightfall I had 7 men and my own diging Petatoes Jon Bryen gave me ye weight of ye butter 1:2:7 for wch my sistr Ireland owes me I sent her six shirts to be made and a shirt to cutt them out by
20th att home reading Jon bought a fatt Cow from Jon Haly for 2l:13s:0 I gave him in Parte 1l:4:4d
21 I went to Courte from thence to Limerick wth an actt ye Mr Vere Hunt[17] killd one of Mr Hartstonge's black dere gave 2d att ye boat[18]
22d I gave 2s:2d for a pair of gloves pd 7s:2½ quitt rent pd Mr Melsop 10d I ow'd him and pd 6 for Sidr and a dram and spent ye day in town

---

16 Cyprian Purcell, who did building and other work for Nicholas Peacock. He owned a farm at Ardnavolagh, near Ballygeasy and Kildimo. Indenture between William Bury and Thomas Bury of Corrobridge 3 January 1738. Dunraven papers D/3196/K/6/3/1; D/3196/K/6/3/2.   17 Vere Hunt of Friarstown, Co. Limerick, eldest son of Henry Hunt. Married Anne Maunsell of Co. Limerick and had three sons and two daughters. The Hunt seat is at Curragh Chase, Adare. Burke (1958).   18 To cross the river Maigue.

23 aboute 11 I left Limerick gave 3d for a dram 2d to ye turnpike so to Courte gave some directions so home att night

24th I sent John Bryen to ye fare wth 4 sheep but he did not sell he pd John Haly 1:2:9 on acct of ye cow I sent a roasting pig to Court to be carried to Limerick I had 2 men diging Petatoes

25th I went to Mr Jacksons to advise wth him aboute Mr Hartstonges dere from thence to Robin ormonds and agreed for trees att 3½ each So to Mr Quins aboute 1 home I spent 11½ I had 2 men diging petatoes aftr night fall I markd 3 large dishes 4 a size less 3 small ones 11 bowls 3 punch bowls 4 old dishes 1 old black bowl 18 new trinchers 3 old I put my new Sheets on ye bed

26 I went to Courte to snuff 6½ 27th to Church

28th ye men att Courte began to trash wheat I got a chees vatt a dish and 3 little bowls I pd ye turners and cleard wth them

29 I gott 10d ned owed me I had a barrill of Mault ground

30 one of Mrs Hartstongs cows died

xbr ye 1st 1743 Jon Bryen came to me and let me know yt he had finishd ye sowing my wheat in ye Corcas I sent home ye above wooden ware and a midling bowl he brought me one of ye new Shirts Sistr Ireland is making I orderd him to kill my pig in ye morning I pd M Cosgary 2d John Borrowd ye fare of Ballingarry[19] 2 and 3 ye men trashing 3 4th I put on ye new Shirt wch did not fit

5 and 6 ye men trashing I orderd a wether to be killd another died 7th I sent ye Motton to Town and pd 2d for pepper

8th I went to Killgabon to borrow 8 or 10l but did not gett itt

9th Jon Bryen gave me an acct yt he sould 3 barrills of barly for 1:4:0 he gave Den 6 and ye remaindr he gave me I sent Mrs Hartstonge 5s:5d to pay for ye mending my clock

10th all ye men loading a boat wth turf wood 3 chears 2 hogsheds of table drink a Salte box a bed and cadow for ye Palatine one of ye calves died I gott 2:2 for ye skin

xbr ye 11th 1743 I sent Phill Palatine to Limerick to stay there

12 13 and 14 att Courte Jon Bryen sould 12 bushills of barley for 12s I pd Jon Haly in full for ye cow 6:7½ for hops 1:1 he sould three sheep for 1:8:0 and gave me ye money I lost 1d and lent ~~Norse 6½~~

15 aboute 2 I finishd trashing all ye Corn att Courte and went to Limerick I pd 1:1 for oysters &c. 1:1 to ye maids

16 in town gave Mr Hartstonge ye 2:2 I gott for ye calves skin and 7:19:7 ye Mrs Howard fitzGerald sent by Mick Cosgary I lent her 3 att 11 4½ and 8:8 lent Mastr 6½ to snuff 4d to Rum 2d to brandy 2:2 to powdr 1:4 to glasses 1d

---

19 A market town 16 miles from Limerick, on the road from Rathkeale to Charleville.

December 1743

to a stick 6½ wine 8 sidr 4 to Honr 2:2 to a wig 7s 0½

17th left town aboute 8 pd turnpike and a dram 4½d so to Courte and sent a barrill of Sidr to town gave out peas and beans to be sow'd did some other jobs so home safe blessed be God

18th I agreed for 180 two year old apple trees att 3d each got 2 of my new Shirts home and got my cow killd

19th I sould ye hide for 6:6 and a sheep for 7s I sent John to Limerick for salte wch he pd 1:8 for and 3½ for 14 shirt bottons turnpike 1d he brought me ye things I bought in town a Friday last and Ms sent me some snuff Madm sent me 4:11 to ye men for making a dich in ye derepark

20th att home very unwell wth a cold I got my wheat remaid Jon gave me acct how he laid out 6:6 he gott from sistr McGan and 3:3 he got for a pig from Thos Gibon

21 I went as far as Suppels to go to Ballynorte but finding myself so unwell I turn back but pd Supple 4:11 to pay ye men I spent 3d

22d my Clock came home and ned broke ye pendaly spring

23d I agreed wth Jon Pursill to wach my turff for half a barrill of Oats I am to give him att Candillmas I paid him for 5 hurlds 1:8 I gave John Bryen 5:5 on acct of his wayges

24th to Courte stayd there till 2 so home I gott my mare and gelding removed I gott ye 6½ I lent Norse

25 att home

26 I walk to see my Sheep spent ye day claining my gons I gave Den: 6½ on acct of His wayges to Suger 10d to Steel 1½

27th ye Priest and others came to me for their share of Christmas and went away in good ordr

1744

28 [December] I had 9 schore apple trees came home I went to Courte where I stayd to **ye 10th of Janr** a seeing flaxseed trashed and claind I sent 9 barills to town I pd 1 10 10 towards bags 12d for 3 sives 2d turnpike 1 to ye men yt planted my trees to buy snuff 1s:1d to Jon Bryen to pay for more sives 1s:1d to him out of wch he pd 4d turnpike to sales 5 sulpher 1 and gave me ye change I pd 2 for an almanack I gott a letter from my brother[1] to desire I would attend at Rathkeal a thirdsday

11th att home ye men diging petatoes

12 I went to Rathkeal dined att my brothers but did no good so home to turnpike and snuff 1½d

13 att home gott my peas putt in

14th I went to Courte gave some directions so home my sistr Ireland sent 4 shirts but no money

15 att home unwell

16 gott ye gardin beans putt in

17 ye plow fallowing ye Peas gardin

18 to Courte gott a barrill of flaxseed clain'd and ordrd it to town so home ye horses putting out dung I had ye young gelding harnissd to ye truckle

19 ye horses putting out dung Tim Callahan putting new Posts to ye bounds gate ye has 2 day more mending ye truckles I pd 8d for 2 new Sives

20 ye horses puting out dung

21 to Courte orderd a motton to be killd and sent to town so back calld to se Sam Harding so home ye men finishd puting out ye dung

Janr ye 22 1743/4 att home unwell

23 I gott my old sow killd went to Courte orderd a ten bushills of Mault to ye mill so home and sent 4 buishills of my own mault to ye mill D Harragan and my own men a spreding dung I gott my peas winnowd

24 D Harragan and my own men trashing beans I got some Plants yt Paul griffy sent me I took phisick and one of my fore teeth came out I had my Sow salteed

25 I had D Harragan and my own men diging for Cabidge and beans I sent a chine of pork to Courte to be sent to Limer and wrote to Mrs Hartstonge to send me a pound of hops I bottled two dowsin and 4 quarts and 4 pints of Sider I gott in my lining I gave to be washd last week and ye Lord make me thankfull

---

1 Husband of his 'sister' Ireland. Because of Nicholas Peacock's habit of referring to his in-laws as 'brother', 'sister', 'mother', it is difficult to say whether this was his real brother or not.

I have 8 table Cloaths 5 napkings seven new Shirts Seven yr old and 12 old Shirts 24 stocks 6 towells 5 pair of new sheets a fine pair a course pair and an old 11 pair of yarn stockings 7 pair of wosted three of cotton 3 of thread 2 of silk 3 old 3 new handkerchef

26th D Harragan and my own men diging and setting of beans

27 I gott ye mault brewd and ye hops from Mr Hartstonge for wch I owe her and directions from her to gett 8lb from Ned ffitzGerald wth other directions I wrote to Mr Harding to borrow two ginies wch he lent me about 10 to Courte D Harragan and men trashing oats

28th att Courte

29th I sent to Killgabin but did not get ye money I went to Tough but did not meet Mr fitzGerald att home

30th I sent david Bryen to Limerick and sent fitzGeralds letter to Ms and wrote a letter to send to tough in ye morning

31th I went to Limerick where I stayd till 12 then in company wth David Bryen and M Daniel spent 1s:4d to Elinr and Nick Mcgan 1:1 so to Courte sould all ye skins to Mr daniel fo 40s and agreed wth him for 3:3 for each skin yt will be taken down till sheere day I gott 5:5 earnest and in parte for ye skins 17:4 I orderd Tim to kill a motton we supt on ye inside so to bed

febr ye 1st 1743/4 I sent 3 quartrs of ye motton to town 3 torkes and a pig I sent ye skins I gott a bus: of peas and 6 bus of wheat claind I sent home my truckles

2 I sent 3 salte samons to town got ye drink tun'd and gott a roast torkey from home

3d as I was seing staves reconed to Day I had an express for me and ye horses to go to Dublin for Mr Hartstonge was very ill I gott to town by 2 orderd everything to be gott ready against ye morning I hired a boat to go for ye Staves for 11s:11 earnest 1:8

4th mrs Hartstonge sett off aboute 10 I stayd in town till 12 pd severall bills wch are in acct so to Courte to a dram in 3 snuff 5

5th I was impatient till 2 then I gott an acct yt Mr Hartstone was better

6 I sent David bryen wth 2000 staves to town 2 barrills of drink and a bus of peas to be left att ye Docters I spent ye day in ye barn seeing ye barly winows

7th att ye barly till diner as I was att it I a letter from ye Doctor wth ye malincoly acct of Mr Hartstongs departing this life ye 4th inst I gave Jon Pursill 1:1 and sent him wth a letter to Standish Grady aboute secureing ye effects

8th Mr Grady came and seald some of ye doars and ye chest of drawrs I went wth him to Limerick where we found ye goods seased and canted after some advice wth Mr Grady I came to Courte and gott 53½ att 1:1 and a stick of wax att wch D Bryen sent me I sent 6d

9th I sent ye Dublin mare to Mr Burys and wrote to him George fosbery and Jon Hartny for grass for some of ye stock wch they gave but finding yt there

February 1744

was spys out I sent 47 hed of ye black cattle to Elltown[2] and gave P Daniel 7:7 to bare their expences and spent ye day devideing ye cows I sent 8 cows and 7 calves home 4 cows to Corroheen tenants four cows to ye tenants here and 6 to Ballygeasy corroh and gave 1:1 for taking care of them att night I sent a desk sea-chest and a trunk coverd wth lether to Shannongrove to tim Borks a chest of drawers a bed and garde vine to Cosgary a bed to P Daniels a bed to Tho Hanrahan a bed a box wth a hat wigs furniture and some thred in itt

10th I sent 2 sconce glasses wth 3 pr of scroles a small glass and 2 hooks to Cosgarys I pd 8:9 for claining ye flax seed I gott a letter from mr Mellsop I sent 5 picters to Tim Bork ye men came from Eltown ye Cattle got safe there ye pd turnpike 2:3 for watching ye Cattle 13d their expences 4:3

11th I gave old Palatine 6½ for making ye bags I sent Connr for one of my own cows I went to Ballygeasy where Mr Jacson tould me I was betray'd for ye People at Adear knew where ye Cattle and things were as well as I he gave me a lettr to Vall Keating[3] and desird me to send ten of ye cows and to send ye picters and sheep to him I swore ye men and sent them wth ye Cattle and Cosgary wth ye Picters I wrote to my brotr for 10l and orderd John Bryen to go wth it

febr ye 12th 1743/4 in patient for a letter but gott none I sent ye black colt by Patt Daniell to his Brotr in law and gave him 6½ I brought home all ye things I sent to ye tenants a thirdsday T Hanranhan came home no letter by him

13 Jon Bryen tould he did not get ye mony from my brother I sent Mrs and Mr Hartongs beds to Corcamore[3a] I gave John 2:2 to pay ye men att home D Bryen gave me an acct of ye money he gott for staves wch I interd in acct

14th I gott 2 letters from Mr Grady wth directions to go to Mr fosbery wth one of them wch I did and having waited for him till 6 I showd it him wth wch he sent me to Tough Jon fitzGerald[4] took a coppy of itt and aftr some advice I came away and gott to Courte about 9 no Dublin letter

15 I gave Mr Grady an acct of what I did yesterday I wrote to doctr Martin for some advice and aboute 30c in ye eavening he calld and aftr some talk he gave me 2 ginies and went away I pd ye high Counstable 11s:2½d in full for Country charges

16 I pd Law Bork 1:2:0 in full for ticking I sent Jon Bryen by Connr Quelly 1:2:9

17 I wrote to Mrs Hartstonge but did not get a letter to Salte 3½ to 300 of plants 6½ young Rahily came for Mrs Quins tythes but I did not pay him

18th I gave P Haly 13d I borrowd of his son I he sent me 3 he o

---

2 Elltown/Elton, seat of the Grady family. 3 Valentine Keating was a son of Jeffrey Keating of Bay Bush and Mary, daughter of Thady Quin. He married Sarah, daughter of Patrick Creagh of Tiervon, Co. Limerick. He emigrated to France and settled in Poitou where he died in 1780. Burke (1958). 3a Corcamore, Co. Limerick. 4 John Fitzgerald of Ardlahan, Kildimo, Co. Limerick. He is probably the John Fitzgerald of Kildimo, son of Maurice Fitzgerald, who entered Trinity College on 23 November 1723 aged 17. His wife was Honor, and his sons Terence, Patrick, Edmond, Maurice and Michael. His will was dated 16 July 1790 and proved 1 November 1793. *Irish Genealogist*, 1 (April 1937).

19 I had a letter from Ms yt they were as well as cod be expeckted
20 P Haly mended ye harrow
21 I wrote to Ms and pd 10d for peaper 1d for blew I got a peice of lining from home wch I cutt into 24 towells ye Priests are taking
22d I sent to see Mr Quin to ink 2d ye news 1d
23d I finished my towells and got a letter from ye doctr for Mastr Harry[5] Mare and yt Mr Quin was in a bad way
24 I gott a letter from Ms to keep close watch I wrote to Jon Bonfield for yards of black sarge and 3d worth of thred wch he sent and I owe for Mr Quin is better I drue up an inventory of ye goods and stock to send to Mrs Hartstonge I had in all 4 barrills of Oats from home
25 I gott a letter from Colonell Taylor aboute flax seed and an other from Ms and another from Mr Grady to desire I wod meet him in town a thirdsday
26 I wrote to dublin and gott a letter from Ms to stop ye plow for ye lands wod be sett I orderd ye Plow tackling into ye wheat garrott to nails 3d to a padlock 13d
27 I gott up early to see if ye nights tempest had done any damnage and found ye wheat Garrott stript in several pleaces and ye windows much shatterd I orderd ye men to thatch ye garratt and then to dig petatoes Mr Quin worse
28th and [2]9th on gard

March ye 1st 1743/4 tho a bitter morning I went to Limerick and aftr some talk wth Mr Grady we came to a resolution not to do anything in ye sad unsettled afairs wthout Counsiller Minions[6] advice and appointed Sunday to talk to him ye gates being shott I was oblidged to stay in town till 2 I borrowd Ed Mcgans Mare calld att adear and Port so to Courte I spent 17d
2d I wrote to Mr and had ye Pleasure of a letter
3 I had a taylor mending my coat and weastcoat to him 13d I gott a letter from my Brotr and 5:0:0 wch wth 3l he sent me by Jon makes 8l:0s:0d I lent D Bryen 13d
4th I gott up att 3 wrote to Ms aboute 6 sett off for Limerick and borrwd 4s:5d from Paul Haly I ad oblidged to stay in town till 2 to know Minsions directions att last he and Mr Grady came to a resolusion to have ye Stock getherd and given up to ye Sherriff expences 2:9 lent David Bryen 1d
5th I wrote to Mr Grady to have Courte and Corroheen advertized for wch I pd 5:5
6 I sent for letters and got one yt Mr G Mellsop brought to ye news 1d nothing in it but a proclamasion against ye Roman Catholick clargey[7]

---

[5] Henry Hartstonge, who succeeded his father Sir Standish Hartstonge in 1751. He married Lucy, daughter of the Revd Stackpole Pery in 1751, and became MP for Co. Limerick. Burke (1985). [6] Matthew Moynham, attorney. See above. [7] A proclamation published by the lord lieutenant offered £150 reward for the successful apprehension and conviction of any 'Archbishops, Bishops, or any other Popish ecclesiasticals' and for 'suppressing monasteries, friaries, nunneries and … Popish Fraternities and Societies'. *Faulkner's Dublin Journal*, 28 February–3 March 1743/4.

March 1744

7th I sent away ye beagles and satt on gard making pens I sent a hondr to Mr Mellsop to send to Dublin

8 making more pens att 12 Mr Badham and Thornill[8] came and desired I wod ride abroad wth them wch I did we came back aboute 4 and att 10 they went to bed but I did I was oblidged to lye in my cloaths for want of a bed I wrote to Ms no letters but ye news wch cost 9 an allarm of wars[9] I spent ye day wth Badham and company

10 Badham still here I took Killmoreen

11 I wrote a Minnitt and gott one parte signed they all went away to Ballyengland

12 I went early to Ballyengland gott ye other part signd aboute I came back Mr Mellsop came to see me

13 I gott a letter from Ms I wrote to Mr Bryen and sent him my rent by Jon Mr Mellsop went away to sope 4d to ye news 1d

March ye 14th 1743/4 on gard

15th I agreed wth a carrier to carry hogsheds of Sidr to dublin att a giney each I sent Cosgary to Mr Jackson for ye picters and orderd him to give them to Mr Mellsop to him 6½

16 I drue ye Sidr and gott it ready I gott a letter from Ms and an ordr for 2 barrills of flaxseed I was obligeed to give 2 Quittrent drivers 13d for not driveing to a stick o Sealing wax 6½

17 I sent ye carrier away 18 on gard 19 on gard

20th I wrote to Standish Grady aboute ye Quittrent I gott a letter from Ms Snuff 2d news 1d

21 I bought a yearling from Harnedy for 14s and am to pay 2s in cash ye rest in corn

22d I walked to Killmoreen and veiwd it I wrote to Ms and to John Johnston aboute his white mare I bought a cow from Tho Hanranham for 1:14:0 am to pay att Midsumr

23 I sent for letters but got only one from mr Johnston I got a letter from my brotr butt no money I gave Mr Johns mare to James Melane for a cow a yrling and 7s in mony

24 I bought 3 heffers from Jon quain for 7:10:0 Jon pd earnest 13d then went back to my garrison

25 no letters I spent ye day on duty

26 I went home and bought 2 cows from Tho Harragan for 4 ginies 2 heffers from Jon Quain for 4:16 I went to Killmoreen look'd aboute so to Courte

27 I wrote to ye Collectr about ye Quittrent no letters to peaper 5d to Salte 7d

---

8 Richard Thornhill of Rockfield, Co. Cork married Sophia, daughter of Brettridge Badham. Will of Brettridge Badham, 30 July 1744. IGI; NLI GO 139; NLI MS 4177. 9 The War of the Austrian Succession had been declared in 1740. Alliances were formed between Austria, Britain and the Netherlands against Prussia, Bavaria and France, and in February 1744 the British fleet fought a Franco-Spanish fleet off Toulon.

28th I agreed wth Martin fitzGerald then went to Mellon and bought 2 cows and 2 heffers from charles Carty for 9:16:0 I dined att Jack Hartnys so to Courte

29 I went home and bought a heffer from Jon quain for 1:8:0 I carried him home and gave him my note for 14:3:11 and my note to Tho Harragan for 4:10:8 then fitzGerald and I signed ye agreement so to Courte then Carty brought ye cows and I gave him my note for 9:16:0 all ye notes payable at midsumer God enable me to pay them honestly

30th I gott a very dissatisfactory letter ye drivers came and drove 4 cows for ye Quittrent and put them in pound I pd them 3:4 and they went away

31st ye appraisers came and valewd ye cows and gave me a bill of apraisment I owe them their fees

Aprill ye 1st 1744 I got a letter from Judy I sent 5l:2s:0 to pay ye Quitt rent

2d Jon Johnston and Rob: hewson[10] breakfasted wth me aftr wch they went away and D Shea came I gave him some drink so away

3d Jon sent me ye Quittrent receipts I got a letter from Mr Badham to snuff 4d

4 I sent ye lennets an a cage to Adear **5** on gard

6 I gott a letter from Ms wth acct yt ye land wod not be sett this year

7 David fitzGerald came to let me know yt ye Sheriff wod come but he did not[11]

8 I sent a letter to Ms and one to Mr Badham I gott one from Ms

9 I wrote to ye doctr for a giney wch he sent I wrote to Mr Grady to know when I shod send for ye stock ye sub Sherriff came and seased ye asses 1 cow 3 reeks two stacks of hay and ye beans so away[12] Tho Jackson gave me a snuff box Mr and Mrs McGan call on me

10th I wrote to Ms and went to Rathkeal and by much ado devided my sistrs effects so home no letters to snuff 6½ expences 1:5

11 I went to see Patt: Kennys cows but cod not agree

12th I went to Adear talkd aboute Courte affairs so back twintey of ye cattle came from elltown

13 Mr Mellsop and his wife dined wth me then home I got a letter from Judy to onion Leek and radish seeds one and a penny news

14 I sent tho Jackson and 2 men to bellvidere for ye cattle and horses my horses harrowd ye Peas field

15th ye cattle came from Bellvidere I gave Aherly 5:5 to pay for Mr Badhams Saddle &c

---

10 Robert Hewson (*c.*1703–?) younger son of George Hewson of Ballyengland, Co. Limerick and Katherine daughter of Nicholas Peacock of Barntic, Co. Clare. He married Lilian Lees who died in 1780. Burke (1976). 11 Sheriffs were appointed annually for each county by the lord lieutenant. Their names were put forward by county magnates and were drawn from leading gentry families. They were the moving spirit of the courts, and were essential in organizing its business. Through their sub-sheriffs, they summoned juries. Garnham, p. 94. 12 Sub-sheriffs or under-sheriffs were chosen by sheriffs to carry out much of their work. Garnham says that by and large under-sheriffs assumed most of the sheriff's duties in relation to the function of courts and the implementation of their decisions. Garnham, p. 95.

May 1744

16 ye subserriff came and seasd 57 head of black cattle and 7 horses I sent to Baybush for ye cows and Cosgary for ye Cattle yt tyred ye Plow and 2 harrow going

17 ye Plow and one harrow finishd ye peas ye Sherriff seasd ye cows yt were att Baybush and ye sheep I wrote to Ms but got no letters my mare foled cosgary came and pd 3:3 for his and ye Cattles expences new and turnpike 2

Aprill ye 18th 1744 ye 2 harrows at Cloriphest Mr Grady and Jon Mallsop came

19 went to Limerick for keys but did not gett them to peaper and a dram 13 aboute I gott back and sent for ye Mare and other things to Shannongrove

20th ye Malincoly cant began in favour of ye fitzGerald I made ye hearth mony man seas ye glasses and gave him my note for 2l a Courte of enquiery calld to prove ye Childerns cattle Mastr lost his heffers no letters

21 ye Cant going on I bought 153 boards for 2:9:0 half ye timber aboute ye house for 1:5:6 ye sleats att 6s a thousand they all went away

22 I wrote to Ms and Mr but no letters

23 I counted 2429 sleats and made them be put in and all ye loose sleats aboute ye house then employed ye men getting ye timber I finishd sowing and sent home my Cow and horses 2 Slabs and a block of timber

24 I sent for letter, but got none to bread 6 sope $9^{1/2}$ snuff 4

25 aboute 9 Mr Grady and Jon came about 10 Sherriff they broke open ye Clossitts and spent ye day and most of ye night in debates I bought 3 hackles for $11:4^{1/2}$ a fish for 3d three books for 1:9 having Patt up all night I lay on ye bed but was oblidged to gett up imediately

26 I bought 9 pleats an old dish for 1:5 two ale barrill 3:9 three ankers for 1:6 an ash tree for 10s ye turf 15 ye mault 7l:0 then having finishd we dined and ye Company went away

27th I saw Mr Gradys Shear of ye beer draw and sent away Mr grady and Jon went ye two fitzGeralds stayd no letter I went early to bed

28th I wrote to Doctr martin and got an answer and began to prepare for Dublin I gave out 4 dow and 4 bottles of them I had and gott a letter from Mr Harding yt my dogs had killd 6 ewes and 6 lambs of his I wrote to him that ye dogs should be killd and he shod be paid

29 I left Courte att 5 breakfasted in Limerick dined att ye Nenah and supt in bur[13] and being fitagued to bed early

30th I dined at Tullamore and lay att Johnstown[14]

May ye 1st I gott to Dublin aboute 20 where I stayed till aboute 12 ye 4th and lay at Monstereven[15]

5 I breakfasted att Mooreboro dined at Rogray[16] and supt att ye mines[17]

---

13 Birr, King's County.   14 Tullamore, King's County and Johnstown, Co. Tipperary.   15 Monasterevan, Co. Kildare, 30 miles from Dublin.   16 Maryborough, Queen's County and Roscrea, Co. Tipperary.   17 The Silvermines are 15 miles north-west of Limerick, and 3 miles south of Nenagh. They had been worked since the 13th century, and possibly before then. The mines were

6 I gott to Limerick aboute 8 and having deliverd ye Messages and letter I gott home by 2 having spent of my own 1:8:5 and 11:4½ yt Mrs Hartstonge gave me and 4:10½ of Ms Nancys money and have nothing to shew but a pair of pumps wch cost 4:4 and a pr of buckles wch cost 10d when I came home I found all in confusion ye horses turned of off ye lands and ye doars shott

7th I sent for ye smith and made him strip ye little mare yt was tendr in her feet then went to Tough to know what was ye matter all was made of a reporte being made yt some body had administerd[18] and yt whatever was found on ye land would be seased

8th I went wth ye desk chest of drawrs sea chest and trunk to ye high Sherriff[19] where I breakfasted we came to faha[20] and he calld on me att Courte and having drunk and glass of ale he went away and I walked hom

9th I gott my sheep washd

10th I walked to Courte orderd ye men to dig in ye gardin and fence ye fields gott my mare shod and Tho: Bury came home wth me we drank 5 bottles of Sidr and he went away

11th I went to ye dere Park and walked aboute then I bought 9 hoggetts from Harry Supple att 6:1 each wch comes to 2:14:9 gave him 13d earnest wch helped on me so home and sent ye men to wash then so loyterd aboute till night

12 I took a list of ye cows I bought wch is 22 cows for 46:14:4½ 2 yearlings 1:20 15 Sheep 4:5:9 then I settled out my roome and gett 3 lambs I gave James Madin a dowsin and half of broad hoops

13 I went to Killdeemo Church from thence to Courte to try for a lettr but gott none then I came home and aftr supper I gott a letter from Standish Grady about ye Dublin Mare wch I answerd so to bed

14th I gott up early and sent my 8 yearlings to ye fare and aboute 6 followd them and a little aftr I got to ye fare I sould them to Mr Jon Bury for 8 ginies then I drank a mug of ale left Jon Bryen to gett ye money then I came home to a mug of Sidr on ye road 4d John came home and tould me he gave Martin fitzGerald on acct of his cow 2:1:0 to Dennis 4:4 expences 13 to me ye remaindr

May ye 15th 1744 I went to Courte and sent home ye joyners binch a levill and some boards then I gave will Ronan 2 10 0 in full for his cow and to James

---

acquired by Col. Henry Prittie as part of a large piece of land to settle arrears of pay for service in Ireton's army. In the eighteenth century they were only mined for lead. They were about 5 miles wide from east to west, and the summit of the Silver Mine Mountain was between 1274' and 1607' above sea level. As well as lead and silver, other ores were gold, zinc, copper, sulphate of iron, and sulphate of baryta. Thomas Dineley, writing in the late seventeenth century, said that the 'melting houses and Mill ... hath a large Water Wheele by whose motion a Great Forge bellows is lifted up and blown'. Its dismal aspect reminded him of Spenser's description of a workhouse for melting ore in the *Faerie Queen*, Canto VII. It was a regular stop on the Limerick to Dublin road, and there were two inns there in the seventeenth century, one called 'the sign of the Holy Lamb'. Dineley, pp 272–4; Gleeson, 101–15. 18 Meaning that the estate of a deceased person was being managed by its executors or administrators, which is conferred by letters of administration. 19 George Fosbery of Clorane was high sheriff in 1744. Lenihan, p. 744. 20 Faha, the seat of the Tuthill family, three-quarters of a mile north of Court, close to the river Maigue in the parish of Kilkeedy. Lewis.

McCoghlin 2 6 0 in full for his cow then I agreed wth ye tenants to make a neck trinch att 12d a perch then I shewd them a place to cutt turff aboute two Robin hewson came and I agreed wth him for ye gras of 40 bullucks att 6d a quartr each and I agreed wth him for 8 acres of medow for a giney an acre then I gott a letter from Mrs Hartstonge yt Ms was sick I pd for ye letter 4d new 1d things for ye Mare and snuff 13 then John Bryen came to me wth a complaint yt John Nash made an assalte against him so home

16 I sent John for a warrant for Jon Nash wch he gott I had my sheep shorn and pd 1:8 for itt

17th I sent Jon wth ye wolle and butter to Limerick ye wolle he sould for 6:6 a stone and ye rise till 24th of June it weighd 10s 3$^{1}/_{2}$lb wch came to 3:6:5 he pd for sope 2:5 oyle and tar 9$^{1}/_{2}$ wayage 3 Costom 3 expences 4 I gave him for himself 8:1$^{1}/_{2}$ and sent him wth ye 2 ginies I borrowed from mr Harding he pd Mr Johnston 1:18:10 on acct of ye mare

18 I went to see two Cows of Mr Burys from then to Killmoreen to took a list of ye greasing Cattle vewd ye deary cows so back to Porte and Marked 46 sheep of wch 19 is wethers 1 ram and 26 sheep and I marked 24 lambs I gott a letter from Judy wch cost 4d news 1

19th I went early to Courte gott some drink brewed and lay att Courte gott a letter from Standish Grady aboute ye mares

20th att Courte sent for lettrs but gott note I wrote to Judy I sent 20 sheep to Court and sent ye colt to

21 I went to Killdeemo to ye vestory whereafter a stiff argument we allowd 4:5:0 to be raised on ye Parish then I came home

22d I went to Courte sent for letters and gott one from mrs Hartstonge wch cost 4d then I gott ye drink tuned left a letter for to send wth Dublin Mare to Persons here and a shilling for ye groome so home Jon Bryen gott 5:6 from Coghlin Grady I gott 7s from J Melane for Mr Johnstons mare got my lining washed

23 I finishd ye 5th pair of my new sheets I sent John to ye Court to appear against Law Cleery he pd summons 4 entry 4 distringuis 6[21] sewing ye distringius 4 no atturney being at ye Court yt all he cod do till next Courte day ye little mare took ye horse

24th I send ye gray mare to ye horse but she wod not taken him I went to Courte gott ye drink stopt and gott my mault measured and have thirteen big barrills I gave Jon haly 1:2:9 to give Edward fitzGerald I borrowd of him to Carry me to Dublin then I came home and spent ye day walking aboute I gave Charles Slattery 4 bus oats John agreed wth Ballycraheen[22] men for liberty of a road att an acre of medow a year

25th I gott my oats winnowd and send four barrills to ye kill and gave B Halloran a bushill

---

21 Distringas, a writ bidding the sheriff to distrain goods to recover money. 22 Ballycraheen/Ballycahane, a townland a mile south west of Pallaskenry, covered 199 acres 3 roods. The former proprietor was Francis Barkley and it was granted to George Peacock. Barry, p. xx.

26th I sent Jon to Limerick to try to sell ye butter but it was so bad He could not sell itt he pd 2s:2d for 2 pd of hops kepd 1d turnpike he gave me 2:2 he gott for some of ye sheep ye dogs killd I allow [illegible] B Halloran for Martin fitzGerald 7d Den: 2:2 Patt 1:5 and she gave me 3:6 wch comes to 7:8

27th I wrote to Mrs Hartstonge and to Ms Nancy and sent her 3:9$^{1}/_{2}$ I laid out on ye Road and spent ye day walking and reading

28th I sent John to Limerick aboute ye butter and gave him 7:2 to give Mr Johnston I wrote about ye clock wch cost 2:8$^{1}/_{2}$ and I could not gett it to go ye butter my much ado he sould for 20s he did not give ye money to Mr Johnston I accted wth Shea and do owe him 17s ye 2 boys broges included J nele att work

29th I wrote for Cyp Pursill when he came he cod do no good to ye Clock so home Mrs Warne and her son come to treat aboute greasing but as I wod not come into her measures she went away in a huff I gave Dick Dilane 4 bus of petatoes J nele at work

May ye 30th 1744 I went to Courte and stayd there yt night men began ye neck trinch

31 I came hom in ye morning but I had scarce put up my cloaths yt were washed when I was sent to Courte I went and agreed wth mr John Bury for 8 bullucks and David Divier came and brought 21 bullucks I stayd att Courte

June ye first I sent for letters and gott one from Ms and another from Rob: Cox I sent home some timber to mend my truckles and gott ye horses shod and sent home 18 quarts 7 pints and 2 large stone bottles a tongs poker and 2 quillting freams

2 I wrote to Ms and sent Ms Maney to carry her when she went 2:14:10 parte of ye pryce of her bus luck[23] and owe her still 5:2 wch I gave P Daniel to give Mr Olivers groom I send 2 coach geldings to Mrs Perry so home and got my drink brewd Tim Callahan yesterday and to day puting ye truckles in ordr Jon Nele trashing yesterday and to day

3 I wrote to my Brother for money but got none but a very disatisfactory answer Patt Daniel came home and brought ye mare but did not pay ye groom he spent 19$^{1}/_{2}$d

4 I began my turff and had 4 truckles of my own there of Mr Hartings 1 of Tho Harragans and 3 horses wth basketts from Courte and E Danels horse att ye bog P Boyle and J Nash Mick Daniel a tending T Harragans horse

5th att ye turff and finishd aboute 1 ye same men and horses and Jon Newman then I sett ye men to trash

6th I went to Courte sould 3 bus mault att 20d a bushill and gott 3:4 so home got my drink tund and my six new stocks Ironed

7th Ned Mcgan and M Daniel came I drank wth them till drunk I do not know how they scaped but I slept ye day and night

---

23 Luck money was paid to seal a bargain; a deposit.

June 1744

8 I was craw sick so loyterd aboute

9th I went to Courte gave Jon Mellsop a barrill of mault so sent home 2 ankers of drink so home Jon came home and brought me a bottle of Shrub he gave Mr Johnston 8s he left my briches in town to be mended I gott a letter from Mrs Hartstone and another from Mr Taylor wch I answerd

10th att home museing and reading

11th I went to Croome to know what my brotr intended his answer was a bill in chansery I spent 1:3 so home

12th I went to tough and advised wth John fitzGerald who tould me I could come by my mony no other way than by fileing a bill so home and gott my Oats winowd I pd Harnedy for his yrling and he owes me 2:6½

13th I went to Courte and stayd there till aboute two and gett ye gray colt shod rid so home Jon came from ye Court but did no good about 7 Paul Haly came and tould me his house was burnt I went wth him to Court

14th I gave P Haly 4 dowsin tevanes and 16 rafters so home Jon gave me 13d he gott from Maddin I gott my two geldings shod Tim Mahony and Killbreedy men came and settled accts they owe me 7:2 J nele and P Boule trashing

15 I gave Joan Daly 13d on acct of her wayges I loyterd about J nele trashing

16 I sent Jon wth Martin fitzGerald to town he sould 2cwt 3 14 26 lb of butter att 18 a hondrd wch comes to 3:13:8 he pd Mr Johnston in full 9s:2d allowd Martin 2:4 he gave me 2:2:2 I gave Martin a note for ye bullock Joan Daly came to live wth me

June ye 17th 1744 att home reading and [illegible]

18th to Courte 19th att Courte and devided ye beer wth John fitzGerald to bred 3d sugar 3½

20th I gave half my bear to Geo. ffosbery I lent David Bryen 13d

21 att Courte a triming my horses

22th att Courte 23d I gott 3:3 from Jon he got from B Halloran 13d from E Daniel

24th to Church and back to Courte

25th I sent my two gelding to ye fare but sould nither to a bridle 2:2 expences 2:2 John sould my two heffers to Tho Harragan for 3:9:0 in parte of ye money I owe him

26 att Courte waiting for a letter but gott none ye big gray mare took ye horse

27th I gott 13s 6½ from David Bryen and gave conner 2:8½ ye dog boy 10 10 so home and gott my corn winnowd

28 att home unwell I gott 13d from B Halloran ye men making Killmoreen Bounds

29th ye men att Killmoreen bounds I gave Wat tobin a bushill Oats

30th I went to Courte gott my horses shod to carry to ye fare wth Gods blessing a Monday I gave Richd Harnedy half a barrill mault I wrote to Mrs Hartstonge aboute 2 home and was wett

July ye 1st 1744 I left home and went as far as Charlevill and there lay to ye Childern 7d

2 I went to ye fare and sould ye big gelding for eight pound 5s and ye Mare for 8l aboute 3 I left ye fare and gott home by daybreak

3 I sent John wth 3 yrlings to ye fare and cast up my expences every way wch amounts to 10s:7½d I made Will Hickey cutt out briches for ye 2 boys I gave Tho Hanranhan on acct of his ~~sheep~~ cow 1:2:9 Jon came from ye fare but did no good I gott a side of Kid from Hen Supple

4th I sent 3 att 1:18:10 wch comes to 5:16:6 by Jon to John Quain in parte of his cows and sent by Jon to Harry Supple 2:5:0 on acct of his sheep he brought me 7:2 in full from ye Clerys I gave Brig: Halloran a peck wheat to tabaco and snuff 1d

5th I paid Patt Day and fitzGerald 7s0½ in full for their appraisers fees and gott maddin to mark a butter ferkin to send to town in ye morning I gott a letter from Ms

6 I sent John Bryen to Limerick wth my own and ye deary mans butter he sould my own att ye rate of 17 itt weighed 0:3cr::5p neat wch comes to 13:6 stopt for a wrong tear 1:7 so yt I gott but 11:9½ he pd 3s for making my wig 1:2 for gon powdr and 5d for snuff he gott 1:2:9 from Ned Mcgan in parte of ye mare

7th I went to Courte and walked aboute so home and gott from martin fitzGerald 2:17:6 his butter weighd 3 1 5 I gave him 6½ wch wth 1:7½ stopt on acct of his cask makes 2:2 I owed him so yt we are cleer on all accts to Charles Slattery half a barrill of barly

8th to Patt Hanrahan on acct of his wayges 2s:2d I gott a letter from Ms wch ocationed my going to Courte in ye eavening

9th I made 26 bottles of wine to be packed and carried them to Killbreedy and gave P Hanrahan on acct of Carriage 13 then walked ye dere park then in Company wth Supple and Brian Spent 13d so to Courte I gave Supple 1:7½ to pay for mending ye locks of ye dere park

July ye 10th 1744 I bought my and Mr Hartstongues two thirds of ye tyths and gave my Notes to Doyle for 10s my own and 40s Mrs Hartstongs and gave him 5s:5d fees

11th att Courte Ned mcgan gave me 6:17:3 in full for ye mare I gave him 2:2 in lew of a bridle he was to gett wth her I gave Mick Cosgary 5:0:4½ I borrowd of Ned Sheehee ye 23d of March last I reconed ye boards and have 139 wch wth 18 I sent home makes 157.

12 I had ye men att ye hay and agreed for dressing Ms Hartstongs flax I pd Man: Rahily 4:0:0 in full for last years tythes and gott my notes

13 ye men att ye hay and ye flax going on I sould ye two ye two Mr Westrop[24] 16 acres of Medow att 17:3 an acre ye two thirds of ye tythe included

---

[24] The Revd Cecil Westropp was the son of Thomas Westropp and was born and educated in Limerick. He entered Trinity College, Dublin in 1730, took BA in 1734 and became prebendary of Ardcanny from 1757 to 1788. He married his cousin Hannah, daughter of Ralph Westropp of Caherdowgan. Leslie.

August 1744

14 ye hay and flax going on I came hom in ye eavening to Ned Naghtin by Jon Bryen 1l 2s 9d in full for his sheep Jon has 2s 10^{1}/2d in his hands

15 I went to Adear in ye aftrnoone I gott an act yt Mr Badham was dead in ye back to Courte I spent 4d

16 dined att Killdeemo to Sidr 8 I took Mr weights parte of Mrs Hartstonges tythes and gave my note for 2l then I came home in a frett yt David Bryen forged my name

17th I devided parte of ye timber wth tutthills men I got a letter from mrs Hartstonge wth directions to go to town Jon Bryen brought me my frize wch being spoild I ordered him to carry it to Elener Hines yt spoild it and sent her word to pay me for itt

18th I went to Limerick served Law Nihill and ye tenants wth a coppy of a reversary decree[24a] I gott 15s on acct of ye wolle I pd for ye hire of ye Pillion 8s for Iron 5 to peaper 2 expences 1:8^{1}/2

19 20 21 ye men att ye hay

22 I gott a letter from Mrs Hartstonge wth directions to send ye horses for her

23 I wrote to Siss Westrop for ye Pryce of ye hay and gott but 18s:1d I sent a barrill of Mault to Mill

24th I gott a letter from Ms Hartsgon and pd 2:2 for a poun of Snuff for her

25 I gave T Honranhan 18:1 and sent him wth ye horses to Dublin then I went to Tough from thence to ye dere Park and spen 10 in drinking many happy birthday to Mar. Hartstonge so back to Court where I mett a company of tag rag and bobtail yt came to Celebrate ye day they envited me to make one of ye Company wch I refused

26 I got ye drink brewd

27 to bread petatoes and ye news 5^{1}/2

28th I finished gathering in ye hay I gave ye man on acct of dressing ye flax 1:7^{1}/2

29 I went to Mr Quin who is very weak I spent in Sidr 7^{1}/2 so home and gave Dennis earnest for his heffers 13d to P Hanranhan 13d to Joan and Jon to buy tabaco 6^{1}/2 so back to Courte

30th I sent P Daniel wth Ms Hartstongs horse to Cnockony[25] fare and gave him 2:8^{1}/2 I heird Mr Quin was dead[26]

31 I sent for letters but got none to ye news 1d

Augst ye 1st 1744 aboute 10 I went to Mr Quins funirall and gott a hatt band and gloves[27] I spent 20s on me and ye People of Courte

2 and 3d nothing materiall

24a At some time in the future, when Lawrence Nihill died, the lease reverted to the Hartstonge family.
25 Possibly Kilknockan, west of Adare. 26 Valentine Quin asked in his will that his body be buried in Adare 'in my family burying place in a decent manner according to the usage of the Church of England whereof I am a member, without any vain or unnecessary expense'. Will, 12 March 1743/4; probate granted 15 September 1744; copy in Dunraven papers D.3196/K/1/4. 27 Valentine Quin had left £10 for mourning to his son-in-law William Barker, £10 to his son-in-law Perse Creage, and £10

4th Tho: Hanrahan came home I gave ye mowers 11s:4¹/₂d

5 to Church and back again

6 I bought warners[28] Parte of ye tyth and gave my note for 20s pd ye Procters 2:8¹/₂ fees

7 I begn to make Sidr to news 1d

8 I had 8 of Courte men att Port reeping wheat

9 Geo fosbery took away his sheep but did not pay

10 ye men att Porte I went to Limerick to meet Mrs Hartstonge but did not expences 2:1

11 ye men att Porte

12 Harry Supple brought a lace of patridges an a cowple of Rabitt I satt reading all day

Augst ye 13th 1744 ye men att Porte Phill geany took his cows and pd 3:6 wch wth 10:6 I allow'd him on Connrs acct makes up 14s ye quartrs grass I gave ye boy yt rid fitzGeralld 3d

14th I went to ye fare of Rathkeal sould my horse for 6l:10:0d I gott 4:0:9 I sould Ms hartstongs heffer for two pound and six sheep for 1:16 and gott a giney and half of ye money wch is 1:14:1¹/₂ I borrowd 2¹/₂ ginies of Simon Daniel I pd Charles Carty 9:16:0 in full and took up my note expences 8:3 so to Courte

15 Jon Bryen gave me an acct of ye hands employ'd pulling my flax vis: 33 men from Courte 12 hired men 17 women from Courte and 20 other women I gave him 5:5 to pay 20 women and pd Courte women 3 Shillings I sent Simon Daniel 4s:11d

16 nothing materiall

17 I went to Limerick and found Mrs Hartstonge had been in town since ye 15 I went wth Mastr to Adear back to town so to ye ffery[29] then was obliged to go round thro Adear and gott to Court very wett expences 1s:0

18 I sent some butter and fowl to town

19 I went hom and brought ye two new peices of linin to Courte and gott 1l:19s from Jon Hartney for ye grass of his sheep and he took them away

20 I sent ye horses to Limerick and the Linin I brought from home yesterday ye gave ye mowers a giney and half wch wth half a giney I gave them before ye greasing makes 2:12:6

21 I went to Adear to Mrs Hartstonge from thence to ye dere Park orderd a dere to be killd expences 1:2 to bread 3 to thred 3¹/₂

---

each to his nephews, Valentine Quin, Thomas Grady, David Barry, Valentine Keating and £10 to his kinsman Edmund Morony. Will, loc. cit. 28 The Revd Simon Warner was vicar of Kildimo from 1742 to about 1757 and prebendary of Ardcanny between 1731 and 1757. The son of Thomas Warner, he was born in Co. Cork, entered Trinity College, Dublin in 1700/1; he took BA in 1705. Leslie, 'Diocese of Limerick'. 29 The ferry over the river Maigue was certainly in existence in the period when Lord Protector Carew (1555–1629) built a castle there to protect the crossing. Charles II granted the Hartstonge family income from the fees from a ferry between Killcollum and Contaghlagh. In 1792 a bridge of three arches was built over the river. Sheehan, p. 204; Limerick papers NLI PC 876/3/19/20.

September 1744

22 I sent to town for ye horses and sent P Daniel and Tho: Hanrahan wth fitzGerald to Killdarrary[30] fare and gave them 2:8½
23 ye Priest took ye horse yt J Geany had and only pd 5:5
24 P Daniel came home and sould Ms Hartstongs horse for 4:0:0 and gave me ye money
25 I went to town gave Ms ye Price of her horse and mastrs illness hindered me from doing any good so came away expences 1:6 ye **26** nothing material
27 11 men att Port aboute 11 Halpin came and fell to dressing sleats
28 D Bryen brought me a loaf of bread wch I owe for
29 I was woke by ye violentness of ye Storm and gott up but I bless God I found no harm done but a prodigious pile of Apples thrown down ye storm continuing I was in dread of an overflow and sent ye men to ye banks
30 I sent Cosgary and his horse to Limerick and he brought home 500 of laths
31 I sent Cosgary and horse to town he broght home a hondrd of hoops and a hogshed Divier came for his bulluck and pd for them I sould 3½ acres of medow

7br ye 1st 1744 I went to Limerick gave Diviers some mony to Mrs Hartstonge about two I came away to sope 4½ gloves 2:2 turnpike 3 expences 1:6
2 I went to Porte and accted wth martin fitzGerald and gott 13s:2½ wch I gave Den and Patt on acct of their wayges so to Courte and brought all ye thread of Madm yt I had wth me to bread 3d
3 I went to ye vestory and consented to have 6:7:2 raisd then back a shouldr of motton for supper
4 nothing materiall
5 I sent Tho: Hanrahan wth a ronlet of Sidr to town ye Sleaters finishd they have 9 day att 18d a day Mr Sis Westrop pd 16:11 wch wth 18:1 he pd before makes 1:15:0 ye greasing of his sheep to a knogin of oyle 4
6 I had an acct of two Mss were ill **7th 8th** nothing material
9 I was sent for to town where I got just as ye gates were opened[31] aftr some talk I gave Mrs Hartstong ye money Mr Bury sent for ye grass of his bullucks and ye 16 11 I recd from Sis Westrop in all 3:11:1 so back to Court to snuff 6½ to N Mcgan 6½ to other expences 6½ ye **10** nothing materiall
11th John Bryen came to give me an acct of what he did att ye fare and Limerick he borrowd 1 6 0 off Martin fitzGerald he gott for butter of my own
12 ye remaindr of ye Pryce of ye woll 10s for wheat 5s in all 2:13:6 of wch he laid out for 10½ bandills of frize 7s:10½d to Soles for ye 2 boys and Joan 13½ Salte 1:7 Costom turnpike and wayage 5 he gave me 2:2:4½ he changd ye dishes and plates I bought at ye Cant for 11 new plates to ye news 1d

---

30 Kildorrery, Co. Cork, is 27 miles north of Cork on the main road from Fermoy to Limerick. Lewis. 31 The walls of Limerick in the late 17th century were wide and had a paved walk, as in York. Even as late as 1760, there were 17 gates still standing in the walls. By order of the mayor and corporation, the gates were opened at 4 in the morning in summer and at 7 in winter, and were closed at 10 pm in summer and 9 pm in winter. Dineley; Lenihan.

7br ye 12th 1744 I gave Harry Supple a bill on ye high sherriff for 4:0:0 being on acct of ye greasing of his sheep

13 Martin fitzGerald came I gave him a receipt for 3c 1qr 3lb of butter he sould for 18s a hondrd wch comes to 2l 18s 10½ I pd him ye money J Bryen borrowd ye 8th inst I gott a letter from ye Doctr to offer 9:2:9 for Suppels horse for Mastr Hartstong I wrote to Supple and got an answer he wod not sell ye horse undr 9:10:0

14 I wrote to ye doctr and sent Suppels letter

15 I spent 2d stilling whiskey

16 mostly reading I gott 2 shirts from home

17 I sould an acre and half of medow for 18s

18 I went to Limerick sould my 20 sheep for 9:10:0 then I took up cloath for a rideing coat and triming for a coat and weastcoat from Jon Bonfield then I pd for mending my whip 1:6 expences 9d So to Court and wrote to ye two westrops for money and to Robt Hewson then I went to Tough wth a letter but did not meet Jon fitzGerald so back to Court and did not gett a penny of money

19 I wrote to Robt Hewson by John Bryen and to Har Page and sent 3l then I went to Limerick stayd to dine and gave 1:10 for two handkircheifs to peaper and expences 1:10 so to Court no money from Rob. Hewson Page refused to take ye 3l

20 being treatned to be drove I sent to ye westrops to no porpose then I borrowd 6:10 from Patr Sheehan Mastr Hartstong dined wth me and agreed for Suppels horse on my giving my note Payable on demand wch I did for 9l ye tealor cut out a rideing coat a coat and weast coat for me ye hackler began Ms Hartstongs flax then Mastr rid his new horse to Shanongrove

21 I went to Port and sent Jon Bryen wth 9:14:8 to Page then I sent all ye flax I had of Mrs Hartstong to Court and gave Moll Pallatine 4 pound and Mastr Sawn ~~and Mary Sawn~~ 4 pound more to spin about 2 Jon Bryen came back and brought Pages receipt then I took my tythes and gave 2 notes one for two pound ye other for one pound I gott my rideing coat home

22 I went to Limerick gott Mrs Hartstongs things put in ye boat to beverige and other expences 1:7½ so back to Court

23 I got home my Coat and weast Coat but did not pay for them ye forenoon I spent reading ye aftrnoone a seeing ye things in I pd for ye making my cloath ye 31st

24th ye hackler finished Ms Hartstongs flax She has seven stone tear 1 stone Shorts 4s:5lb of to ye hackling comes to 9s I pd him 4s:6 I gott 14[?] and 10 quarts 4[?] 5 pints from Limerick Jon Moran pd 4s:8d in full for ye grass of his colt and took him away I sent 2 dowsin small beere and 3 bottles Sidr to Limerick

25 Rob: Hewson came to me and gave me a sow to news 1d

26 I sent for ye sow and gott her Jon Bryen tould me he agreed for 14 calves att 9d each from Martin Quain ye money to be pd att midsummer ye hearth mone[y] collecter came for his money I gave him a bill for two pounds 4s on ye westrops

October 1744

27 nothing materiall

28th I sent 2 dowsin of small beere to town I broke one bottle I wrote to Mr Hewson and westrop for money but got no other but yt Mr Westrop pd ye hearth money collectr. 2:4:0

29th I sent 7 stone of flax to Ms Hartstonge and gott 3:4 from Patt Daniel part of ye Pryce of ye pad I walked to Killmoreen veiwd ye calves so back and diverted me as well as I could

30 I went to Church and back to Court

8br ye 1 Mounty westrop[32] gave me my note ye heart money collecter had and ye receipts and began to draw his hay and I began to still

8br ye 2d 1744 I gott 13d of ye 2:8$^{1}/_{2}$ I gave Jon Bryen a Sunday so yt I must charge to his acct only 10$^{1}/_{2}$d I agreed wth Martin fitzGeralld for 19 calves att 9s:6d wch comes 9:0:6 to be allowd in his acct I sent smalle beer and appels to town to snuff 10 I gott 5:5 from Harnedy he owed me for mault

3d I had my truckles from home and began to draw oats I sent a pig to town ye Clock was clained

4th I finished ye still and bottled 6 dowsin and 2 bottles of small beere I gott 3:3 from G Pursill and 3:3 from w Staff I sent a torkey and a bit of bacon to town ye men finishd ye Oats and began ye hay

5th ye hay drawing I send some Sidr and petatoes to town and spent ye day a drawing up my acct wth Mrs Hartstonge

6 I had ye favour of Mastr Hartsgons dining wth me I began to put in ye hay but ye wettness of ye day hinderd me

7th I spent reading

8th I began wth ye hay att 2 I was oblidged to stop ye day being wett Mast dined wth me I pd a mower for mowing ye walks 1:1

9th ye men att ye Sidr claining ye yard garding & medows Cosgary and mare in town I gave Connr Quelly on acct of his wayges 1:7$^{1}/_{2}$ to ye mower for moing ye back garding 1:1

10th I sent ye horses for Mrs Hartstonge and a dowsin of Sidr I gave ye chimney Sweeper a tickett yt he swept nine chimnies I went home and brought 2 pr Course Sheets and a pr of fine and a pillow case I gott 7s from James all in ye grass of his horse and sent him away

11 I sent I sent Patt Sheaghgan ye 6:10 I borrowd of him ye 20 of last month and then I gott things in readyness to receive Mrs Hartstonge about 3 she came and we spent ye Eavening malincoly enough I gave tim County 13d to 8d for Mattys coat

12th I went wth Mastr Hartstonge to hunt but hereing Mr Bury intended to dine att Court I left them having given Mastr 13d and ye huntsman 13d aboute

---

32 Mountifort Westropp was a son of Thomas Westropp of Mellon, Pallaskenry, Co. Limerick and Elizabeth Bury of Shannongrove. He married Martha, daughter of — Roberts of Britfield, Co. Cork in 1744, and they had two sons and three daughters. Burke (1958).

3 Mr Mrs and two Ms Burys came and stayd to dine ant till 5 so away to ye new Id

13th I gave ye Gleasier 6½ and he went away I pd Patt hure 5:5 on Mastr Hartstongs acct aboute 10 Mrs Hartstonge went away and I gott ye things clained so rambled aboute

14 satt reading

15 I sent 2 dowsin small bere 9 bottles Sidr and 4 pints of Rasbery wine to Limerick ye boy brought back two dowsin pleats 8 dishes and 7 dowsin bottles

16 I sent ye bay mare to ye horse wch she refuseing big sow was brined Tim Hearlyhy brout me 3:6:0 wch wth 2:13:3 he sent me by Cosgary makes 5:10:3 wch is in parte for medows

17th I sent 4 horses to town I gave John Quain on acct 1:2:0 and gave him a token to Jon Bryen for another giney

18 19 nothing materiall 20th I recd a letter to go to town in ye morning

21 I went to town and dined there about 5 I left it so to Courte delaid at ye boat till aftr nightfall lost my handkercheif to snuff 1:1 to expences and turn pike 4:6

22 I sent home for 2 bushills of wheat and about eavening they came ground

23 I gott a letter from Tim Mahony for Rent and sent to Jon Bryen to make up some money I gott a letter from Rob: Cox[33] to stand Godfather to his daughter I wrote to him and sent a giney and gave ye messinger 13d

24 I went to Limerick to bring up Mrs Hartstong sent away ye things about 3 came away wth out her expences 4:6

25 to Limerick and brought her to Courte and her Childern

8br ye 26th 1744 Jon Hogan gave me 1:14:6 for ye hay and father Nicklas[34] 17s:11d wch I gave Mrs Hartstonge I gave lame Connr 12s in full for this and last years hire then John bryen brought me 2:3:8½ and 1l 18s 10d I borrow'd of Martin fitzGerald wch makes 4:2:6½ then I aded as much to it as made it 5:10:3 and orderd him to carry it to Tim Mahony in ye morning lost att cards 6d

27th nothing materiall

28 to Church and back again

29 nothing materiall but yt Ned fitzGerald send a surrendr of ye ffery

30 I had Jon Donevan making a doar for ye turff hos and sent David Bryan to cutt lathers to Coraheen I went wth Ms and Mastr to town and back expen 8d

31 I went for them and gott 10:1:6 from Ned Merowny wch was all he had of Mrs Widenhams money then I bougt a quatr of beefe for 7:7 to hogsheds 1:2:9 snuff for matty 7s to ticking 13 brogs and stockings for her 1s:10d to bread and tape 1:7½ a lock 2:2 to straps and buckle 1:1 oysters wine drink and turnpike 4s aboute 2 we left town gott to Court by nightfall

---

[33] The Cox family seat was at Ballynoe, Ballingarry. Bence-Jones. [34] Fr Nicholas Morony was priest in Adare but was hunted from there by Southwell and fled to France. He returned and was appointed to Kilkeedy. Begley, p. 600.

November 1744

9br ye 1st 1744 I gave Tim Nele 1:2:9 Garatt Pursill 1:2:9 James farrill 1:2:9 Jon donevan 2:5:6 all wch is in acct of money Mrs Widenham owed them then I gave Mick halpin 11:11 in full for ye worke he did here last 7br aboute three our family went to Shannongrove and I came back imediately

2d I sent ned fitzGerald wth a letter to Ms Bastables[35] and gave him 1:1 to bare his charges and spent ye day roving

3d nothing materiall

4th I went to Shanongrove from thence to melan Church back to diner in ye eavening we came home to ye servant 1:1

5 I went honting and gave ye boy 13d

6 nothing materiall

7th I brought some chees sidr and other bottles from home and 8 books I gott 15s 5d from Jon Bryen and 12s from Tim Bork for his hay

8th I wrote to ye subsherriff to come to open ye things I gott 3 shillings from Jon Pursill for his hay

9 I went to Limerick pd my Quittrent and bought a hogshed and severall other things wch I pd for I got to Courte by diner aftr wch I pd morroh 12 in full for ye hire of his boat brining up ye goods then David Bryen came home and fell att me

10th I went to Castletown[36] and brought home ye things to 2 drams 6d pd ye hackler 3:4

11 we prepared for company to open and cant ye goods aboute 11 ye subsherriff and company came ye things were valewed and Mrs Hartstonge pd ye money ye sherriff and fitzGerald went away and Mr Bury stayd till 9 and Mr Jackson all night I pd for bread 9 wine 6 beefe 7s 1$^{1/2}$

12 Mr Jackson took a list of ye goods to be devided then went away I got 5s for ye gras of a yrling att Killmoreen

13 Burn came aboute his money he claimd 1:2:9 but on ballencing accts there was but 2:5$^{1/2}$ due to wch I gave him

14th I gott 1s6 from Slattery on acct of ye gras of a horse and pd 6:6 for knives and forks 2 2 for cards and 9d for bread

15 I accompted wth ye men and bought a yrling from Mastr Hartstong for his Mama for 9 6 and pd him 1:7$^{1/2}$

9br ye 16th 1744 I pd for wine fish tea whiskey &c 15:4$^{1/2}$ and pd Mastr Hartstonge 7:7 in full for his yearling I gott 13s greasing money

17 I went hunting and aftr fitaguing ourselves and horses killd nothing to bread and starch 15d

18 att home wth Mr Pery and both ye Mrsrs Quins[37]

---

35 Arthur Bastable who died on 7 April 1773 aged 75, was born at Castlemagner, Co. Cork, the son of Arthur Bastable (who died in 1727) and Anne Mansfield. Grove White, vol. 2, p. 117; IGI and Vicars. 36 Castletown Manor, 2 miles from Pallaskenry, was built by John Waller in the parish of Kilcornan. Lewis; Burke. 37 Wyndham and George Quin.

19 Pery went away I brought from Port ye 3 suts of furniture and some books and prapard for our familys going away

20th I gave my accts to Mrs Hartstonge aboute 12 they all went away and I spent ye day lonsome enough and taking an acct of ye things

21st Robin Hewson gave me 12:5:7 on acct of greasing I spent ye day a stoping ye Sidr and gave Connr Quelly 1:2:9$^{1/2}$ wch 2th 1:6:2 he got at sevreall times before makes 2:9:0 all wch is in acct of his wayges since Mrs Widenhams death

22d I went to Tough from thence to Ned fitzGeralds and gott a bill on Will Bork for 8l wch I got and gave to Mrs Hartstonge then I pd John Bonfild 2:6:8d in full of all accts and gave him 2:6d for 3 handkircheifs and he gave me another then I gave Harry eavans 1:2:9 for a Saddle then I came away wth an intent to come home but ye wettness of ye night made me turn back and I lay att my sistrs expences 2s:4d

23 I went earley to Mrs Hartstongs and gave Mastr 1l:5s:od in full for his 2 yr old heffer then having breakfasted I gave or lent Ms 1:2:9 they wod not let me come away and being unwell I stayd all night and gott a sizers from Mrs Harts

24 I came away before any of them were up I came away having pd 1:1 to ye osler and made 12d for my horse other expences 2:4 snuff 13d so to Courte and foun some of ye house stript I sent 2 dowsin of Sidr to town and spent ye day idley

25th before I was up I had an acct yt ye coach was coming to town and directions to go down I left Courte att 10 and got to town before ye gates were opened wch cost me att Lynchs 9 and 5d turnpike I dined at my sistrs and spent 10d then went to lodgeing and calld for a dram and sugar wch cost 4 norse Judy and I drink it in a bottle of sidr then ye Ladies came in and we prepared for ye jurney in ye morning I bought an ounce of snuff wch cost 3d and 3d for apples so to bed wth Mastr Hartstonge

26 I got up early and went to see ye horses and got them fed then I pd Ned Mcgan 7:5 in full of his bill to Mrs Hartstonge then I gave 1:1 for havana snuff then I gave Ashton a mug 3 drams wch cost 8 then I pd for 1:3 for drams for ye men then I gave Ms Nancy 2:8$^{1/2}$ then we sett of and stop att Tolloh where I pd 2:2 then we went off to ye mines where we lay and I lent David Bryen 2:8$^{1/2}$ to tom Hanrahan 6$^{1/2}$ to ye maid drawer and ostler 1:4 for my self then abote 9 to bed

27th I gott up earley got ye horses ready and breakfast ye bill pd and att 7 sett off and went them a mile then calld att ye in and pd 3d for a dram gave ye ostler 6d on Mastrs acct so away breakfasted att Tolloh wch cost 2:3 so to Limerick and gave 6:6 for a hatt pd for turf 2:10$^{1/2}$ for a knife 5:7$^{1/2}$ to snuff yt Mastr did not pay 1:6 to his washerwoman 1:1 then I came to my Sistrs and eat my diner and spent 2:8$^{1/2}$ I pd ye whole day 7d turnpike so to Courte and got there by night fall then being tired I went to bed att 8

28th I sent Cosgary and horse to town for 2 tables of Mrs Hartstongs and Tom Bork for ye chair aboute 8 they came home and pd 11d for peaper 3d for bread I gave Jon Bryen ~~2 ginies to buy a beefe~~ and 5d to Matty

December 1744

29th young Tutthill[38] came for 4 cowple of ye Timber I breakfasted wth him then went to see calves to Ballygran but did not like then then I lent Mick Cosgary 3:9 to norse a penny for blew

30th I had ye china dug up and ye pigs brought from Porte and ye apples picked I gott 6s from ned fitzGerald for his hay I pd ye man yt broad ye flax 2:5 in full Jon Bryen brought ye books &c

xbr ye 1st 1744 I sent 5 hogsheds of Sidr in ye boat to be sent to Limerick and from thence to Dublin No 123 whitemos 45 russidin I wrote to Rob Hewson for money but got none

2d nothing materiall

4th [3rd] I went to Limerick pd Costom 1:8 Porters 3s boat 5:5 watch 4 to finch 1:2 turnpike 2 to me David and Cosgary 2:2 So home att night fall

5th [4th] ye Quttrent drivers came and I was oblidged to give them 4:4 fees then I sent home a peice of a barrill of Sidr in ye eavening I bought 12 calves from Ned Sheehee for 11:6 each wch comes to 6:18:0 wch I am to pay ye 3d of May next when I came back Harry Evans brought my new Saddle and a Saddle Cloath Ned Mc gave me yesterday

5th I had Halpin mending ye house I sent to Hewson and Westrop for money but got none Cosgary gave me 2:8 of ye money I lent him ye 29th of last month I marked ye sidr

6 Ned Mcgan came to Courte and brought me a letter from Ms for wch I pd him 4d and gave me a pocket peice of ye vallew of 2:9

7th I wrote to Jackson for advice aboute ye tenants Quttrent and to Rob Hewson for money but got none

8th I sent Mick Cosgary wth 6:18:4 to Limerick and wrote to Ned Mcgan to make it up 10 4 0½ wch he did and I owe him 3:5:9 to bread 3d

9th to Church and got Mastr Hartstongs knife from Mrs Bury so back sent a letter to Dublin but got none

10 I got up early misd ye torkey Cock found him eat in ye barn Syprian Pursill came to make a case for ye glasses

11 I sent Cosgary wth ye glasses a box wth cheeses in it and a carpett to Limerick to be sent to Dublin and I sent Thos. Bork for money to Mr Hewson John Bryen came to let me know ye flaxseed was clain and yt he had but four barrills wch I orderd him to carry to town tomorrow Pursill went home aboute 2 a violent Storm rose wch continued till 8 and stript a grate dale of ye houses no money from Hewson I gott a letter from thornell

12th I sent again to Hewson and gott 4l:3:8d I made ye men geather up ye slates and cutt reed and gads

13 I sent Cosgary wth a letter to my brother and had four men thatching ye houses yt was stript

---

38 Possibly a son of George Tuthill (1693–?) of Kilmore and Faha. Burke (1958).

14 I went to Limerick and deliverd my flaxseed I had four barrills and half and half a barrill I gave 12:6 for these I pd ye porters 1s then I came to my sisters pd reconing 2:2 to Jon to buy Salt and other thing 13:3 then I pd for 2 board 25 half pd of glew 2½ sope 1:1 snuff 1:1 4 drams and turnpik 1:2 to Jon and Martin 1s then tom: Hanrahan gave me a letter from Ms and some needills then he and I came away and left Jon in town to deliver ye bags to to Court by night fall I gott a disfactory letter from my Brother I pd Ned Mcgan ye 3:5:9 I borrowed ye 8th to pay ye Quittrent

15 I gave my cloaths to be washd and aboute 11 came home and gott 6s from Sulivan for ye grass of a cow then I fell to settling my things being att Courte since ye 17 of July last about 6 Jon came home and brought me flax bill and an act of ye 3:8 I gave him yesterday to Salte 1:6½ to his expences 1d½ returned 10d

16 I gott 2s:4d from Thos Collins on acct of tresspas I gott 6 ginies from Rob: Hewson of Thos Bork I spent ye day mostly reading

xbr ye 17th I gott up early went to Ballylongfort and bought of John Quain 4 heffers for 8 pound from Tho: Harragan 3 cows a heffer for 8:2 and four calves for 1:6:0 from Mick McTomas 3 cow for 6:0:0 then I went to Killmoreen walked about and found one of my lambs dyeing and I had ten 51 sheep and lambs I bless God I have 50 sheep 47 cows 57 calves 4 horses and a fole four pigs and 3 young calves when I came home I gott a letter from Mr wch I answerd and sent a drake to Court then Doctr Martin sent his man for wheat 4 bushills claind for him and spent ye night a settling my accts I gott 7s from Dennis Hartney on acct of ye grass of his horse att Court

18 I sent 4 bushills of wheat to ye Doctr then I gott a stack of wheat putt in aftr wch I sent 11 cows to greas att Courte and had Ned Mcgans mare and Red Robin brought up

19 I went to Tough to settle accts but could not so went to Adear to see Mrs Quin but she was gon to town and only saw Mr so home by 12 to a dram 3d I gott 10:10 from Jon Bryen and 6:6 he gott from Garrott Pursill then Jon gave me and acct yt he sould as much medows as came to 5:7:4 and recd 1:13:4 greasing money wch comes 7l:0:8 Patt Hanrahan 2s:1½ D Bryen 2s Paul Rahily 6:6 to ye taylors 1:2 ticking 9:0½ W Welch 1:1 hops & salte 4:2 to weaving 77 bandills of Linin 3:2½ to meat Court 2:3:8 in Limerick 13 all wch comes to 6:1:0½ ye remaindr wch is 10:7 he forgets what he did wth itt for fere of a mistake I mention ye People he recd ye money from viz: W Covard 1:11:3 M Edigan 12:6 Den Gorman 12:6 Patt Hanrahan 12:6 Jon Nash 12:6 James Mcmahon 18:9 Ed Bahan 6:3 Jame Covard 4:9½ Law Daniel 13:6½ fitzGerald 9:0 Will Boyle 3s tho Halloran 3d he like wife gave me an acct yt Tim Callaha had a peck beans for 1s P Boyle a bus of beans 2s two bed of petatoes 10 Mick Pursill two bed of petatoes 2s o So to bed

20 I went to Ballyengland for money but gott none aftr breakfast I came home

21 I had my cow killd and having but little tallo I sould ye hide for 8:4½ and gave 8d for 2q:14lb of tallo I cutt up an old sheet to 4 pillow cases and gave ye rest away

1 James Henry Brocas, *Old Baal's Bridge, Limerick*, reproduced courtesy of National Gallery of Ireland, NGI 2559.

2 British School, *Mrs Valentine Quin, née Mary Widenham (1682–1776)* (private collection).

3 British School, *Valentine Quin (d. 1744)* (private collection).

4 George Petrie, *Kilmallock, Co. Limerick 1826*. Visited by Nicholas Peacock in November 1745. From J.N. Brewer, *The Beauties of Ireland* (London 1825).

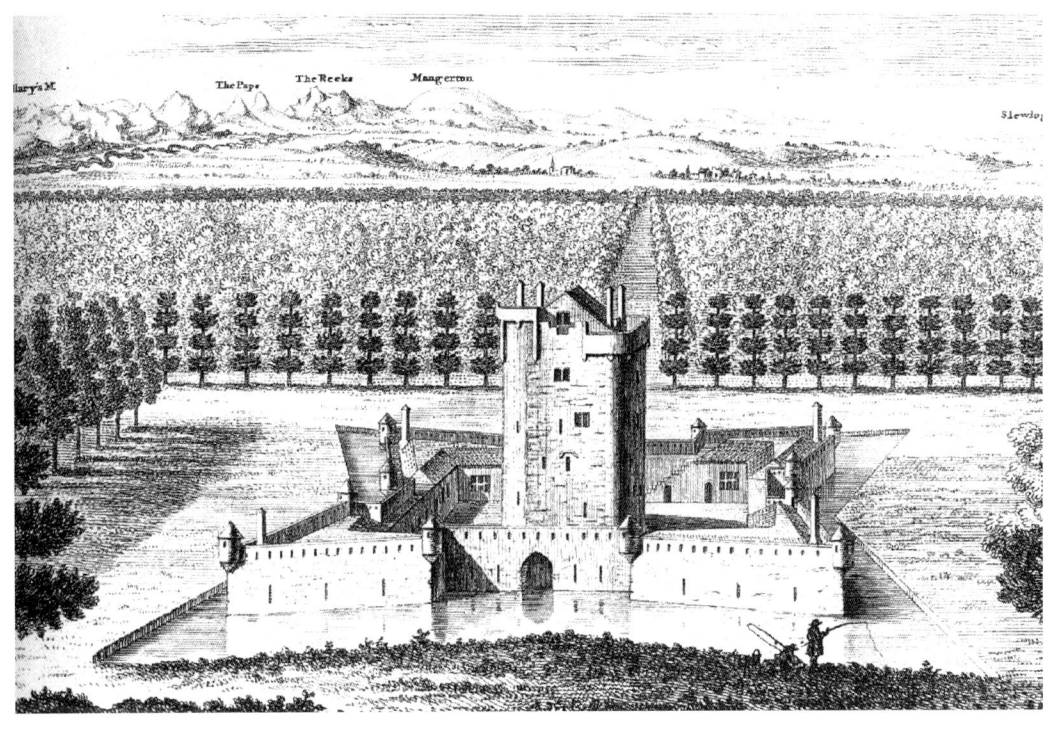

5 *Lohort Castle, near Mallow, Co. Cork* c.*1738*. Visited by Nicholas Peacock in August 1747. Engraved by J. Toms.

6 Itinerant spademen holding 'loys', used particularly for potato cultivation c.1820. Ulster Folk and Transport Museum.

7 George Edward Pakenham, *Irish road scene* c.1737.
From the Journal of George Edward Pakenham, 1737–90.

8 John George O'Brien (Oben) *A View of Adare, Co. Limerick* 1793 (George Stacpoole collection).

December 1744

22d I sent Jon Bryen wth 5 busl of wheat and a small cask of butter and some petatoes to mis Listr then I went to Court and calld for ye linin I left washing and on reconing them, I misd a sheet then I packd up my own things and sent them home and gott hom along wth ye boy and took an acct of my cloaths and only misd an old pr of thread stocking then I gave Joan on acct of her wayges 10 bandills of bandlecloath about 7 Jon came home and gave me an acct yt he sould ye wheat at 10d$^{1/2}$ a stone being 26 stone comes to 1:3:3$^{1/2}$ wch he gave me he sould my Sistr o 2qr:21lb of butter att 1l:1s:6d a hondrd is 14s:9d$^{1/2}$ of wch he gott 10s to whiskey 1:6 wayage costom tear and turnpike 6

23 I went to Killdeemo Church and gave James Mctamos 10s:10d on acct of Mowing att Courte so aftr service home and spent ye remaindr of ye day reading I gott a letter from Tim mahony for ye remaindr of my own Rent and ye Rent of ye dere park I gott a hare from supple

24 I gott up early and gott my cow Cutt up and salted then I bought 2 cows from John Quin 4:4:0 from Ned Naghtin 2 cows 3:6 0 pd earnest 1:1 I gott 6 6 from Mick Moachan on act of greasing

decmbr ye 25th 1744 I spent att home reading in ye eavening I gott a letter from Mr Thornill and another from Ms wch Ned Mcgan pd 8d for

26th I gott up aboute 7 and went to Killbreedy and pd Tim Mahony for ye use of Mr Bryen 5l 10s 0$^{1/2}$ wch wth 5:9:6$^{1/2}$ I pd him him for Mr Bryens use ye 26 of 8br makes 10:19:9 ye may gale then I went to ye deere park and walkd aboute then to feighs where I spent 1:7$^{1/2}$ to 1 garland 13d to tabaco 4$^{1/2}$ so home

27th Nurs and David breakfasted wth me in their way to Charlevill then I wrote to Mrs Hartstonge and Mr Thornell and sent them to Court to be sent to ye office and spent ye day variously Dennis Bryan[39] came to me to stay att his service

28 I spent looking over accts and accted wth Martin fitzGerald and he owes me 4:1:2 besides 3:6:4 he ave me I gott 18:11 on acct of greasing from Darby feeneen

29th I bought a cow from Jon Moleheen for 1l:12:0 and a bushill of barly then went to Courte and veiwd ye things there then home Jon sould a cock of hay for 10s I gave Cate Caffow 12:5 on acct of her wayges

30th I spent reading

31st nothing materiall [sentence crossed out about a debt paid]

---

[39] May be Dennis Brien, a farmer of Ardnocrony whose will was proved in 1769. Phillimore, vol. 3.

1745

Janr ye 1st 1744/5 I send five cows to Courte and pd Jon Molleheen 1:12:0 in full for his care and Cate Carty 6^{1}/$_{2}$ in full for claining flaxseed I gave Matty an old Shirt and brought 4 calves from Court and 10 from Caparoe one of my sheep died
2d I had 32 dowsin of Candills made Mr Harding sent to me to borrow some money I sent John Bryen to him wth 4 ginies I sould ye sheep yt died yesterday to Darby Harragan for 5 days work I had a letter from Standish Grady for ye yrlings yt greas'd att Court about ye pryce of ye Beer Jon fitzGerald had ye 19 of June last and aboute his note he passed to ye Sherriff for Mrs Hartstongs horses
3d I went early to Courte sent an answer of ye letter ye 5 yearlings and gott 1:13:0 for their grass then I orderd Ned fitzGeralds Cows to be seasd for fere of stealing them away having remov'd some of his goods so home and sent John to Adear for Steel hooks and eyes wch cost 5d I wrote to Mrs Hartstonge and left it att Courte to sent to ye office in ye morning
4th I wrote to Jon fitzGerald for ye Pryce of Richd Borks beere but did not gett it I sent Jon to ye forge and gott two spades made
5th I went to Corraheen to try to buy 2 cows but did not then I calld att Courte to try for letters but gott none I lent David Bryen 1:1 so home and in ye eavening I bought 2 heffers from Jon Quain att 1:18:0 each comes to 3:16:0 was invited to dine att Hary Supples to morrow
6th being ye only day I dined abroad I had att H Suppels a rump of beef pork and petatoes a puding and roast torkey aftr wch we drank sidr and whiskey till late I gott home by night fall
7th I gave John Bryen on acct of his wayges 5:5 then went to Courte and had 7 men diging petatoes I gott 15s from Ned Sheehee on acct of greasing att Courte
Janr. Ye 8th 1744/5 att Courte sent for letters but gott none to bread 3d molds 1^{1}/$_{2}$ news 9 men att ye petatoes
9th a violent storm but I bless God no harm done
10th I had 10 att ye petatoes and finished them I gave Mary Mangadan 6^{1}/$_{2}$ for picking them I made 4 stocks and two pair of pillow cases
11th I sent for letters but gott none to peaper 6d then I gott a letter from Jon ffitzGerald to desire I wod go to Castletowne to morrow and pay for ye things I bought att ye Cant then I came home and gott ye 4^{1}/$_{2}$ ginies I lent Mr Harding ye 2d instt and borrow'd 1:18:6 from him then I made up ye Sherriffs money and gott 3:0 from Jon ye recd on acct of greasing
12th I went to Castletown and pd 10:2:0^{1}/$_{2}$ I owed for ye things I bought att ye Cant ye 26 of Aprill last and got my note and 3s:2d for a small desk Mr ffos-

bery canted for Mrs Hartstonge then I came away and spent 4^{1}/$_{2}$d wth Pursill and daniel so home Betty Grimes send her son for money I only gave him 13d for himself I was invited to dine att Mr Hardings to morrow

13th before I was up Jon brought ye tresspas sheep wch I made pay 4s:2^{1}/$_{2}$d tresspas I did not go to Mr Harding but spent ye day mostly reading Pattr Daniel carried ye 6 sheep I sould him ye 14 of augst last I bless god I have a young lamb

14th David Bryen came to me and I gave him on acct of a calf 3:3 wch wth 4s:10^{1}/$_{2}$d I lent him att severall times makes 8:1^{1}/$_{2}$d

15 I wrote to Mrs Hartstonge and sent Jon to Lim att night fall he came home and brought me a letter from her he gott 5:11:8 for ye flax seed he was obliged to pay 10s10d for ye moneys he gott 4:9^{1}/$_{2}$d my sistr owed for ye butter he pd 1:4 I owed her for letters 8d of wch I must I must charge his son wth he pd 1:7$^{1/2}$ for Salte 1d for ye news. I gott 62 trees from Cypriane Pursill

16 I sent John to Killmoreen to Plant ye trees I went to Court pd 1:17:1 in full for ye countery charges of Courte and 1:6:7^{1}/$_{2}$ for my own

17 I had 8 men trashing Oats att Court I sent Jon to Pallas markett he pd 3s:4d for frize for ye boys and 3:1^{1}/$_{2}$ for broges for them

18 I had 7 men att ye Oats I gott a letter from Mastr Hartstonge I gave ye Palatine 2d

19 7 men a trashing and finishd ye Oats and I finishd 2^{1}/$_{2}$ pr of pillow cases

20th I went to church dined att Mr Hardings and there gott a letter from David Bork and one for Arther dohig inclosed to me from Mrs Hartstonge I came home and wrote to Tom Bork to go wth them I found my 2 pigs killd

21 att home gott my pigs Salted and sent a chine to Mr Harding I pd him ye 1:18:6 yesterday I borrwed ye 11th instt and Jon brought 17lb of hare to make ropes but do not know yett what I am to pay for it I gott 7 0^{1}/$_{2}$ from Tho Halloran on acct of greasing I gott a letter from Sistr Mcgan to go to town aboute our dispute wth my brotr I finishd 3 stocks wch wth five I made before makes 8 new ones

22d I gave Jon 15:4^{1}/$_{2}$ to pay for ye hare but Mr Harding wod take no thing for itt he gave his son dennis 3:9^{1}/$_{2}$ on acct of his wayges and sixpence for making ye boys briches I went to Limerick and advised wth Bryen Mcmahen aboute my brothers affair then pd Ma Daniel for dressing my buckskin 4:4 to mending my whip 6^{1}/$_{2}$ to peaper 10 snuff 1:1 reconing and turnpike 1:4 so to Courte and stayd there

23 att Courte and gott Cosgary to settle ye corn

24 I gave Syp: Pursill 3:3 to buy things for sashes left a letter to be sent to ye office and came home

24: 25 att home nothing material

26 att home gave Jon 1:1 and Joan 6 I gott a flesh fork schimer and pot hooks from Tho Daniel

Janr. Ye 28th 1744/5 I sent 70 bandills of Canviss to be wove accted wth Dan Shea and give him 13d I gott 4s from Man Killbreedy for greasing and 4s from Tim Day on dito acct I accted wth Jon nele who had in medow 6:3 in greasing

February 1745

5:4 in all 15:7 for wch he pd 3:7 days wth 4d a day 12:4 so yt I owe him 9d I sould him a bed of petatoes for 5s

29 I went to Killmoreen and walked aboute and offerd Dan Shea ye northwest 50 acres for 6s an acre then I went to Courte and gott a letter from Mastr pd 3d for bread 30th att Courte doing nothing

31 had some of ye Oats winowd and gave ye hackler in full 13d

Febr ye 1st 1744/5 I had ye remaindr of ye Oats winowd about 5 I left Courte and went directly home and sent John to look for money but he got none he gave me 8s he gott from Geo Russill[1] for greasing 2:5:6 he gott for 42 stone of wheat and 12s he gott for 3 bushills of beans he pd 1:3 in all expences

2d I sent Jon wth wheat to ye markett I gott 11:4 from Mick Molleheen for ye gras of his two cows yt greased here a quartr I had a man sent for me to Town on ye affair wth my Brotr I gott 1:18:10 from Martin fitzGerald wch I gave him creditt for aftr nightfall Jon came home but did no good ye Marketts being fallen he brought me a letter from Ms and another from Mr Hartstonge and one from Mr Thornell

3 I went to Limerick to Signe and Swear to an administrasion and am to pay 1:6:8 for itt I spent every way 3:0 and lent my Sistr Ireland 3s:11d then I went to Courte and gave David bryen to give James Bork 1:1

4th att Courte gott ye mault winowd wrote to Ms and Mastr Hartstonge and to Mr Thornill then I gott ye remaindr of ye Pryce of ye medows from Mr Westrop it being 1:15:3 then Mr Jackson came and carried ye wrought bed and blew bed 3 hogshed and 2 tubs one of ye sheep died

5 I came home early and got 16 3 from Will 2s:2d from Law daniel 4 from Tim Herlihy for Tor Cenedy 1:6 from Will Hickey and a penny from Tho: Harragan I gave ye mowers 15:9 and spent ye day making a calf skin westcoat

6 I sent Jon wth half a barrill of wheat and 2 busl of beans to Limerick he sould ye beans for 8:8 and ye wheat he left in town for 1:10:0 he got 3d a pound for ye lard wch came to 5:9 ye wheat he carried to day he left in town

7th I gott 4:4 from Math Sawn and 4s from Jon Conway both for greasing I pd 6½ for church rates for ye dere park I sould John Heas a bed of petatos for 8s

8th I wrote to Mrs Hartstonge and gott a peck of beans sowd and finishd my calf skin weastcoat

9th I sent Jon to sell ye wheat he left in town he sould 50 stone att 11d a stone wch com to 2:5:10 he pd 1:6 expences and 3d for needills I gott 3s from ye pyper for greasing I had 3 men diging for plants and parsnips

10th att home reading I gott 5:5 from Ned Naghton on acct of greasing

11th I went early to Court and drue off a pipe of Sidr yt I found leaking

12 I went to Ballyengland and gott a bill on Colonell Taylor for 6l 4:11:3 of wch he owed me then I went to ballynorte and signed sealed and deliverd my case pd 11l in cash ye bill 6l and was allowd 2:5s:0 for ye appell trees then aftr

---

[1] George Russell may have been a son of William Russell and brother of Philip Russell; see above.

diner and seare of a boule of punch I came to feighs spent 7½ so to Courte aftr nightfall and gott a letter from Mastr Hartstonge

febr ye 13th 1744/5 att Courte a doing little send all ye wooden ware I had to Rob: Hewson wch I must charge him for I send him a little pig

14th I came home and found ye Plow going and got in full for hay 5:7 I made a winow sheet

15th I gave Bryen Boyle a bushill of barly on acct of weaving ye Canvis ye Plow going men trashing for ye straw I gott a letter from Mrs Hartstonge I made 2 bags and another winow sheet

16 I spent ye day mostly wth ye Plow I sent a few peas and beans to Court and gott my linin I gave to be washd a Monday

17th to Courte from thence to church and in company wth Mr Harding spent 3d so to Courte

18 wrote to my sistr to buy 8 pound of steel hemp wch I gave her for sent a pig to Mrs Quin and pd 1:10 for Salte and 1d for tread then orderd ye pigs to be killd then went to Jon Mounsill and gave him my note for 6:10 and brought ye colt to Courte and loyterd aboute

19th sent for letters and got one and ye news then Cyp: Pursill came and wrote proposals for ye ffery then ye glasier came and I orderd him to putt ye pains into ye new frames then gott ye pigs salted so home and gott from John 8s yt Jon Hogan pd for greasing

20th ye gleasiers being done att Courte came to me but wod not gleas my wondows undr 7d a fott wch I wod not give then I pd them 4s 8d for ye work att Courte I sowd 10 busl oats

21 I got some oats winowd and sow'd 10 busl I gave on Rob: Hewsons letter to his boy 5:8 and a note to my Sistr Mcgan for 3d then I gave Henry Supple a barrill oats and sould a bus for 2s

22d I gott a yrling from Garrott Pursill for 7s in full for what he owed me and in parte of a barrill Oats I gave him this day then David Bryen came to excuse Tho Hanrahan and Patt Daniel to acus them both I sow'd 10 bus oatts and gott some winowd

23d I finishd plowing for Oats and sowed but aboute 7 bus I found a penny coming from Caparow

24 I went to Courte and att Killdeemo spent 5 I gott 2 letters from Mastr Hartstong Patt Sheehan gave me a proposal for ye North east parte of Courte

25 Mick Hartny came and vewed ye lands yt P:S bid for and having drank and mug of sidr went away it began to snow 7 men att Port

26 I gott up early and found a pritty deep snow I orderd in ye horses and sent Sheehans Proposall to Mrs Hartstong and answerd Masters letters I gott a letter from Ms and ye news wch cost 2d to bread 3d

27 ye snow continued but what ye son melted

28 I had a letter from for beans att 2s 3 ye barrill an a letter from Jon Monsell for Oats att 1s 5 ye barrill I lett them both know I was content wth ye price I gave Patt Hanrahan 8 bushill of barly on acct of his wayges I gott a proposall from M Hartny it began to very hard

March ye 1st 1744/5 I sent Hartnys Proposall and a lettr to Ms I sent for seeds Saltepeter pepper sugar wch cost 7 [?] to snuff 13d peaper 10 I got my briches home ye snow is so deep I put in ye sheep and calves

2 I gave Jon Supple half a barrill of barly ye snow so deep I could not stir out so diverted me reading ye persian tails[2] 3 I sott reading ye snow begins to thaw

4 I sould a man from Rathkeal a stack of hay for 2:17 0 I lent Peter Daniel 13d I recd some money from greasiers att Court and 4s from J Bryen one of Killmoreen greasier he did not pay for ye odd days

5 I gott a lettr from Mrs and another from mastr Har to bread 3d I sould 2 bushills barley for 5s

6 I sould 2 busl of barly for 5s 2 busl mault for 5s

7th I gott up att 3 and went hom and gave Jon Bryen three bushills of barly and 2:8½ on acct of his wayges I sent 5 barrills Oats to Courte for monsill and went to deliver it but ye boat being aground I could not I spent att Killdeemo 5d

8th Monsells men came for ye Oats I sent 5 barrill at 1s [?] a barrill and 2:15 wch makes up 6:10 ye price of ye horse and I gott my note I had 2 women claining my barly I gott ye remaindr of ye Price of ye hay and gave a receipt for it

March ye 9th 1744/5 I sould Dan Shea 1 bushill of barley for 2s 0 wch he pd me and I owe him 7d I sould J Doil 3 bushill for 7:6 I gave Patt Daniel a barrill of barly I sent Jon Pursill to Limerick for Plants he brought 600 of wch 300 are for me he pd 7d a hondrd and 12d for nails 7d for seeds I gott a letter from Mastr and ye news 10 I satt reading

11 I gott ye Oats att Courte Riddled and gave M Cosgary five bushill of barley and 3 he gott up att home he gave me 16:3 I sould 200 of hay for 3l aboute 2 I came home

12th I gott some drink brew'd I gave by dennis Bryen direction to Elenor Hius 2 bushill of barly Jon gave me yt he got for 30 bushills of barley 8 bushill of Oats and a peck of peas 4:14:0 wch I sett down I gott ye male of 5 barrill of oats from ye mill last night I gott a letter from my Brotr to ask time for my money for a year and yt he wod give his bond and utter I sent him some ellefry wch tho a present cost 7d I sould a bus of barly for 3s

13 I got my Male sifted and rambled aboute

14 I went to Tough but did no good so home and pd Rob: Hewsons man 3:0:1d in full of ye 6 pound bill I gott from him on Colonell Taylor for ye 12 of febr last then I pd M Bryens man 1:2:9 on acct of Rent then I pd Patt Hanrahan 14s:5 in full of his wayges and he pd me for a bushill Oats 2s 6d then I gave Den Bryen 2:8½ on acct of his wayges then I gott a letter from my sistr aboute

---

2 Almost certainly François Petis de la Croix, *The thousand and one days: Persian tales* (London 1714).

ye Law sute wch I answerd I got some Oats winowd and gott a pair of new shoes from Dan Shea

15 I gott ye remaindr of ye Oats I had trashed winowd

16 I came to Courte sent Jon by his son David 10s:10d and 2:2 to drink their Patricks pott I sent 6d for Herrings then Mr Mounty westrop gave me 4s:4d of a months gras on ye mash for his horse

17 to Church back to Courte and dined on a hering gave ye familly 2 coppers sidr gott a letter from Mrs Hartst wth an answer of Pursill and Hartnys Proposalls

18th I went to Hartny and tould him ye answers thence to Warners pd him 2s 0 I owed for Mrs Hartstongs Tyths then back to Courte spent 5d I gott a letter from Jon Mellsop to give his mare to Doctr barkly[3] wch I did

19 I gott a letter from my brotr aboute our lawsute I sould a hondrd of Hay and got ye beacon resalted Donevan painting timber for ye ware

20th I had 8 men and Donevan setting down ye ware I sould two hondrd of hay

21 ye above men att ye ware sold a 100 of hay

22 went to Limerick got ye Administrasion pd 1:1:5 for it Pd frank yeamons[4] 3:9:7½ for Mrs Hartstong breakfast at Mrs Quins took up a coat from Jon Bonfield pd 1:3 to Jon finch he pd coopers when sending ye sidr away I pd 6½ for mending my bridle and girth to snuff 17½ expence 2:2 then back to Court by 6 a clock

23 att ye ware **24th** reading

25 I went home pd Joan Daly on acct of her wayges 1:2:9 so back to Courte and gave Hen Supple 1s2 wch wth a barrill of Oats I gave him ye 21st of febr and 4 bushills barly I gave him ye 2d inst makes up his years salary aboute 5 Thos Bury Willm Taylor came to me and agreed for aboute 9 acres of medow att 2s3 an acre (they to agree for ye tyths) for 5 years I to take up ye medows ye first aprill they to mow them before ye 15th of August and pay for them and draw them to ye fodering place by y3 1st of 7br I to give them 2 acres of Caparow to foddere on they to fence it up and if I should till them 2 acres they to fence in 2 more and so on during ye 5 years

---

3 Possibly Robert Berkeley DD (1699–1787), a son of William Berkeley and younger brother of the philosopher George Berkeley, bishop of Cloyne. He married Anne Elizabeth Dawson in 1734 and they had four sons and four daughters. His will was proved in 1787, when he was living at Ballincurra, Co. Cork. Another candidate is David Barclay, vicar of Ballingarry and Corcamohide. He took a BA at TCD (as Berkeley) and died in 1761. He appears to have not been made Doctor of Divinity, however. Miss Barkeley of Kildimo, niece of George Fosbery, married Christopher Waggett. *Cork Evening Post*, 11 May 1770.   4 Francis Yeamons was a merchant 'at the Golden Bottle in Irishtown', Limerick. He had 'linen of his own manufacture, worsted and thread stockings, fore Men's and Women's Wear'. He also sold caps, hats, handkerchiefs, claret, white wine, Frontiniac, Lisbon, Port, Rum, Whiskey, German and English steel, Iron, powder and shot. He imported Riga flax seed. He and his wife Mary had two children at his death, Mary and Elizabeth. His will was dated 9 August 1755 and was proved 27 July 1757, leaving everything to his wife, save 5s. to his daughter Mary. *Munster Journal*, 23 January 1755; will, 9 August 1755, proved, 27 July 1757; NLI GO 139.

April 1745

26 ye men and Morrohs Sloop att ye ware I sould 200$^{1/2}$ of hay I had a letter from Hen: Page for a giney for drawing my Case wch I paid and gott ye Case Mastr. Hartstongs gray horse died this morning

27th ye men and Sloop att ye ware I sould 300 of hay ye hearthmoney Collectr came I pd him 4s for myself 2s 0 for Mrs Hartstonge and 2 for betty grimes I had 6 sticks cutt for ware att ye grove

28 ye men and Sloop att ye ware and finishd and cought a salmond[5] I pd ye sloop 8:11

30 I pd 4s for a teirce 6s for 2 bus of lisburn Salte to making my coat 4s sope 5d no letters I gave Betty grimes 3 bus of barly in full of ye half giney Ms Hartstonge desird I wod give her sould Mr Connill 9:2:23 of hay he pd 1:3:10

31 to Church and back to Court

Aprill ye 1st 1745 I sould Connill 23:1:5 of hay furnill 20:3:0 severall others 13:0:0 Studders and Richd Rose[6] lay at Court

2d Rose and Studders went away sould furnill 11:1:0 of hay to severall 21:0:0 others go to Conill 3:2:23 sent a letter to Mrs Hartstonge and gott one from her to bread 3d

3 to Conill 15:–:27 to severall 18:0:0 Geo. ffosbery Supt wth me I sent a letter to David bork gave ye man 1:1 I had a letter from Mr Barker and 2 horses to greas here while he was in ye Countery Master Quin died

4th to Connill 1:2:21 to Severall 18:0:0 I pd 2:7$^{1/2}$ in full for countery charges for ye dere park

5 to Severall 36:1:14 of hay pd 4d for a letter from Mastr

6 to Severall 41:2:15 lent tim Bork 1s:8d

7 spent chefly reading lent David Bryen 1:7$^{1/2}$

8 Sould 5:15:2:7 caught a salmond

9 no letters sould 3:2:0:5 10th sould 1:16:1:0 of hay

11 sould 2:14:1:12 12th sould 2:6:0:19 to a letter 4 a rope 1s:4d

13 sould 4:1:1:11 / 14th gott Studderds bond from my brotr

15 went to Limerick pd Merowny 4l pd for Iron 12s:3d to gloves 2:2 snuff 2:2 expences 3s I spoke to Studders who Promisd to pay ye money imediately so home

16 sould 500 of hay to a letter 4d bread 3 sope 10 blew 4

17 18 nothing materiall

---

5 There was a salmon weir near Court on the river Maigue. *Civil Survey of County Limerick,* vol. 4, p. 348.   6 Thomas Studdert (1696–?) was the eldest son of George Studdert of Ardlaman, Co. Limerick, rector of Kilpeacon and Rathkeale 1685–1738. Born in Rathkeale, Co. Limerick, Thomas Studdert entered Trinity College 12 May 1716, aged 19. He married Madam Cusack of Kilkishen House, Co. Clare. His seat was at Bunratty Castle, Co. Clare. Richard Rose was a son of George Rose of Rathkeale and Susanna daughter and coheir of Richard Stephens of Newcastle, Co. Limerick and widow of Edmund Burgh of Newcastle. Richard married Mary Anderson, daughter of John Anderson of Foxhall, Co. Tipperary. He died in 1762. Burke (1958); *Alumni Dubliniensis*; Pedigree of Rose of Rathkeale NLI Mic 8305.

19 I went to Shanongrove and allowed Mr Bury for Charles Slattery 2:4:4½ and gott ye remaindr of ye pryce of 6 barrills of beans and aftr breakfast to Court pd Ned Sheehee in full for 12 calves I bought from him last wintr 6:18:0

20 went to Limerick and found my Sistr in an uprore and going to Port then pd Mrs Melsop 1:10:0 pd Mr Bryen in full for 7br rent 3:19:0 then I pd 5:5 for making my briches and gott a pair of gloves in then I offerd my own and Mrs Hartstongs quittrent but it wod not be recd till July then I pd Jon Bonfield 1:17:0 in full for my coat to garters 4d and coambs 4d then came to my Sistr and patchd up a reconsiliation wch cost me 2s in Lisbun wine so to Courte to Den 6½

21 ye fox beheaded a lamb wch I gott dressd for my diner we cought 2 salmonds pd 1:1 for gelding my colt

22 ye men second diging ye petatoes a Salmond

23 ye men att ye petatoes Jon Bryen gave me an acct of Corn and 4:1:2 I gave him for tabaco 9½ to B Grimes a bus barly

24 ye men att Dito I bought a mare of Ned Sheehe and pd him 8:5:5 in full for her I gave Norse 7d I had keeping for her son Jack and 13d I owed her son Harry for my chip of ye Pig they killed in 9br last

25 ye men finishd ye Petatoes we caught a sallmond and sent that and 2 trouts to Shanongrove

26 I went to Killmoreen and gott 11 lambs thence home and gelt 3 yrlings and 3 calves then I swapt a cow wth Syprian Pursill and gave Jon 1:8 for Soles for ye boys then to Courte and gott 1:4½ from B Grimes wch I am to give her corn for to a letter 4d news 1d

27th I gott a letter from Brotr yt he was taken by Jon Morphey and desird I wold lend him 12l I wrote to him I had not it and to Morphey I wod see ye 12l pd him by mid-sumer Syp: Pursill put up ye sashes in ye hall ye Dublin mare took ye horse I gave Pursill 1:6 for making ye cases ye glasses went in to Dublin

28 I spent reading I lent David Bryen 2:2

29 Jon. Johnston and Son came to Courte ye later eat all my chees and wod eat ye Davill if chees I went to Connill for money but gott none to Sidr 3d so back to Courte

30th I spent prepareing to acct wth ye tenants and measurd ye neck trinch it is 237 perch from one sluce to ye other Garrott Pursill made 13 perch and was pd for it so yt ye men have but 224 to be allowd for I pd 4d for a letter from Madm and 8d for seeds

May ye 1st 1745 ye forenoone I spent accounting wth ye men then went to Clorane to Meet Geo: fosbery but did not spent 6 so back to Courte and recd from Severall for greasing att Coroken 2:1:1 I gave Tim 4 bus Oats and charged it to his acct I gave B Grimes 3 bus barly for wch she pd me last week

May 1745

2d I gave conner Quelly 2 bus barly for 8s wch I chargd Mrs Hartstong wth wch wth 1:6:2 he gott before ye 21 of 9br 1744 and 1:2:9 he gott yt day 3s allow'd Darby morihy and 1:4:0 in ye grass of a cow all make 4:0:11 he gott since Mrs Widenhams death

May ye 3d 1745 I went to ye fare and gave Jon Bryen on acct of his wayges 6:6 I lent David Bryen 1 7 9 and spent 5:5 wch I gott from Jon Bryen on acct of greasing lent Conner Quelly 6:6

4th I gott 5:8 from Norse She owed me 3:3 for a bus barly and spent ye day loytering aboute

5 I spent reading ye ware went away

6 I had ye men looking for Sticks ye parte of ye ware yt went away floated up wth ye tyde

7th ye men preparing to sett down ye ware I gott a letter from Mrs Hartstonge and another from thornell and owe 4d for them

8 ye men att ye ware I sent to Limerick for Iron and Salte wch cost 5:9 I gott a bottle of shrub from my Sistr

9 I had ye men att ye ware I pd Tim Mahony 13:3:9 in full for ye rent of ye dear Park and 3 10 0 for Hen Supple I wrote to Studerds and I lent Mr Thornill a bill of 40l on him then I cast up ye acct between Madm and me and have 6:9:9 of her money and 4:4:6 of my own

10 I sent a letter to Madm but got none I gott Supples receipt for 9:10:0 he gott on acct of his salery to ye news 1d

11 I went home and John gave he gott for beans and Oats and 2 pottles of peas 2:2:3 he pd 2 8½ for services of ye little mare to oyle 6 barm 2 for daging ye sheep 4d then I went to Killmoreen and gott greasing money 7:18:7 I gave him on acct of his wayges 10:8 and 11:7 for cate 5s for his son Dennis and 4:8 for ye weaver I gave amonst them for tabaco 1:2½ to sidr 2 then I recond my sheep and lambs and have I bless God 20 sheep and 18 lambs so to Courte I lent tim Bork 11s:4½d wch wth 1s:4d I lent him before makes 12s 8½ **12** I spent reading no letters

ye 13 I gave Will Purtill on acct of work a bus barley to Norse 2 bushills barly she gave me 2s I pd Patt day and E Cheehan 15:7½ in full for church rates due on Courte Killcollum and Corohee in ye year 1744 then I gott 7:6 from ye Palatine in full for Petatoes

14th I went to Coroheen took a list of ye Cattle there so to Court spent 5 No letter I lent david Bryen 1.1 to pay for Cantelons horse[8] we had an acct in ye news yt ye allyd army were beat[9]

---

8 A David Cantilon, jeweller of the city of London, leased Killgobbin and Lissurlurran land in 1709 from Francis Garvan of Killgobbin, Co. Limerick. Declaration of Gerald Fitzgerald, 13 June 1722 Dunraven papers, D/3196/K/5/16/2. 9 At the battle of Fontenoy, fought on 11 May 1745 in the War of the Austrian Succession, Britain and its allies, Hanover, Austria and the Netherlands, under the leadership of the Duke of Cumberland were beaten outside Tournai by the French, commanded by the Marechal de Saxe; 7,000 allied soldiers were killed and Cumberland was forced to retreat towards Brussels.

15 16 17 nothing materiall to oyle 6½ news
18 I bought 2 yrlings from tim Bork for 1:14:1½ then I gave him 2 bus of barly for 8:8 wch wth 1:8 I lent him ye 5 of Aprill and 11:4½ ye 11th inst makes in all wth 5:1 0 he owd me for ye grass of a horse att Killmoreen half a year is in all 1:7:6½ ye ballance is six shillings and 7d
19 nothing materiall
20th aboute 7 I was served wth an injuncion by ye Studderds and went to Tough aboute it so back to Courte and prapard for Dublin wth Gods blessing in ye morning and do carry 9 att 1l:2s:9d att 9s 22½ att 2s:8½d one att 3s:4d 36 att 6½d and 43 att 1:1 all wch comes to 20l:5s:0d I gave matty 6½d to buy Sand and sope I gave out 5 dowsin Pleats and 16 dishes
21 I left Courte att 4 calld att Limerick baited at mines and lay att Rosgray
22 I breakfasted att Borros dined att mareboro and lay att Monstereven
23d breakfasted att neas[10] and gott to Dublin by 12 where I stayd to ye **2d of June** and pd for a watch 4l 15s to Mr Wallis[11] for a writt and other expences 4:11:0 to Mrs Hartstonge 2:17:9 to Shoes and buckles 7s 0 lent Mr Mellsop 3:11 my own expences 3:7:0 I gott a sillver cup from Mrs Hartstonge

June ye 2d 1745 aboute 12 I left Dublin dined att neas and lay att monstereven
3d breakfasted att mareborah dined att Rosgray and lay att ye mines
4th aboute 12 I gott to Limerick stayd there till 4 then calld att ye Sherriffs and gave a box wth wigs in itt and 2 pairs of buckles then I calld att adear and was Promised 3 horses to send to Dublin and desird to go and chuse them in ye morning so to Courte
June ye 5 1745 I went to Adear and brought to Courte two of ye old horses but wod not take more then I sent Patt Hure by Jon Griffy 5:5 so to Courte and gave Norse 2 bus of barly She gave me in parte 2:2 I gott 1:16 0 from David Recd for greasing while I was away and allowd him he pd for 2 letters and 1½ he gave P Daniel going to try ye Dublin mare I had ye sheep washd
6 I sent a barrill of Mault to ye mill had ye men and 2 women weeding in ye garding
7th I sent Matt Haly to Lim pd for 2 hogheds 16s:3 two stone sope 9s 3 pound hops 5s to bread 6 [sic] I spent yt day making up my acct ye men and women weeding
8th ye men and women weeding I sent Cosgary to town for lath and nails ye 1000 of lath cost 7:6 to ½ of 6 nails ½ a hondrd of 12d nails and a thousand of 4d cost 35 then I pd Morrah 8:1½ for his boat and sloop putting down ye ware
9 I went to church but no clargyman so back to Court and dined on roast lam aftr wch I went to ye dere park wak'd aboute aboute 6 back to Courte spent 13d

---

10 Naas, Co. Kildare.   11 John Wallis, a lawyer in the Court of Exchequer, Dublin, 1734. *King's Inns Admission Papers.*

June 1745

10th very cold wether I gave James Mctomas 5:5 to buy syths ye men in ye gardin and drawing lime I bought a yrling filly from Paul Haly for 2:9:0 and gave him 1:1 earnest will fosbery breakfasted wth me and I sent a man wth him to Corcamore

11 Halpin came to mend ye houses and fell to dressing Sleats I gott a barrill of Mault brew'd

12 halpin att ye Sleats I gave 19 blankett and 2 large ones to be washed and sent hom what barly I had at Court and Jon tould me they finishd drawing my turf

13 I sent a barrill of Mault to ye mill and rov'd aboute I gott ye sillk blanketts a course blankett and rug washd

14th I went to Limerick and gott a letter from Madm then went to hunt for dohig butt miss'd him then back to town sent home a pottle of oyle wch cost 2:2 a pottle of tar bladder 8½ 4 pd of pitch 8 3pd reding 4½ expences 13 so home and sould my cow for 4l

15 ye big gray mare foled nothing materiall I gott in my lining from ye wash halpin att ye sleats ye men in ye garding

16 I wrote to Madm but no letters I send hom 4 yrlings I had a violent fitt of eague

16 [17] I sett ye men to Cutt turf Jon Bryen came to me and gave me an acct yt he Recd from Darby feenen 1l:6s:4d from D Donevan 9:3 James Madin 3s:3d M fGerald 2s Patrick boyle for trespas 19½d for butter 1:1:7 all wch comes to 3:5:6 of wch he laid out 7:4 and gave me ye remaindr then I pd Paul Haly 2:6:11 in full for ye filly I bought of him ye tenth inst I gave Jon 4:4 to buy cards I gott Mrs Hartstong sheep shorn and pd 10d we began to mow Syp Pursill came to mend ye bedsteds Mr Taylors cow and bulluck went away

ye **18** ye turf and mowing going on Pursill and Halpin att work J donevan making a car for ye truckle I sent some Shrimps to Mrs Quin and sent for letters but none

19 ye turf mower Pursill and halpin going on Ned fitzGerald came and pd me 1:19:9 on acct of ye Rent of ye fery and gave me his note for ye reymindr then I pd Doyle my and Mrs Hartstongs tyth money and took up my notes I had watr drawn for brewing

20 ye men att ye hay I gott 4l for ye cow I sould ye 14 and sent her away hallpin and Pursill att work

21 I went home vew'd my yrlings and orderd 20 of them to be carried to ye fare then Jon gave me 1:17:2 he gott 1:0:4 of butter then I went to Killmoreen and gott 16:9:9 from martin fitzGerald he gott for 2cwt:0qr:2lb of butter and gave him a receipt for itt then I took a list of ye greasiers and there is 73 my own three cows Dens 2 heffers and Slatterys callops includ then I came to Courte and made up 22:0:1½ gave it Jon to give Rob: Hewson on acct of ye Rent of Killmoreen then I gott a letter from Madm and another from Mastr

22 Jon went wth ye money and brought me a message yt Mr Hewson wod not receive it till he got ye whole then I sent home 7 heffer yrling to be sent to ye fare and so loiterd aboute ye men att ye bog and hay

23d to church and back to Courte gott 14s for 61 cows of James Russills for a month and 8s for a weeks Profitt att ye fery Harriott Blew foled

June ye 24th 1745 I went to ye fare and by much adoe gott 23l:10:0 for 21 yearlings and had 9:6 expences for them and me so back to Courte aboute 8

25 I went to ye dere Park spent 8 so back to Courte and gave Syp Pursill 3:3 ye men att ye hay I gott a letter from Madm wth directions to send ye horses

26 I went to town and gott bank Bills for 45l and inclos'd them in a letter to Mr Thornell then I wrote to Madm and sent away ye horses and gave ye men to bare their expences 1:16:3 then I spent 8d and took a peice of Cloath for skirts from Jon Bonfield for wch I owe him 2:18:4 so to Courte and gott a letter from Hen Page for ye May Rent to bread 6d

27th agreed wth tho Harragan for 10s an acre for caparow and sould him 20 heffer yrlings for twintey pounds and ten shillings earnest 2:8:0$^{1/2}$

28 Jon Bryen and Thos. Harrogan came to me I accted wth Harragan and chargd him wth 1:12:7$^{1/2}$ he had in Corn and allowd him in ye Pryce of ye yrlings 8:18:4$^{1/2}$ wch is in full for ye cows I bought of him last xbr then he gave me 10:3:2$^{1/2}$ wch wth 1:8:9 he owes make up ye 20l:10:0 he was to give for ye calves then I made up Jon Quain acct and charged him wth 2:5:2$^{1/2}$ he had in corn and greasing then I sent him by Jon Bryen 12:2:2$^{1/2}$ and owe him still 3:9:7 then I made up Tho Burys acct and gave it to Jon to carry him and to gett 1:1:0 and give it to Quain Halpin finishd ye houses

29 I went home and made up 13:11:1$^{1/2}$ for John Quain and owe him still 2:8:5 then I gott my notes from him and Harragan then vew'd my Corn and Cattle so back to Courte by 12 gott ye Roses pickd and began to still[12]

30 sent Harriott Blew to Mr Burys hors wch she took gott 9s from ye ffery gave Mr Cosgary 5:5 downs brought ye nett I gave him 4s no letters from Dublin I gott a letter from my Brotr for 12l but had no mony

July ye 1st 1745 ye stilling going on I began to bring my things up stears to have Mastr's roome claind I orderd Jon to carry my yearlings to ye fare

2d Jon came and let me know he sould [blank] yearlings for 10:0:6 he gave me 1:1:4 from Mr Thos Bury he pd 2:2 for hatts for ye boys 1:1 expences then I gave him 2:8:5 to give Jon Quain to give Mick Mctomas 6:19:0 to give Ed Naghtin 3:6:0 to give Joan 2s:2d for him and Martin fitzGerald 12d I sent for letters but got none

3 I spent a stilling and making Pens J Jonston and his two sons breakfasted wth me Recd from David Bryen 1 16 3$^{1/2}$

4th I spent ye day making pens and tending ye still gott ye 5:5 I gave Cosgary a Sonday last I gave Pattr Daniel 5:5 to pay for ye Service of ye two gray mares last year 5:8

5 ye stilling going aboute 3 I went home and gave Jon: Bryen 3l:0:1 to pay for ye last years tythes and to buy oyle 13d so back to Courte no letter to bread 6d

---

12 Possibly Peacock was making rosewater, distilled from rose petals.

July 1745   143

6th stilling I had a letter from Rob Hewson for Rent wch I sent ye 26 of last month to Mr Thornell wch I let him know I sent ye big gray mare to be tryd she refused ye horse

7th I gott 8s from ye ffery I sent Harriott blew to be tryd she refused ye horse I gave Patt Daniel 2:8½ to give Mr Burys groome I had a cow and yrling brought here

8th I gave Mary Sawn 2:8½ and some tread to knitt into stockings I sent E Grimes 3 shirts to be made

9 I had a letter from Ms yt they must make some Delay in town to ye letter 4d bread 6d I gott ye well and little house claind Herlihy came to have me make up bounds wth him ye gray mares fole sick

10 I gott up att 3 went home and found stones breaking to burn a kill[13] ye flax dressing my own two mades at ye frize then I went to Killmoreen and tould Herlihy yt I wod devide ye bounds when ye Herdsmen made it up as I gave it them so back to Courte and loyterd about ye fole still sick

July ye 11th 1745 my much ado I gott ye men to work Jon brought me some new Petatoes and ye new bandle cloth wch I cutt into 2pr of sheets and spent ye day making them

12 I wrote to Ms and sent for bread snuff and letters no letters bread 6d snuff 1:1

13 I went to Killgavin for ye money but got none ye gray mare refused ye horse I left her att Killmoreen I gott two quadruble pistoles[14] 2 att 2s8½ and 4:4 from Mastr fitzGerald he gott for butter I gave P Hure 2:8½ and spent 7s so to Courte and finishd a new pr of sheets

14th Jon Bryen brought me half a lamb and some parsnips and ye Pryce of a hondrd a qartr and 14l of butter att 37s:6d a hondrd comes to 2:10:10½ he gave 2:3 for a bus of salte and 3 for Costom and turnpike he gave me my note to Mr Quin for last years tythes I gave him 15s for to pay for sheering my lambs &c: I gott 8:11½ from ye ffery I dined on lamb and Parsnips aboute 4 Tom: Bork came hom no letters

15th I spent ye day making a sheet and att ye still I lent Tom Bork 4s:4d ye men att ye hay

16 att ye still I finishd two pr of Course sheets Doyle came here aboute ye tyths but I did not take them no letters to new 1d

17 ye men futting ye turff ye wettness of ye day hinderd us from puting in ye hay

18 I sould two cows to Paul Haly for Mr Jackson for 4:18:0 wch he pd me then I went to Killmoreen from thence home aboute 1 back to Courte and satt wth ye men I sent Jon a fivepence to buy a syth stone

13 To make lime.   14 The value of foreign money was fixed by Dublin Castle. In 1745 a quadruple pistole was worth £3 13s. 0d. In 1737 the exchange rate was reduced, and again in 1750 when a proclamation was issued by the lords justices declaring that the quadruple gold pistole, weighing 17 pennyweight 8 grains, was to be fixed from 10 October at £3 11s. 4d., and 'in like Denomination' for their several sub-denominations. Customs collectors were asked to send a return of the number and weight of Quadruple Pistoles in their hands. Minutes of the Irish Board of Customs and Excise, 4 October 1750, CUST 1/49/59; 6 October 1750, CUST 1/49/91-2.

19 I sent to Lim: for letters and bread aboute 3 ye horses came home nothing materiall bread 6 news 1d

20th I sent to see them to Limerick lent Madm 4 ginies lent Mr 2 to Mastr 5:6 pd Bonfield 2:18:4 in full for ye linnin for my Shirts pd him 11:4$^{1}/_{2}$ for a weastcoat 2:2 for 3 handkircheifs gave ye taylor 2:8$^{1}/_{2}$ to ye glover 2:2 expences 5:7 so home wth Pallisere

21 I went to Church and back to Courte sent my mare for Mastr Hartstonge and he came to Courte

22d I went to adear to buy sheep but did not so to Lim bought 10 stone of att 15d a stone a bag 2:2 making 1 to Porters 4$^{1}/_{2}$ expence 2:2 turnpike 7d so home wth Madm Mastr and Ms Ann so hurryd aboute

23d went to Jackson aboute ye sheep but did not like them so home and orderd some linin for me then to Courte and gave up ye keys spent 4d

24 to Limerick and laid out 4:7:1$^{1}/_{2}$ for things for Ma to Iron for my self 2:10$^{1}/_{2}$ so back to Courte and spent ye night in a frett aboute D Bryen

25th to Limerick and brought home Mrs Quin and Harts to turnpike 6d ye Chair 2:8$^{1}/_{2}$ and spent ye day wth ye company till 8 then to bed

26 ye company dined att adear I spent ye day mostly my bed being very unwell

27 ye company at Shanongrove I continu'd unwell and gott from Mr ed fitzGerald on acct of his note and ye arrears of ye ffery 2 att 1:9:3 and 1 att 1:9:3

28 still unwell tho much better I bless God

29 I bless god pritty well spent ye day deviding ye medows yt were sould

30th I spent varriously gave Madm 4:7:7

31 went home and brought most of ye thing I had above and ye swourd and gave them to Madm

Augst ye 1st 1745 made up my accts to give Madm but she had not leasure to look them over I gott 8:8 Jon Sulevan owed me then Doyle ye Procter came I could not agree for ye tyths of Courte but gave him 7s:7d in full for this years tyths of Killmoreen viz: hay sheep and cows in ye Presence of Mr Mellsop

2d our family went to town and I home where I stayd claining out my Roome till 6 then back to Courte

3 I finishd 3 handerkcheifs and have now in all 11 pockett hankercheifs besides white ones ye Lord make me thankfull

4th to Church and Advertized ye tythes of Courte aftr Service walkd aboute Killmoreen so home and dined on beans and bacon then gott 8:1$^{1}/_{2}$ from Jon Molcheen on acct of greasing att Killmoreen and gave Cate 2:6 on acct of her wayges then I was sent for to Courte and unwillingly got there by 7 and aboute 9 to bed

Augst ye 5th 1745 I sent ye horses for Madm and gott from ye ffery 7:1$^{1}/_{2}$ I sould Patt Pursill an acre and $^{1}/_{2}$ medow aboute 8 Madam came home

6 I gave Madm 1:4:4$^{1}/_{2}$ I gott for greasing and att ye fery and so I gott from Jon Hartney Geo fosbery dined atCourte I satt wth him till 5 then stroled aboute

August 1745

7th I began to draw hay I gott ye 4s:4d I lent Tom Bork I gott a pr of stocking home and gave 2:8$^{1}/_{2}$ to Mary Sawn to give her daughter for knitting more
8 I went wth Mrs Hartstonge to town and spent 8:7$^{1}/_{2}$ so home and thought yt I had lost my watch and on a dispute wth Jon was very uneasy
9 I gott up att 4 and went to look for my watch to no porpose but aftr I came back gott it in my briches Mr and Ms Nancy went to town I gave him 2:8$^{1}/_{2}$ ye hay drawing he came back at night and brought an acct yt Hary Quin died so to bed without a super I gott 9:4 greasing at Court
10 Mastr a hunting and brougt Tounsend and W fosbery to diner and gave them a bely full of punch I gave him 2:2 for Spiritts and 2:8$^{1}/_{2}$ for Thos. Jackson I gott 3:8:7 from Ed fitzGerald in full of his note
11th I gave Mr Hartstonge 2:8$^{1}/_{2}$ and went to ye funerall gave Mrs Hartstonge 3:8:0 then about 1 we sett off gott to adear aboute 5 I drank a mug of sidr so to Courte
12 att Courte alone and hem'd ye hattband I gott yesterday I sent home my colt and filly and pack'd up my Cloaths to send them home I gott 7:3 from ye fery and 6:6 from James Rohan for ye grass of his sheep
13 I gott 4:8 greasing and gave thatt and ye money I gott yesterday to Madm then I sent home my horses truckles and Cloths then sent a barrill of Mault to mill gave Mrs Hartstonge more money I gott for greasing
14 I sent Mr Cosgary to ye fare for a fatt Cow for Mrs Hartstonge and gave him 3:0:8 then gott a sheep killd wrote to Yeamons for a bus of oatmeal so home spent 6d found my men and women pulling flax so loiterd aboute
15 being a holyday[15] my People went to ye Patern I gave Cate 3 bandels of flanning Margett 3 and Den two wch wth 9 he gott before make 11 att 5d a bandle comes to 4s:7d to be charg'd to his acct I spent ye day diversly
16 nothing materiall my own people at ye flax
17th I had 5 women and my own att ye flax I was summond to be on Massys[16] jury ye taylor making a coat for little Jack ye boys began to Reep ye Duch barly
18 to Courte from thence to Church back to Court dined there in Company wth Mr Bury and family aboute 5 they went away and Mrs Hartstonge was unwell wch made me stay there yt night
19 I wrote to Ms to come to her Mama and bring ye Doctr and was very unwell
20th Mrs Hartstone and I better
21 pritty well and came home and most of my flax pulld and moste of ye barly reap'd I spent ye day wth ye men reeping and women binding
22d I sent to see Madm and am tould she was worse I gott from Martin fitzGerald for 6c:0qr:3lb of butr he sould yesterday 9l:6:7d he owes me 1d then Tho: Bury came to me and offerd 9s an acre for parte of Killmoreen ye men and women att ye flax

15 Feast of the Blessed Virgin Mary.   16 Hugh Massy (1700–88) of Duntrileague, Co. Limerick. See Introduction and Barnard, *Abduction of a Limerick Heiress* for the full story of the charges against Massy.

23d I went to see madm where I breakfasted then home pulld ye last of ye flax and bound some barly and walkd aboute

24th I sent to see Madm who is better I tryd both old and new Petatoes and cannot brag of ye goodness of ither I sent Jon to trate aboute ye tytthes aboute four he came home but did not agree

25 went to Courte dined there aboute 5 I came home sent my 3 horses and truckles to draw Pauls barly gott a letter from mr Bryen to meet him in town with my Rent so to bed

26 I gott ready to go to town and carried 2:4d peices 1 giney and 9s 9d wth and ordrd John to follow me to town wth a firkin of butter to more morning I sett out aboute 10 and got to town by three where I stayd to ye 31 I gave Mr Bryen two four pound peices and a giney on acct of May Rents lent Mastr Hartstonge 6:8 pd for my lodging 7s:7d to 6s:6d my expences 1:6:0 I left town at 4 went to Courte lay there yt night I got 1:15:0 from John he gott for 1:0:12 of butter I borrowd 5s:5d of my sistr and spent 2:8 1/2 of it

7br ye 1st 1745 I came home for a shirt went back to church dined att Courte and lay there yt night lookd over Bastables[17] acct and wittnesst it so to bed

2d I went wth Mr Hartstonge to Hally Park back to Courte and home att night fall unwell wth a cold I got a letter from Robt Hewson for money for thornill

3 I continud unwell had 2 horses drawing hay

4 I had Tho Harragan his horse 2 boys J Newman Mick Molcheen and Ned Naghtin and my own men boys and horses att ye hay and finishd I had a letter from Hen Page for Rent and Promisd when I gott ye price of ye medows I wod pay him

5 I went to Courte lent T Bork for Mr Hartstong 2:8 1/2

6th att Courte sent home for my horses and eleven barrils of lime I did not go to ye buck but was inployed mending and alltering locks att 9 our family came and brought Ms Stanford Mrs Sillver Mr Bury and Mr Mansill att 12 to bed

7th I gott 6 men from Madm and sent them home to Reep oats aboute 10 I follow'd them home and spent ye day wth ye men and women I sent six shilling by John to pay griffy for ye torkeys

8th att home my Servts att ye well

9th I sent Jon to Rob: Hewson wth bills for 45l for wch he sent me a receipt I had six men from Courte wch wth my own men and women we finishd Reeping binding and stooking ye Oats

10 I went to Courte and stay'd there seeing ye hay in till ye 13 so home

13 I gott from John he gott for medows 12:12:3 viz from Mr Willm Taylor 10:4:9 from Will Covard 1:5:2 from David Welch 11:6 from Mar Edigan 10:10 all wch I carried to Askeaton and pd Hen: Page on acct of May Rent so home spent 9d I had 3 men from Courte and my own men and horses drawing Oats

---

17 A member of the Bastable family of Castlemagner, Co. Cork.

September 1745

14th I had ye above men and finishd my Oats

15 I was sent for to Courte to Appels 3d

16 I went to Ballynegolla and Grilla[18] to veiw them I dined at Bruff[19] overtook Mr Hartstonge Mr Bury and their company att Clampitt[20] spent 2:8½ wch ~~I owe Clampitt~~ then in Company wth them went to Courte where I stayd to ye

22d so home having got from Mrs Hartstonge 10:10 to bear my expences to ye County of Cork

23d in Company wth Pattr Danil I began my jorny to Castlemagner[21] and gott there by 4 in ye aftrnoone

24 I viewd all ye lands wch I think very bad we sup'd att Chattmans[22] so back to Bastables att 10 to bed

25 aftr breakfast I left Castlemagner and bated att Charlevill gott to Courte aboute 6 where I found Mr Dawson and ye young Docter[23]

26 I gave Mrs Hartstonge an acct of my jurny and fell to Racking sidr Mr Bury and his family dined att Courte I gott from M fitzGerald 5:3:27 of butter

27 att ye sidr and made up 5 hogsheds gave them a cag of Salmon a box wth bacon and a baskett wth bottles to ye carriers and saw them tye up their loads I lent Mr Hartstonge 18:3

28 I wrote to thos Studderd and waited for an answer till it was to late to come home to snuff 1:1

29 I sent home for a shirt and a goose aboute 10 I gott them and went to Church from thence hom and had a goose boild and another roasted for my supper

30 I sent to Courte 2 large stone bottles then accoumpted wth Jon Nele and do owe him 10:6 he had 54 days att 4d a day wch comes to 18s and 9d he had due to him he gott six shillings in medows and 4:6 worth of beans then Jon gave me 20s he got from Nele and Mau for medows and acctd wth me for for 3d he gave

---

18 Ballynegolla, 3½ miles west of Adare. In 1840, the remains of an old church stood on the eastern slope of a hill. There is a cromlech of some size. Grillagh is in the parish of Tullybrackey. Ordnance Survey Letters, f. 349.   19 Bruff is 11 miles from Limerick and named after an earthwork in a grove thickly planted with elms. The de Lacy family built a castle here, which they held under the Desmonds. The castle, on the north bank of the river, came into the ownership of the Hartstonges. A major battle was fought here in 1641, in which the English forces lost. There is a memorial to the Hartstonge family in the church, reproduced in Lenihan. Ordnance Survey Letters, f. 98. Lewis.   20 A George Clampitt came to Ireland in about 1650, and obtained a grant of land in the city of Limerick. His grandson Isaac Clampitt, was high sheriff of Limerick in 1738 and mayor in 1739. Isaac's son George was a clothier and admitted as an Honorary Freeman in 1753. Some of the family were Roman catholic, as one of this generation became a Franciscan priest and another a student in the Irish College, Rome. Lenihan, p. 105.   21 Castlemagner, Co. Cork is 3½ miles from Kanturk. Its name comes from the family of Magner, who built a castle here. This with its lands were granted to the family of Brettridge, and were inherited by the Hartstonges. Lewis.   22 The Chapman family of Lisdogan, Co. Cork were to be Nicholas Peacock's in-laws. A George Chapman was born in about 1705 in Cork, and in 1731, George Chapman married Jane Roche in Cork. George Chapman of 'Lisduggan' promised to ride in Lord Egmont's independent militia troop, and a George Chapman proposed to surrender his lease of Clonribben and was in arrear with his rent to the Percevals in north Cork in 1746. Lists of 8 May 1744, 12 Sept. 1745. BL Add. MS 47001 B ff 66 138 v; f 87; 47002A, f 43.   23 Probably Dr John Martin jnr, son of Dr John Martin of Limerick.

ye weaver 1:3 for tucking 1s:7½d turnpiks then I allowd Will Covard 3:7 for Dennis acct then I gave him 8:2 in full for ye binders hire all but C Carty

7br ye 30th continued then I gave John 3:9 to give Margett in full of a quartrs wayges and orderd him to hire one in her place then I allow Patt Boyle on Cates acct 6:2½ then I orderd my Plow to Courte and gave ye men 2:2 to drink and 4½ for tabaco I gave Joan Daly on acct of her wayges 19:6

8br ye 1st 1745 I came to Courte Sett ye Plow going
2 ye plow going
3d I was sent for to town aboute Studderd I got by ten but did not good but apoint a meeting ye 10th att Rathkeal I gave 3:4 to Mills as Visitation fees I pd my sistr 5:5 I borrowd att ye Sizes then having din'd I pd for a bottle of wine and a sneaker So to Court and lent Mastr 6½ to give ye Plowmen
4 ye Plow going lent Ms N 1:1 to Mastr Harry to give ye boy yt brought jett 1:1
5 ye day was so wett I ordr ye Plow to be untacled and sent a letter to Hen Studderd[24] **6:8:9** ye Plow going I came home
10 to Adear and Mett Ed McGan att Clampits where we breakfasted and pd 1:5 for it I pd Clampitt 2:8½ I borrowd ye 16 of last month so to Rathkeal and did not good but spent 7:2 so home and Ed McGan wth me
11 12 13 14 att Court ye family began their jurny **15** Court
16 I went to town to meet ye Studders but did not 2s:9d
17 I was sent for by ye Studders but did no good but a Promis to pay ye 3d of 9br then Mrs Hartstonge let me know she intended to begin her jorny in ye morning wch made me stay in town I pd for Thos Jackson 1:8:7 to flin[25] 18:6 to Mastr 4:6 to Ms 5:5 to Ms N 6½ to oyle of Marsh Mallows[26] 2:8 aboute 2 to bed
18th we sett off aboute 12 baited att tollo lay att he mines
19 I went wth them parte of ye way then D Bryen Ed Mcgan and I breakfasted att ye mines dined at tollo lay in Limerick expences since I left home 18:10
20 to Court **21** began to Still ye Plow going **22** att dito
23 ye plow finish I sent them home to bread 6d new 1d sope 2:3 to a coatt for Matty 7:7 Jon Bryen gave me he gott for 1c:1qr·5lb of my own butter 2:0:0 Rob Hewson came to me for Rent and to know if he can gett reed
24th I gott from Martin fitzGerald 9:16:1 he gott for 6:0:26 of butter and gave him a receipt then I wrote to Mr Bury for greasing of his bullucks but did not gett it ye still going **25 26** ye still going news 1d
27 to Killmoreen veiwd ye sheep and calves gott 6s from thos. FitzGerald for ye grass of a 2 yr old so to Church then to Court gave Ma: 5:5 I had a goose from home
28th I went home gave out 67 pound of wolle and six stone of flax to carry to ye fare I gott E Daniel and T Callahans shear of ye giney I pd for liberty of a

---

24 Henry Studdert was the second son of the Revd George Studdert and Thomas Studdert's brother.
25 Flin may be James Flyn, glover of Main Street, Limerick.   26 Marshmallow (*Althea officinalis*) grows near salt marshes.

November 1745

road thro Ballycraheen wch is 6:2 Jon pd for costom and turnpike 4 I gott 8d M Edigan owd for ye hay I gave 2:8½ to C Slattery on acct of dressing ye hemp I gave 18d for Schallops then to Court and gott a letter from Hen Page to allow Mrs Taylor 1:15:0 wch I did and charg'd it to my acct wth Colonell taylor 29 I sent ye men to Porte Geary[27] and Day came aboute ye Church rate wch I apllotted to spurs 1:1

30th 31 ye men att Port to eggs 2d

9br ye 1st 1745 I wrote to colonell Taylor and Ms I gott from Darby Day for ye grass of 6 heffers att Killmoreen half a year 1:16:0 2d ye men att Port 3d reading

4th Clampitt came to me for Mastr Harttstonge bill I pd him in full 11:6 and got his receipt I pd him for Conr Quelly 16:4½ ye horses came from Dublin and brought me a byble yt Ms sent me and a Schrew yt Palliceir sent he men att porte

5 6 and 7 ye men att Porte

8 ye men claining out ye turf house Donevan 6 days mending ye truckles I gave anstice 2:8½ on acct of knitting then I pd Ter: Mcmahon one pd for John Mellsop to ye news 1

9 my 2 ye smiths Cosgarys and M fitzGeralds horses drawing turf I pd M Halpin 13s for his work in June last to Matty 10d to buy Stockings to bread 6d then Jon gave me 1:17:3 he gott for wolle from James Mcmahon 3s from Law Daniel 12s M Mctamos 13s:1d J Mctamos 6s in all 3:11:4 then Jon Donevan gave me acct of his note 5:0:5 then I made up 30:16:2 for Mr thornill

9br ye 10th 1745 I wente to Ballyengland and pd Mr Hewson for ye use of Richard Thornill Esq 30:16:2 wch wth 5:5: I gave Tim Herlihy for ye use of Bretrige Badham Esqr makes 31:1:7 and gott a receipt from mr Hewson[28] then Breakfasted so to feighs spent 9d so to Courte gave 10d for half a goat and had it dresst but ye tooth ake wod not let me eat any of it H Evans came to make some bridles &c

11 ye men and horses att ye turf I pd M fitzGerald for 2 days his mare was drawing 2s and sent her away I sent ye Quittrent by thos. Bork and he brought me Receipts for it and a bottle of Shrub my Sistr

12 ye men and horses att ye turf and about 12 finishd I sent hime ye horses truckels and 2 roulets cidr Jon gave me 1:19:0 he got from thos. Haragan for

27 John Geary, farmer of Ardnavolla. Indenture between William Bury and John Geary, farmer, 2 April 1745 Dunraven papers D/3196/K/6/3/3. 28 Peacock was settling the lease of land, probably in Co. Cork and probably through relatives of the Hartstonges, Richard Thornill and Brettridge Badham with Robert Hewson acting as an intermediary. 'Use' was a transaction by which a man who owns land covenants in consideration of a blood or marriage relationship that he will lease it to his child, wife or kinsman for life, in tail, or in fee. As Blackstone so felicitously puts it: 'Here the statute executes at once the estate; for the party intended to be benefited, having thus acquired the use, is thereby put at once into corporal possession ... without ever seeing it, by a kind of parliamentary magic'. Blackstone, vol. 2, p. 338.

greasing at Killmoreen I gave him 3d for tabaco and to buy Iron 3:3 I got from James Slattery by John Geary 5:11½ on acct of greasing att Killmoreen

13 ye men trashing and Pressing I gave H Evan 3:3 on acct of his work

14 finisht ye sidr and barly ye men puting in timber

15 I went to Limerick pd frank yeamons for Mrs and Mastr Hartstonge 1:6:3½ to Ned Mcgan for Batt Linerd 10s:11d to Jon Bonfild for linard 1:6:11 to him for cloath for a cloak for Matty 1:8 and 1l:11s:0 for triming and a Rug weascoat for my Self I pd for wine 29 for a pr of sisers 8 sugar 10 dram 3 reconing 2:9 and 2d turnpike I lent Dan O Bryen 5:5 so to Courte lent David Bryen 10:10 then having supt on 7d worth of elles and drank a glass of sidr and to bed

16 I wrote to Jon Bonfield for ye Pattern of a coat of Cloath and triming for David Bryen and lent Patr Daniel 9:9 and gave him 2:6 to by Iron for me then I left ye men att ye hay and walkd home it being a frost

17 spent ye day mostly reading I got 7d of ye 3:3 I gave Jon to buy Iron ye 12th

18 I spent mostly settling my accts and gott 10s:8d from Jon Williams for greasing att Killmoreen and Coroheen

19 ye taylor came to makeing Cloaths

20 I gott a letter from Jon Pordon aboute Studders I gott 2:8 from Jon Morphey I gott a lettr from Mrs Hartstonge and wrote to David Bork about his Rente spent ye day variously

21 I went to Courte gott Borks answer and wrote to Mrs Hartstonge and sent her burk letter

22d Sent to ye office news aboute 3 I came home and gott from Jon for Jon Mcmahon 9s James Nash 5s and from Tim Day 8 all greasing att Killmoreen I gave Jon on acct of his wayges 1:2:9 I gott two lined girths home and gave one to Thos Bury aboute nine Hary Evan came for me to meet ye Studders and brought ye other gyrth

23 I went to town but did no good but spent 4s in my way home I had a quarill att Clampitt and lost ye handle of my whip so home

24 aboute 11 I went to Clampitts dined there went from thence to Killmallock[29] lay there and pd 17:9 for our supper &c:

25 I went to ye Array but none of ye Lord Southw. Troop[30] being there I came away dined and lay att bruff pd 15:8 so to Limerick and waited for ye Studders to ye **29** but did not meet them so to Courte where I lay

30 I wrote to Mrs and Mastr Hartstonge and to Jon Mellsop and to Ned Mcgan and sent his 11:4½ 6:1 of wch was his bill while I was in town and 4:4 I bor-

---

29 In the eighteenth century, Kilmallock, Co. Limerick, was a fortified, walled town 15½ miles from Limerick, on the small river Lubach which sent two MPs to parliament. Its condition was such that in 1760 Edward Willes told the earl of Warwick that admirers of ruins need not go as far as Balbec or Palmyra. In 1775 it was observed that 'the town is entirely composed of old castles joined to each other ...' Lewis; Ordnance Survey letters, f. 349; Willes, p. 54; Seoighe, p. 162.   30 Thomas, 2nd Baron Southwell of Castle Mattress (–1766) married Mary daughter of Thomas Coke of Melbourne, Co. Derby. Peacock was not among those named present at the Limerick militia array raised by Lord Southwell in 1756. This raises the possibility that he was dead by then. NLI GO 608.

rowd of him then I gott all ye sidr yt was fitt stopt so home gott from Jon he gott from Mau: Donevan 12s from Jon Nash 5:4 for greasing att Killmoreen then I gave Jon 4 bandels of flaning and Joan 12 bandels on acct of their wayges

xbr ye 1st I spent mostly reading
2d I gott from Jon 12:3 on acct of greasing att Killmoreen of wch he gott 5:8 from Mick Daniel and 7:6 from James Harragan then I gave ye taylor 12 in full for making my and ye boys Cloaths so to Court
xbr ye 3d 1745 nither letter or news to bread 6d
4 I pd Jon Moneen for a busl and $^{1}/_{2}$ a peck wheat 5:7$^{1}/_{2}$ I ordrd ye men to thatch ye hay
5 ye men att ye hay Mr Bury gardiner att ye trees
6 ye men finishd ye hay ye gardiner ye trees I gave him for his trouble 2:8$^{1}/_{2}$ I pd 4d for eeles[31] 6d for nails news 1d I gott a letter from Thos Studders wch I sent his brotr I lent norse to pay Carty 2:8$^{1}/_{2}$ I we for $^{1}/_{2}$ a pd snuff
7th nothing material **8** I spent reading mostly
9 I spent wth Martin fitzGerald and Tho Harragan
10 I went to Killmoreen to see Harragans heffers but did not meet him so back to Court sent a cow home to be killd and ye Pole ax then got ye barly winowd no letter to news 1 so aboute 5 I went home I pd 2:2 for a busl of Salte aboute 9 to bed
11 I had my cow killd sent ye hide to Killcorly and gott 9s 10d itt weight 4s: 13lb I sent Jon to ye fare to he did no good nor I much
12 I bought from Jon Quain 1 cow for 2:14:0 a heffer for 2:3:6 from Thos. Harragan 4 heffers att 2:5:0 each wch comes to 9lb and one heffer for 2:8:0 I spent ye day wth them a a drinking Quain gott 1:1 earnest
13 I sent a Stone of flax to Doctr Martin and sent for letters I gave ye little taylor 2:2 and 2so to ye women for binding I gott 9:5:0 from Mar fitzGerald horn money and 4:16:5 he gott for 3:0:2 of butter in all 14:1:5 ye boy came and I gott a letr from Mastr Hartstonge to bread and news 7d
14th I pd Thos. Harragan on acct 8:14:0 and pd Jon Quain 4:17:6 in full then went to Courte intrd ye Labourers work vewd ye sellers and aboute 1 came directly home and pd Cate Carty in full for ye days She had binding spining and pulling flax 3:3 then when night fell I litt a candill and satt down to wright my will ye Allmighty God give me grace to mend my life
15 spent ye day reading
16th aboute 6 I went to Courte and in ye eavening ye entire roofe of ye Kill house fell
17th I gott a letr from Wallis wth a supena[32] for Thos Odle[33] and directions to drive D Bork I gott ye barly piled and winowd and gave ye women 12d for claining ye barly to bread and news 7d

---

31 *Civil Survey of County Limerick*, vol. 4, p. 362. 32 A writ issued to order the appearance of a defendant in court to answer charges against him. 33 Thomas Odell (−1761) of Shannongrove, eldest son of John Odell and Anne Fitzmaurice. He married Constance Fennell, daughter of Robert

18 I wrote to Mr fox for 2 bottles rum and 2 of brandy for wch ~~I owe him 6d~~

19 in Company wth D Bryen I went to Ballylongford bought a cow from Thos Harragan for 3:5:0 and a cow from Jon Quain for 2:17:0 gave each as earnest 6½ I gott 5:4 from J Hickey greasing att Killmoreen then having dined and drank pritty hard I ordrd Jon to go for Rum brandy and Sugar and gave him 8½ he got from Mr Crotty to pay for them he pd 7s for them 6d bread 1d turnpike then I went back to Courte to expences 1:1

20th I gave 18:9 for 31 yards of ticking to new 1d

21 I sent Da: Bryen to serve ye supena and sivere to ye service I gave him 5:6 I gott a note from Patt Sheehan to give Lowry 4 of Mr Bury wethers

22d gott a letter from Jon Longane to deliver his cow wch I did and spent ye day reading

23d I went to David Bork aboute his rent he sd yt if I drove he wod replevey³⁴ so I came back to Limerick and stayd yt night and pd frank yeamons in full for mending my whip 2:2 then came to my Sistrs playd cards aboute 9 to bed

24th I went to ye office for letters but got none then I send Madm ye affidavid of ye service of ye supena and and gave her an acct of Borks affair then I gave 1:4 for gon powdr 13d for snuff 5d for wax and then dined so to Courte ~~4:11 in debt to my sistr~~ news 1

Debr ye 25th 1745 I came home and aftr diner gave my little family a boul of punch and 2:8 amoungst them for their Xbrmas box

26 I spent wth Geary Supple and day

27 I veiwd all ye Cattle att Porte then wakd to Killmoreen and veiwd ye cows and sheep so home and bought a cow of Tim Calahan for 2:17:0 gave him no earnest

28th att home spent ye day wth A Clampitt R Flanigan and S Henry they gave me a hare

29 in ye Eavening I went to Courte

30 we had a dinr and Company

31 I went home and had Company my Sistr Mcgan lay at porte I gott directions to drive Bork and a lettr from Wallis aboute Studders affair we all drank pritty hard

---

Fennell, of Curraghbane, Co. Cork and d.s.p. Burke (1958). **34** Replevin, an action by the tenant to recover goods which have been distrained, on the debtor's agreeing to have the case heard in court, and to return them if the case goes against him. The tenant would have to lodge a claim with the clerk of the peace who could order the sheriff to replevy, or return, his distrained goods. De Moleyns, pp 615–16.

1746

Janr ye 1st 1745/6 my sistr went home and I to Courte gave ~~Ned Megan 13~~
2 D Bryen P Daniel and I went to Bruff and drove Da Bork he replevied ye distress then I went to Limerick expences att Bruff 5:2 and 13d for a coppy of ye Replevin
3 I wrote to Madm and sent a coppy of ye Replevin I wrote to Wallis aboute Studders affair so to Courte I gave 5s for a whip and aboute 3 I came home a little aftr I had an express to go to town to meet Studders I borrowd 10:10:9 from Thomas Harragan and gave him my note for it Sam: Harding lay wth me
4th I went to town but did no good as usuall I lay in town and got a knife and fork from Whistler for wch I owe ~~1:8~~
5 aboute ten I left Limerick calld att Courte home and Martin fitzGerald gave me yt he gott for 2 2 2 of butter 3:9:6^{1}/$_{2}$ he owes of ye Pryce of ye butter 11^{1}/$_{2}$d
6 Jon and I left home calld to se Rob cox gott to Charlevill aboute 10 where I baited and spent 2:6 I left it att 12 calld att Lisscarroll[1] to ale 4 and gott to Rockfield[2] as they were att diner aftr wch we drank till y then went to bed allmost dronk
7 att Rockfield and sett my leas to be drawn
8 att Dito ye leas in hand
9 we Perfected ye lase to Commence from this day gave my bond for last Mickelmas Rent and a note for ye Rent yt shall become due ye 25 of March next and a note for 3:17:8 I borrowd of Mr Thornill to give his lady then I gave ye Atturny and Clark yt drue my lase 1:11:5 gave ye buttler 2:8^{1}/$_{2}$ ye made 2 8^{1}/$_{2}$ ye cook 2 8^{1}/$_{2}$ ye boy 6^{1}/$_{2}$ ye groom 2:8^{1}/$_{2}$ to ye man yt went to shew me ye way to Castlemagner 13d Bastable not being att home I went to Chapmans where none were att home but his daughtr[3] who desird I wod stay yt night wch I did
10 aboute 9 I left ye Ladyes gave ye made 13d and ye boy 13d then breakfasted wth Bastable gave ye servt 13d aboute 11 I came away and att Liscarroll pd for a mug sidr 4d att one I gott to Charlevill where my mare was taken ill butt recoverd so well yt att 3 I left itt and came directly home pd 2:3 for wine and a dram
11 I sent for a horse to Court and aboute 12 gott there and gott a letter from Mr Hartstonge then I vewd ye house gave some directions so home

---

1 Liscarroll, Co. Cork, is 4^{1}/$_{2}$ miles from Buttevant, on the road to Newcastle. 2 Rockfield, Liscarroll, Co. Cork. The Thornhill family lived at Ballyheene, a house which was deserted by 1840. Lewis. 3 Catherine Chapman, Nicholas Peacock's future wife.

12th I spent mostly reading

13 I went to Shanongrove from thence to Court where I lay and was desird to go to town

14th I went to town aboute Mr Studders but no good only yt he has given ordrs to rais money and cant 300 sheep then I pd 1:8 to Ned Mcgan for a knife yt he pd Whistler then I drank a bottle of wine and aboute 2 I left town so to Courte expences 3

15 I gott some peas sow'd aboute 12 I came home and accted wth Martin fitzGerald he has 22lb of buttr above his complyment for wch I allowd him as by his acct appears and do owe him 2s:7½d then I gave him a receipt in full for last years acct

16 I gave Martin fitzGerald 3 cows and 3 heffers wch wth 34 cow he has makes up 40 then I accted wth Den Bryan and on ballanceing accts do owe him 4½ but did not charge him for ye grass of his heffers only till 1st of 9br then ye hackler finishd ye hemp and flax I have 2 stone 4lb of hemp and 2 stone and 6lb of flax I gave him in part 7s:1d I gott my gardin beans and peas in

17th I gott my beans and some of ye peas trashd and gott 23 dowsin candills made

18th I gott ye remaindr of ye Peas trashd and they and ye beans winowd and 21 pecks of beans ye Peas are to foul to be measurd I gott a letter from Madm wch I answerd no directions aboute ye Shirffs Court

19 I went to Church from thence to Killmoreen shewd M fitzGerald some of ye orchard for a gardin then to Courte no letters to bread 3

20 I went to advise wth Mr Jackson[4] aboute Burks affair his advise was to go to town and wait for ye next days Poste if I gott directions to by them if not to drop ye affair I went to town lay there and lost 18d att cards

21 I gott from Ned Mcgan 3:5:7 part of Studers money then I pd Ned Mcgan in full for money lent me and my expences to this day 15s:7d then I gave ye drawer and made betwen them 13d then I gave Mrs Mellsop 1:10:0 then I lent Mrs Johnston 5s:5d then to snuff 13d wine 9d dinr 12d no letr or direction aboute 3 to Courte

22 att Courte ye men diging for Onions

23 ye men diging Petatoes

24 ye men att ye Petatoes I gott 19lb of Iron from Mr yeamons wch I owe for no letter to news 1

25 Parkers[5] barn burnt ye men att ye petatoes

26 Jon brought me a shirt and busl oats for ye mare I gave him 13d to buy Plants and 1½ for tabaco I got a letr from Bastable yt Chapman could not give his daughtr half ye fortune I expected

27 I went to Limerick to meet Studdert but did no good I lay in town

---

4 Joseph Jackson, an attorney at the Tholsel Court. In Dublin, the Tholsel Court appear to have been occupied with minor offences. Ferrar; Garnham, p. 83. 5 Possibly George Parker, freeman of Limerick, 9 September 1748. Herbert, p. 119.

February 1746

28 I gott a letter from Wallis wth ye execution and one from Mastr then I gott 9:2:0 from mr Studdert and lent Ned Mcgan 1:2:9 pd Matt fox 6:4 I owd him to boots 11:4½ news 1 snuff 6 bread 6d needills 4 to expences 2:10 so to Courte

29 att Court gott my mare shod

30 I sent Thos. Harragan by Jon 7:19:6 in parte of ye money I borrowd of him ye 3d instt Jon gave me he gott from Jon Haly for greasing 13:6 I wrote to Madm and Mastr Jon Johnston breakfasted wth me

31 I wrote to Ned Mcgan for a 1000 of 4d nails a 3d loaf and ½ a hondrd of Oysters all wch I gott and must allow him for them I finishd diging ye Petatoes

febr ye 1st 1745/6 ye men in ye garding

2d I came hom ye cow I bought of Jon Quain calved

3d I gave 8d for 400 of plants and sett 390 of them Jon gave me he gott for greasing 8:8 I gave him to buy Iron to make a new Duch Plow Iron 11s I bought a heffer from Martin fitzGerald for 2:9:0 and allowd me 9 shillings he owed for her grass so yt I owe him but 2:0:0 I gott ye two cows I had att Court brought home, and one of my bacon hogs Salted

4th I bought a cow from Jon Quain for 3:6:1 Jon went to Limerick for Iron and gave 9s 11d for itt I went to Court sett ye Sleatr to work so home no news nor letter

febr ye 5th 1745/6 Jon went to ye forge to gett ye Plow iron made I gott some Oats winowd and some flax seed claind ye boys trashing

6 Jon att ye forge ye boys trashing and diging petatoes I went to Courte gott 3:11 from D Bryen and pd ye Sleater 6s so home and pd 6½ for gads and 3d for trashers then I walkd out to see my cattle to ye smith 1s

7 I began to Plow for Oats I had my windows glaisd and pd 6:5 for them I gave Doyle a bridle and stirrop lethers one of ye cows calved I gave Jon 1½ for tabaco

8th ye Plow and flax seed going ye men puting out dung for beans I sowd some Parsnip seed I gave Jon on acct of his wayges 13d

9 aboute 12 I went to Court gott a letr from Madm and another from Mastr I pd C Lynch 3d for picking petatoes att Courte so home and gott 8s from Matt Grady for greasing then walkd aboute my cows

10th I had ye Plow going and 2 horses from Court harrowing I sowd aboute 10 bus of oats and gott beans putt in and all ye oats winowd I pd Jon Quain by his wife 10s:10d on acct

11 ye frost was so hard yt yc Plow cod not work till 12 then they finishd ye gardin south of ye house but we could not harro itt I gott some beans trashd for seed

12th ye frost hindrd ye Plow and harrow I gott some beans set in ye orchard

13th I finishd ye beans in ye orchard then I wrote to Madm and Mastr to Thos Studdert for money to Mick Hartny aboute maulting ye barly att Courte to Mr Thornill aboute Killmoreen to Bastable and Ned fitzGerald to borrow a 100lb for Mastr Hartstonge to Ned McGan aboute ye news to David Bryen to send a way ye letters and to Thos. Bury for ye greasing of ye sheep to Pay supple no money from Mr Bury

14th it pleasd God to send a thaw and I had ye Plow tackled and sowd betr than a barrill of beans ye men diging ye hed ridges in ye Oat garding I gott an answer from Ned fitzGerald yt he cod not answer Mastr Hartstongs expectations

15 ye Plow att ye beans ye harrow finishd ye oat gardn I gott a letr from Mick Hartny aboute ye barly Bryen Boyle brought home 47½ bandills of blanketting I did not pay him for weaving

16 att home being a very wett morning I sent Jon to look att some cows one of ye wethers att Killmoreen died

17 I finishd sowing beans and sent ye Plow and horses to Courte went to see ye cows yt Jon saw yesterday and offerd 10 10 0 for ye four so away and in my way home calld att Tho Burys drank Shear of 3 bottles of wine so home and spent ye Remaind of ye day examining my maids aboute porke wolle flax and oats &c: ye night I spent museing on ye want of one yt would take care of my house

18th I sent Jon to Courte to Plow and sent 5 Limerick barrills of oats and went down in ye eavening

19 to Courte ye Plow going I gott 6:3:0 from Pattr Sheehan in full for ye gras of Mr Thos Burys Sheep then I pd Doyle 2:11:6 for gleasing ye Church

20 I sent 4 barrill and 3 bus of Barly to Mick Hartny to be maulted for Mrs Hartstonge then I went to Mick feigh and pd him Countery charges for Courte Killcollum Corroheen and ye Derepark and 8:9½ for my own holdings then I pd Hen Supple 3:10:0 on acct of his sallery spent 1:6 so to Court and Martin fitzGerald came to let me know yt one of my cows died I tould not mine but his and bad him do what he wod wth her ye Plow going

21 sent to Limerick no letters I pd ned Mcgan for ye nails bread and oysters I had ye 31 of last month 2 11 to bread 6 ye Plow going

22 I came home and found my Serts hat taken ye carcas of ye cow from M fitzGerald I made them cary Part back and take away ye rest one of ye cows calvd ye boys digd for Parsnips Jon came home but did not finish at Court

febr ye 23d 1745/6 I went to Court where I stayd to ye **15 of March 1745/6** in wch time I gott 5 barrills Oats sowd 5 barrills of petatoes 3 bushills peas some beans and gott ye nere wing of ye ware set down

15 I came home in ye eavening and found they had finishd my Oats but nothing elce done

16 very unwell and spent ye day mostly on ye bed

17 I pd Jon Quain on acct 2:5:6 and orderd 3 yrlings and 2 cows to ye fare I went to ye fare pd Tim Calehan in full for his cow 2:12:0 to Jon Quain 1:2:9 to broge for J Rohan 1:6½ to frize 3:9½ to ye Servts 2:8½ to my expences 1:1 then I orderd home ye Cattle so to Court

18 dined att Shanongrove back to Court and Mett Mr Bendons Men who came

April 1746

for 2 cows Mick Lowry put to greas att Court they pd me ye greasing and took ye cows wth them

19 att Court Jon Donevan mending ye little boat

20 went to Limerick took up cloath for a coat pd 6s for a hatt bought some seeds to my diner 3s so to Court

21 I gott from P Daniel he recd from Mr Jacks 7l:15s:2d wch wth 3:12:4 I gott att ye fare makes 11:7:6 ye greasing of ye 70 yrlings he hadd att Court then pd Pattr Daniel 3:8:8 wch wth 1:2:9 he gott before makes 4:11:5 in ye Eavening I sowd some seeds in hott bed[6]

22 gott some beans sett **23** reading

24th I gott ye new Stable claind out and was Servd wth a Prosses for one of ye cows yt greasd att Coroheen wch was drounded a Friday last

25 tho a Holy day[7] ye men were att ye wear but did no good ye foundation being so hard yt we must take up all yt we sett down

26 I wrote to Mr O Bryen aboute ye Prosses he was not come to town

27 I went to Limerick and putt ye Execution again Studders into Bryen Mcmahons hands then Pd ye taylor for trimings and Making my Coat 9:2½ then I agreed for a roome att 12d a night and gave 2:8½ earnest I gave N Mcgan 13d to me 2 11½ so to Courte

28 I had ye Colts put in

29 I came home and found my Pig killd

30th to Church aftr Service home

31 I gave Joan Daly 1:1 and she went away I ordrd Cate to take care of ye Cows and deary then came to Killdeemo and Chose Jon Hogan and James Sheehy Churchwardens[8] so to Court packed my cloaths to be to town wth Gods blessing to morrow[9]

Aprill ye 1st 1746 I got 6 from W Buttler for ye gras of a cow at Court for a quartr to town to a bushill Salte 22 sould my flax seed for 2:2:0 advisd wth Counsill Minsion aboute serving Mr Grady wth an ordr of ye Court then to Court where ye County was calld and adjurnd so abute 8 to my lodgings I gott a letter from Madm and Mastr

2d about 8 to Court was impanilled on Donegans Jury and brought him and another in Guilty aboute 2 ye Jury were discharg'd I dined at Mr Mcdanills then I servd Mr Grady wth a copy of ye ordr and pd 3:9½ for making an affidavid of ye Service so to Court where I stayd till 6 and then went to my sistrs aboute 8 to my lodging

3d I breakfasted and din'd at Rob Peacocks and spent ye day att ye Court till 5 then to my lodgings and wrote to Mr Wallis and Madm Hartstonge I gave D Bryen 23d lent him 22

---

[6] A walled seed bed with fresh manure under the soil, which heated and made the soil ready for early sowing.   [7] Annunciation of the Blessed Virgin Mary.   [8] Lay members of the church elected to assist the incumbent with parish business.   [9] He was to serve on the jury at the assizes.

4 breakfasted and dined att my sisters spent ye Day at court aboute 7 I went to ye Studderts but did not good supt att my lodgings aboute 9 to bed

5 I borrowd from Frank yeamons ~~2:5:6~~ and got 2:11:0 from Ned Mcgan of wch he owd me 1:2:9 so yt I owe him 1:8:3 then to court dined att Rob Peacocks borrwd 5:5 from him then pd Mr Bryen 9br Rent then spoke to Bryen McMahon about Studders affair then pd for my lodging left town about 6 got to Court aboute 7 to expences 1:4:9

6 mostly reading 7th wth ye men got some cowcomber plants from Mr Burys Gardiner

8 Billy fosbery came to agree for ye grass of 20 heffers at Killmoreen he is to be att ye rate of 16s ye collop for a yr to making ye net 4:6

Aprill ye 9th 1746 unwell wth a Cold ye hearth money man lay here but I did not pay him one of Will fosberys sheep died

10 11 and 12 unwell tho I bless God better than I was bread 3

13 to Church and back 14 wth ye men in ye Garding

15 wth ye men no letter to brandy 10d sugar 9 lamb $7^{1/2}$ I got a pinte of Kidney beans sett I got a mare from Doctr Martin

16 and 17 nothing materiall I gave my brotr some petatoes

18 went to Killmoreen gelt 9 lambs 2 of wch are Richds and one lame one and 16 ewe lambs in all 26 so to Court and paid Marget Casy 2s in full for a months hire Ned Mcgan and my Sistr came to see me and stay'd till 6 ye mare I got from ye Doctr took James Hewsons hors for wch I pd 5:5 and 6 to ye groom I gott a letter from Jon Melsop to bread 6 sugar candy 2d

19 and 20 nothing materiall

21 went to Kildeemo agreed wth James fitzGerald to build ye walls of my house att 18d a perch he to find every thing belongin to ye measons work spent 3s so to Court H Evans cut a side of lether

22d I gave Hen Evans 4s:$10^{1/2}$ to buy buckles 9d and sent ye things by him and ye boy I gott a letter from Mr Hartstonge and one from Wallis to serve George Rose[10] Junr wth

23 I gave David Bryen 13 and sent him to serve ye supena

24th wth ye men

25 I gave D Bryen 5:5 and sent him to make affidavid of ye Serving of ye supena and wrote to Mr Hartstonge Mr Wallis and Mr Mellsop I got a letter from Madm wth a bill on David Bork for 90:6:0 and a bill on Arther Dohig for 40:2:3 to turnop seed 2 parsley seed 2 to bread 6

26 I sent a barrill of Mault to ye mill my gray mare refused ye horse to snuff 13d lamb 10

---

10 George Rose was the second son of George Rose who was high sheriff of Limerick. He married twice, first Jane, daughter of Thomas Hickman of Barntic, Co. Clare, and second Susanna daughter and coheir of Richard Stephens of Newcastle, Co. Limerick, and widow of Edmund Burgh of Newcastle. They had one son, Richard who married Mary, daughter of John Anderson of Foxhall, Co. Tipperary. Pedigree of Rose of Rathkeale, NLI Mic 8305.

May 1746

27 to Church and back to Court
28 went Parte of ye way to Borks spent 17d so back
29 to Borks and got 87:1:6 wch he says is his years Rent I gave him a receipt for so much on acct of ye bill then to Limerick gave ye money to young Doctr Martin then pd Frank yeamons ye two ginies I borrow'd y 5th inst and pd him 4s:8d for 2 pd of hops then came to my Sisters and gave Madm an acct of what I did then I borro'd of my Sistr to pay for a bitt and snaffel 6:6 so to Court expences 4:8
30 wth ye men in ye Garding

May ye 1st 1746 I surveid ytt Part of ye bog ye tenants hold and find it 9a:2r:16p then yt part yt P Daniel holds and find it 7a 04 2p then I sett David Bryen a peice wch contains 6a:1r:16p for 6:6 an acre so to Court accted wth Pattr Daniel and spent ye day variously
2d I went to Limerick from thence to Dohigs but did not good so back to town pd my sistr ye 6:6 I borrowd ye 29th of last month and pd Robt Peacock ye 5:5 I borrowd of him ye 5th of last month then Ned Mcgan shewed me $3^{1/2}$ barrills of Oats he had from me and wch we are to acct for then I dined and so to Court
3 Jon Bryen came to me and I gave him 5:5 to give Cate of wch I owed her $2:8^{1/2}$ and ye other on acct of her wayges then I went to ye fare but did no good but drink and gave Jon 1:2:9 to give Thos Harrogan expences 9:9
4 at Court settling wth greasiers
5 I accted wth Thos Harrogan who owes me on ballanceing of all accts 3l:2s:1d then I ordrd Jon to give Jon Quain it being ye full ballance I owe him 1:15:10
6 I sent for letters and got one from Madm aboute Bork and Dohigs bills and directions to send ye horses I got ye drink brewd to half a stone of sope 2:4
7th I sent Tim Mahony on acct of ye Rent of ye dere park nine pounds 18s:5d but he went to Dublin and it came back I pd 13d for gelding my colt and 3 for gelding three pigs I wrote to Dohig and ye Doctr but got no answer
8th nothing Materiall
9 went to Limerick to try for bills but got none aboute 3 I came home but wrote to Madm and Mastr
10 went to Killmoreen took an acct of ye Cattle gelt two yrlings had a plan of ye house and orchard from Pursill so to Port gelt 3 calves vewd ye gardins sent ye sheep to be washd and sent Jon to Henry Page wth $13:5:2^{1/2}$ so to Court and roved aboute to bread 6
10 spent ye day roving aboute
11 sent away ye horses gave tim 1:12:6 to Church and back again got a letter from Madm
12 Sould some Medows so to ye vestory gave up my accts dined att Killdeemo back to Courte roved aboute
13 went to Limerick gott a letter from Madm wrote an answer gave M fitzGerald 7:7 in full for his heffer gave Jon $2:8^{1/2}$ to buy Iron &c so to Court expences 4:6

May ye 14th 1746 I went to Killmoreen bought 2 heffer from Simon Daniel for 4:10:0 and time till ye 15 of august then orderd my sheep home to be shorn to morrow then I gott ye foundation of my house marked out and gave ye men 6 so to Court orderd ye 2 heffers home

ye 15 I went home and got my sheep shorn for wch I pd 1:8 and pd 10 for daging and 7½ in Part for dressing them I gave James Nash a fleece of woll and 10 for last years tyth and a fleice of woll 1:2 for this years tyth then I markd ye Sheep and have 52 of all kinds and 23 lambs aboute 7 I came to Court and rovd aboute a Cow drownded

16 sent for letters and got one spent ye day wth ye men

17 acompanied by Syp Pursill I went to Limerick to buy timber but did no good it was so dear aboute 12 I left town so to Court Jon sould 2 barrills barly for to lamb 7½ expences 3:0

18 reading and walking I got a letter from Madm

19 I pd James fitzGerald 5:5 in part for my house and 1:1 for mending ye Church yard wall I sent my Mare and Colt to Killmoreen

20 I went to Limerick from thence to Tooreen[10a] to meet Dohig but cold not I waited for him in town till 4 then wrote to Madm then to Court and found ye Gray mare foled expences 3:3

21 a carrier came wth some of Mrs Hartstongs goods viz: a bed Boulster bedstid teaster hedpeice vallans &c: a kittle and cover 2 stewpans a cheafing dish and bottom a lamp a colindr a fish kittle and pleat a saucepan copper ladle basteing ladle skimmer a beakfast table a tea table a card table a large table a sea chest lockd I had ye things putt in and spent ye day mostly wth ye men att ye dich

22 I went home and having veiw'd my own affairs I gott in my lining yt was wash'd and got 17s from John he gott for a barrill of barly and 2s parte of ye pryce of two bushills ye other 3s he gave Tim Callahan in full for his cow then I came to Court and spent ye day wth ye men

23 I went to ye dich then not finding ye men att their work I ordrd their cows to pound sent Mick Cosgary to Limrk for Rock lime and other things to Iron 1:3 bread 3 no letters

24 wth ye men nothing material 25 reding & roving

26 wth ye men in ye garding and dug up all ye safforn

27 went to Limerick to mett Dohig but did not I bought a peice of balk for 7s and 9d for sawing itt I gott a letter from Madm wth directions to distrain Dohig aboute 5 I wrote to her and Mastr then back to Court expences 3:3

28 I spent wth ye men and in ye Gardin ye Dublin mare took ye horse I sent T Bork and Cosgary to ye fare

29 P Daniel D Bryen and I went to Ballynegolla for money or to distrain and waited for A Dohig ye best parte of ye day he came about 5 and pd me on acct of ye Rent 48l:5s:5d and Promised to pay me 33l to morrow in town then we came directly to Courte expences 1:8

10a Tooreen, Co. Limerick.

June 1746

30th I went to town and spent ye day att expences but got no money but Batt and Jack Dohig swore they wod pay me att Court by 10 a clock on Monday or give distress att Court I gave ye Doctr ye money I got yesterday so to and got to Court by night fall expences 1:7½
31 I loyterd aboute ye garding to lamb 7½

June ye 1 I spent mostly reading no letters
2 I had all ye men claining for turf bread 6d
3 Mrs Hartstongs men cutting no letters bread 6 Iron 1:1½
4 I had 3 plows att work a wett day
5 ye 3 plows att ye turf I had a letter from Mr Studdart to go to Eniscouse but refused and wrote to him to meet me in town to morrow att 9 I made up Conr Quelly acct and find he has got in all to this day 6:2:3 since Mrs Widenhams death
6 I went to Limerick to meet Studders and by much ado I gott 3l from him I got a letter from Madm aboute 7 I left town and got to Court by 9
7th John Bryen came to let me know ye barn was broak open and some buttr meall and salte taken out I pd tim Mahony 7:13:0 on acct of ye Rent of ye dere Park
8 to Church and back to Court dined on lamb then went to Killmoreen lookd about then went to see my stone horse and spent 4d pd ye church rates 7 Paris taxes 4 so to Court no letters
9th wth ye men att ye bog and garding
10 went to Killmoreen began ye walls of my house gave ye men 6 for tabaco so to Porte for ye administrasion gott from John he recd for 2 bus of barly 5s from Conr Maddin for greasing 2:8½ and he allow'd him for 12 ceelors 11 then I pd Patt Hickey 4:10½ in full for his wayges and 3 day hire and agreed wth him for 40s and a pair of broges a year so to Court and spent ye day wth ye men att ye bog
11 to Limerick mett Mrs Judy and she came to Court wth me I sould 4 barrills Oats for 2l:4s and gott 10:10 from Jon wch I laid out as by acct expences 5d
12 wth ye men att ye bog and aboute ye house
13 I went to Limerick but did no good Expences 2:8
14 I had ye sheep washd ye Carriers came wth some of Mrs Hartstongs goods
15th I pd ye carrier 1:2:9 in part
16 Mrs Hartstonge and hers came home
17 I spent ye day Mostly wth them Jon gave me ye remaindr of ye Pryce but 2s he gave for Salte
18th I went to Killgobin and from thence to Adear and spent 4s and gave my Stone colt and 4l for a gelding and gott Lary for ye little old Mare and sent 1s to ye groom so to Court overtook our own and Mr Burys family att diner and spent ye day wth them and was invited to diner
19 dined at Shanongrove to ye groome 1:1
20 to Courte seeing one of ye cows I sent here killd

21 to I deliverd my Stone Colt to Ned fitzGerald but did not get ye geld spent 1:7½
22 to Church and back again to Court
23d went to Limerick aboute Mrs Hartstonges business but did little good expences 2:8½
24 I gott home ye gelding and gave my Note for 4l provided he getts ye better of his lameness gave ye man 13d and spent ye day wth Mr Bury
25 I went to Killmoreen walk'd aboute back to Courte
26 I sent my black mare to Mr Burys horse and gave Jon 4:4 to buy timber att ye wood and spent ye day wth Mr Quins and yr Company was invited to Dinr by them
27 I acompanied Mrs Hartstonge and hers to adear where we dined about 6 back to Courte to ye groom 13d
28 to Courte was invited to Dinr to Parkers
29 ye family and I went to Clotakey where we dined
30 Jon gave me 2:19:6½ he gott for 2 1 26 pound of my own butter and 1:2:0 he got for 2 barrills oats

July ye 1st 1746 I went to Limerick and bought 8 beams and 2 deal boards for 6:7:3 and pd for them and did some business for Mrs Hartstonge so to Court
2 and 3 att Court I gave James fitzGerald 5 5 on acct
4th I came home and Jon gave me 2:7s:0d he go for 4 barrill and 2 bushills of Oats I gave hm 4s:2d to pay ye men he hired and spent ye day variously
5 I went to Killmoreen walkd about and got 10:12:1½ from Martin fitzGerald he got for 8c:3qr:10lb of butter I allowd him for his brotr 3:9½ Spent 6 gave ye measons to drink 2:8½ so home and pd Tho Harragan in full for Countery charges 8:10½ I gave Joan on acct of her wayges 1:1
6th I was sent for to Court and dined there wth ye family of Shanongrove gave Honer 2:2
7 att Court gave Syprian Pursill 1:2:9
8 I gott a letter from thornill for money
9 I came home and made up 9l:10s:0 and carried it to Mr thornills man and took up ye note for 3:18:10 and indorsed 5:11:2 and ye back of ye other note then I went to Court and spent ye day wth ye men
10 11 12 and 13th att Court ye timber came there
14 15 16 17 at ye Sizes I gott 2:6 and 11s:4½ from David bryen wch I spent at ye Sizes I lent Conr 5:5
18 19 and 20 att Court
July ye 21st 1746 went to Limerick [illegible] gott Bryen Mcmahon to wright for an execusion for Harry Studderts effects so back to Court expences 1:3
22 I went to Killmoreen on a diffre between me and Will fosbery wch Jon Hartny and Pattr Sheehan made up then we went to Carrolls and spent 6d each so home

July 1746

23d att home and sent Jon to ye wood for 2 dowsin tevan wch I owe for he brought them to Killmoreen

24 in ye aftr noone I went to Court to prapare for ye selebrateing Mr Hartstonges byrth I wrote to M Cusick

25 I spent wth ye Company very cheerefull

26 aftr breakfast I came home and spent ye day variously

27 att home reading

28 ye men cutting wattles to make schaffolds I had Char Slatter James farrill and Jon Hogan mowing to tabaco for them 1 1/2 a weet day

29 I was calld for to Courte where I stayd to ye **16 of 7br** in wch time I gave Syp: Pursill 2 8 1/2 and allowd him in Limerick 4:4 1/2 then I came home and gott from Mart fitzGerald he gott for 7c:3qr:11lb of butter and pd him ye money I borrowd of him ye butter came to 9l:4s:5d then I gott from Jon he gott for 4c:0:10lb my own butter 4:16:8 and 4l:13: he got for 11 stone 14lb of wolle so to bed

17 I came to Court and in ye following manner made up 40:19:7 and sent it by Jon to Robt Hewson for Mr thornills use then Robt hewson came wth Jon to me and Promisd he wod bring my note from Mr Thornill then I gave Jon 10:10 and bid him try to by my tythes

July ye 9th 1746	to cash pd Rourk	5 11 2
augst ye 1 1746	to ye grass of 2 cows	0 14 0
20	to cash pd Tim Herlihy	6 17 0
7br ye 17th	to a bill on Mr Bastable	11 14 9
	due to me from Mr Hewson	1 15 3
7br ye 17th	To cash pd Mr Hewson	14 7 5
		40 19 7

18th Jon came to let me know yt I was distraind by Tim Mahony and yt he agred to give 5l for this years tyths and 3l for last years then I sent my note for ye money so rovd aboute

19 I wrote to R P and F Y to borro 5l but did not gett itt to snuff 1:1 ink 6 weafors 2d

20 nothing materiall 21 I came home aftr dinr

22 drawing home Oats

23 drawing hay I gott 6:4:0 from Geo fosbery on act of Medow and from Will fosbery on acct of greasing then I made up 10:19:9 and sent it by John to T Mahony and gott his Receipt

24th finishd ye hay went to Court being sent for

25 Court family went to ye county Cork I pd M feigh in full for his horse 4l and H Supple on acct of his salery 3l so loyterd aboute

26 I sent 13s to Jon to Pay Jon Gerin &c to bread 6

27 Gerin gave me ye Map of ye North east Park of Killmoreen wch contains 33a 14r 20p taylors Medow 5:3:20 Commin 5a Medow 2:0:10 he tould me he gott 5:5 I lent Martin fitzGerald 13s:3d to pay ye Meason

28 I gave simon Daniel on acct of his cows 0:17:6
29 Jon tould me he gave 7s for bandle cloath and to charge his son david wth parte he gave me 7d change
30 I began drawing turf

8br 1st 2 3d att ye turf
4th I had an acct ye my sistr was very ill
5 I went to see my sistr who is better back to Court by 2
6: 7: 8 att ye turff and finishd
9 Jon came to me and gave me 1:0:2 he gott for lambs wolle he pd ye Proctr 6s 6d to his son david 2:2 costom 4 to me 11:2 wch wth 1:9 I gave him makes 12:11 and bid him pay B Hall 3s to E fox 2:3 to A f Gerald 2s to J Lowry for 6 days 1:6 60 H Coagh 6d to Jams Dilane 1:4 to Burn 5 to J f Gerald 4d to M Welch 1s
10 to Limerick did no good to mending my watch 2:8½
11 loyterd aboute to tea 1:8 expences yesterday 4d
12 walkd to church and back to Court I gott a bill of James fitzGeralds in favour of Tim Herlihy for 19s wch I took and Herlihy gave me 12d in full for his medow
13 14 nothing material 15 ye family came home
16 I gave ye horse ridr 10:10 in full for pading my mare
17 18 nothing Materiall I gott a cow killd
19 to Killmoreen and found ye turf out of ye bog so home
8br ye 19th I spent with Martin fitzGerald
20 21 22d a drawing turf [illegible] Carolls horse 3 day J Geary horse 2 days and [illegible] 3 Den Bryen 3d making it up Martin fitzGerald gave me 6:3:9½ he gott for 4:3:2 pounds of butter I had a letter of invitation to go to a Christning of my brothers
23 I went to my Brothers Christning dined drank share of a bowle of Punch so home to ye boy 6d dram 3d
24 ye men heding ye hay I roving aboute
25th I had 91 gallons of Sidr from Thos Harrogan of wch he gives me 24 gallons and ye remaindr I am to allow him for
26 I went to Church from thence to Courte 2d
27th to Courte 28 I came home and got my Stable claind to Sope 12d to Schollops 2:5
29 I sent Jon to Limerick wth a casge of butter wch weigh'd 2:0:13 for wch he gott 2:15:0 expences 6½ I spent ye day Settling my accts
30 I gave to be washd 11 Shirts 10 Stocks 2 weast Coats two ½ half pr of Sheets &c and have in all 26 shirts and 27 stocks 3½ pr of pillowcases five white handkercheifs 12 colourd ones 13 nightcaps 23 new towells 3 cotten weastcoats 7 pair of cotten 3 of thread 2 of silk 6 of wosted 9 pr of yarn stockings 27 pr of all kinds 1 pr of fine pr of large course 2½ pr of new bandle cloath 5 pr of hemp one old sheet in all 10 pr of sheets 8 table Cloaths 5 napkins two corse pillow cases I borrowd 4 ginies of Mar fitzGerald

November 1746

31 I went to Asketon and pd ye May gale and ballanced accts to 5s:6d wch I owe from thence I went to Court expences 1:5

9br ye 1st 1746 to Courte accting wth ye men 2 3 4 5 6 att Court pd 2:2 for two halters 3:3 to Caroll for drawing my turf 2:8½ to Patt Daniel to buy Iron to Shoeing Colts snuff 1:1 lent Andrew Quin 2:8½ Recd from Mick Cosgary 5:5 for greasing att Killmoreen

7th I went wth Money of Mrs Hartstongs to Ned Morony and got back to Courte by 2

8th aftr breakfast I went hom and found ye horse ridr rideing my colts and ye men trashing I had a sheep brought home to be killd 9 at home reading

10 I pd 10d for Salte and loyterd aboute

11 my filly was shodd

12 I sent Jon wth a cask of butter to town wch he sould att ye rate of 24s a hondrd it weighd 1d:0qr:26lb and came to 1l:9s:5½d to frize 3s Iron and steel 2s:0 Salte 1 10 bread 3d turnpike 1½ costom 2 expences 4

13 I sent ye Colt to be Shod and pd Simon Daniel on acct of his cows 1:8:11 I gott 3l:0s:4d from Martin fitzGerald he got for 2:2:2 of butter

14th I gott a letter from Thornill for money but had none I bought a pig from R Harnedy for 6:0 ~~but did not pay~~ I gave Jon 1:7½ to for 3 horses I had drawing turff a cow calved

15 I loyterd aboute 16th I satt reading

17 went to Court Stayd there yt night

18 came home and agreed wth Martin fitzGerald as deary ye insuing year for a hondrd of butter and seven shillings for every cow he is to have 40 cows and his calves and to rare but 20 calves his mare on ye Pasture ye rest of his Cattle in ye west Park

19 I had 200 trees from Syp: Pursill and gott one of my Pigs killd to a shoe for ye black mare 5

20 I spent variously

21 help killd a hare I making a cap

22 a violent storm some of ye houses stript

23d to church dined to Courte so home

24th sent Jon to pay ye Quittrent he pd 14s:5d to thread 2 reding 11 turnpike 1½ one of ye Calves died I gott 4½ barrills of Oats winowd and some barley

25 I gave Thos Collins wife 2 bushill Oats [illegible] full for 16 weeks hire I had ye barly dryd

26 I sent 5 bushlls barly to mill I walk'd aboute Killmoreen so home got 1:8 for [illegible] Calves Skin

9br ye 27th 1746 [illegible] to Wallis aboute ye Studders is [illegible] and to send a justisia[11]

[11] A writ which empowers a sheriff to do same justice in his county court as if it was heard at a superior court.

28 att Killmoreen planting trees I pd 1:4 for Hackling 16 pound of thread
29 ye men planting I gave John 3:9 1/2 to pay Caroll on acct of cutting turff spent ye day unwell
30 I went to Court

xbr ye 1st 1746 att Court
2d att Dito lend Mrs Hartstonge 1:1 Lent David Bryen 1:1 to snuff 1:1
3 I went to Bruff aboute Burk and found him burying did no good spent 2:8 1/2 of Mrs Hartstonges and 2:3 of my own I got home by 7 at night
4th to Courte 5 aftr diner came hoome
6 one of ye Calves died I got my pig killd I gott a letter from Wallis aboute Studders affair
7 att home reading
8 I sent 5 barrills Oatts to ye kill and mill
9 I gott for 7 sheep 2:2:0 for a cow yt dyed at Killmoreen 1s 1 from thos. Harragon 8:2:0 1/2 on acct of Rent for a cow hide and Calve skin 6s 11d I gave Jon Bryen on acct of his wayges 1:2:9 to him he borrowd ye 11th of July last 2:8 to him to Pay for ye bandle cloath for ye little boy 1s:3d to him for a shew for ye for ye black mare 6 I pd Caroll on acct of ye turff 8s I had my pig salted and ye two mares shod
10 ye men drawing Stones to build a hen house
11 I sent Jon to ye fare wth a cow but did not sell he pd Mr Quin and Buckners Procter 3:0:0 in full for last years tythes and took up my notes he pd 3d for a pr of garters for me
12 all ye men att ye mill ye little boy gott a pr of broges and stockings
13 I sent Jon for two buchills for wch he pd 3:8 I got ye male from ye mill I finishd 2 nightcaps
14 I agreed wth peter Hogan to tend two horses to ye 25th of March next for 8:4 and 1:1:8 for rideing ye filly and Colt so to court where I stayd till 11 att night so home
15 pd Tim Leary for Petr Hogan 13:8 pd Richd Harnedy 6:6 in full for his pig I bought ye 14 of last month ye men drawing stones
16 I gott my bulluck killd little tallo
17 I sent ye hide to be sould and gott 8:4 1/2 it weighd 4 stone 8lb ye men trashing
18 I went to Killmoreen to see ye trees planted and removed 47 I had planted before and bed M fitzGerald to allow Syp Pursill 19:6 on my acct so home
19th I gott my shirts dryed and ye male sifted I have 13 bushills clain sifted
20th I went to Courte and pd 4:3 1/2 in full for Church rates for Killmoreen
21 I went from Court wth ye Ladies to Killtenan[12] and left it at 5 back to Court Supt and got home aboute 9 so to bed

---

12 Killtenan is 3 miles south west of Adare in the parish of Croagh. The Pery family owned land at Kiltenan, leased from John Dowdal. Limerick papers, NLI PC 875/2/15/56.

December 1746 167

22 I pd Tim Herlihy for Mr Thornells use 4 gineis one att 1:18:10 1 att 9s 2 att 1 1 and 6d brass in all 7:1:6 then I gave Jon 1:3 and sent him wth a ferkin of butter to town one of ye Cows died I gott 6:8 for ye hide I gave Pattr Hickey 19$^{1}/_{2}$ on acct of his wayges I got 1:5:10$^{1}/_{2}$ for 1:0 14 of my own butter to brandy 3 Rum 1:10 whiskey 1:8 sugar 1:8 Costom and turnpike 3$^{1}/_{2}$

23 I had ye tallo rendrd and gott 20 dowsin of Candels made ye weaver brought home 35 yards of ticking for ye weaving of wch I pd 17:2 gave him some sidr and away wth him

24th I lent gillbreth 2 bushills meal sent a torkey and 2 ducks to my sistr

25 to Church and back again Spent ye day at Court

26 att home wth Gillbrath

27 gave Joan Daly 13 she went to town I to Court where I stay'd to ye 30 so home gott 2:2 for a calfes yt died to snuff 8d

31 att home drinking

1747

Janr ye 1st 1746/7 I accted wth Martin fitzGerald and allowd him for C Pursill 19s6d for John Bryen 10:10 allowd him for his brotr 12:9 allow him for a sow 16s so satt drinking till night I gave torlo farill 9d for 3 basketts

2d I gave John 3:7:9 and sent him to town he gave Cate on acct of her wayges 1:2:9 he pd for a gallon brandy 6s 2 quarts rum 2:10 sugar 2:6 an almanack 2d snuff 1:1 soles 6½

3 I had Jon Gery his wife and H Supple wth me I gave H Evans 2:8½

4th I gave D Bryen on acct of his wayges 3:9½ so to Church then to Courte

5 gave Jon Dilany 10 and owe him 1½ ye family of Courte and company went to town

6 dined at Jon Geerys back to Court

7th att Court lent David Bryen 2:8½ **8th** att 9 to Mrs Hartstonge for bread 1s snuff 1:1

10 ye family came home

11 to Courte and rideng aboute

12 to Killmoreen aboute will ffosberys Cattle and having ended ye dispute came home and found 2 measons att ye henhouse ye sow pigd

13 went to ye quartr Sessions and was Chose way Warden[1] wth Mr Hartstonge for ye Parish of Killdeemo ye Courte adjurn'd to next day so home to a Schrew 13d expences 4s

14 to Limerick was on ye grand Jury aboute two ye Courte broke up I came home expences 3d

15 att home Craw Sick

16 to Courte where I met Robt Cox and wife

17 we rode abroad Mr Cox and his went away

18 Mr Dane and his went away I got ye 2:8½ I sent David Bryen ye 7th inst

19 at Courte was sent word yt T Mahony came for ye 9br Rent

20 came home and borrow'd from Tho Harragan six pound and from Martin fitzGerald 5:2:6 and gave them notes for ye money

21 I sent Jon wth ye money to Tim Mahony and got his Receipt for 10:19:9

---

[1] The parish had a duty to ensure that the highway was kept in good order, and a way warden ensured that 'annoyances' were removed. They could summon local inhabitants in the parish to work on repairing the highway; they could spend their own money and be reimbursed by the parish for work on guide posts and drainage, and with the consent of quarter sessions they could raise a rate. Blackstone, vol. 1. pp 358–9.

22 I cutt up my ticking and sent as much as will make a bed to Ms Hartstonge to sope 1:1

23 I sowd a bed tick

24 I sent John wth a barrill of Oats to town wch he sould for 6s he pd for 54 pound Iron 8:8 for two dale bords 2s 5d costom &c 4½

25 ye wettness of ye day hinderd me from going to Church to Will Hickey for schallops 12d

26 I gave to be washed 11 shirts 11 stocks 2½ pr of sheets 1 white Handkircheifs 1 pr Stockings 3 caps two tablecloaths and 3 pocket handkercheifs and about 2 I went to Court wrote to Wallis for a fiery facies[2] against H Studdert

27 and 28 to Courte 29 I came home and got some of my barly windowd and got in my linin

30 wett morning ye aftr noone more barly winowd Syp Pursill came for mony but I had none

31 a wett windy day a little of ye tyde overflowd

febr ye 1st 1746/7 ye tyde overflowd ye banks in severall places ye Corcas all coverd I gave patt Hickey on acct of his wayges 2:8½ I lent Pattr Daniel a straper to give him milk I made up 40 cows for Martin fitzGerald

2 I gave Martin fitzGerald 2:8½ to pay for my Shear of y tythes

3 Mr Thornills man came for Rent and I settled acct wth him and gott a dowsin of sidr from Thos Harragon wch we drank

1746–7 febr ye 4th I went to Courte

5th to Courte 6 to Courte and gott a supena and execution[3] against Henry Studdert 7th at Courte

8 to church and back to Court 9 to Courte

10th to Limerick advisd wth Bry McMahon so back to Court expences 3:9 11th 12th att Court

13 I came home and sent Jon Mellsop ½ a barl oats and wrote to Thos Bury for a warrant against Dav Quory for Stealing one of my lambs

14 I servd Henry Studdert wth a supena for costs for reviveing ye Judgments 4:9:4 but he did not pay me so I came away and spent 3:9½

15 att home reding Slattery brought a lambe ye fox killd I a quartr of it for supper

16 I cut out 2 pr of lining drawrs I got 13 sheep and 7 lambs home from Killmoreen

17 I went to Limerick and borrowd 3:9½ from my sistr 2:8½ of wch I gave a mastr in Chancery[4] and 1:7½ to Peter Hogan then I left town and came to Courte where I stayd to ye first of March to snuff 1:1

2 A writ of execution of a judgment, ordering the sheriff to raise the sum due from the defendant's goods and chattels. Abbreviated to Fi. Fa.   3 An order to carry out the terms of a judgment.   4 A subordinate judge trying civil affairs.

March 1747

March ye 1st 1746–7 I came home
2d in ye eavening I went to Courte and borrod from Ed Sheehy 3:8:3
3d I gave Sypryan Pursill 3:7:2 on acct
4th I came home and found ye Plow going I sowed beans in ye Rush Corcas and finishd then I gave Peter Hogan a bill of Thomas Harrogan for eighteen shillings 8 1/2
5 I began to Plow for oats
6 Mastr and ye two Ms Hartstongs came to see me I went wth them to Courte where I dined so home about 6 Judy gott a pig
7th ye Plow finishd ye little Corcas att Portacaca
8th I spent reading M fitzGerald got 2 young pigs wch I am to Charge him wth att 2 8 1/2 each
9 a sowing Oats Jon looking for me I went to Thos Harrogans where I stopt aboute 9 home
10 Sowing oats I had 6 men and my own a diging for beans &c Thos Harrogan had a young pig and Den Bryen another I lent my bay mare to Ed Mcgan
11 ye Plow going I had some beans sett
12 ye Plow going **13th** ye Plow and harrow att work I had a letter from Wallis wth his bill of Cost wch amounts to 8:19:0 and an execution against H Studdert for ye Cost
14 ye Plow and harrow going
15 I went to Court from thence to Killceedy[5] Church sent Mr Hartstonge 13d aftr sermond we went to see Carigoginall[6] so to Courte where I dined about 5 I came home
16 ye Plow and harrow going and finishd ye oats then I sett ye men to dig Petatoes I gave John Quain half a barrl oats and Charls Slattery a busl of beans for 2:6 worth of work
17 I gave my family lave to go to see Patrick and bad them drink 3:3 on my acct att Thos Harragans Thos Coagh took a pig for 8 days work I spent ye day at home
18 diging Peatatoes and Thos. Coagh
March ye 19th 1746/7 I had Slattery Coagh and D Bryen and my own men diging Petatoes I sent Jonn to Pallas Markett wth beans he sould a peck for 12d and gave 6d for 300 of Plants aboute 6 I was sent to for ye Quittrent Receipts I went to Courte where I stayd
20 I left Courte at 5 and went home orderd barly be winowd to send to Limerick Slattery Coagh and my own men at ye Petatoes
21 I sent Jon wth a barrill of Barley and 3 bus of beans to Limerick aboute 7 he came home but did not seell expences 6 he brought me word ye sherriff was in town I lent my Irish Sull and Irons to Court

---

5 Kilceed/Kilkeedy is on the river Maigue 4 1/2 miles from Limerick. 6 Carigoginall Castle, 5 miles west of Limerick, was held by the rebels in 1642. They surrendered in 1648 and retook it soon afterwards. They finally evacuated the castle on Cromwell's arrival. It was held by supporters of James II and surrendered to General S'Gravenmore, who destroyed it. Some of the remains were repaired in the nineteenth century. Lewis.

22 I borrowd 3:3 of Thos Harrogan and went to Limerick gave ye two Executions agains Hary Studdert in ye Sherriffs hands dined in town att night fall gott home to snuff 13d expences 6½

23d I sent Jon to gett inteligence aboute H Studdert and he brought but a disatisfatory one to him 6½

24th I wrote to Ned Mcgan and sent him ye black mare to goe to ye sherriff I had a letter from Mr Thornill for Rent ye Lord enable me to pay him washd 11 shirts 12 Stock 1½ pr sheets &c:

March ye 25th 1747 I sent Jon to town to try to sell ye barley or beans ye 3 bus of beans he sould for 4:6 to Costom and turnpike 2½d I gott a lettr from Studdert to meet him a Monday and yt he wod settle my affair wch God grant

26 ye men and Den Bryen diging Petatoes

27 ye men and Tho Coagh and welch diging of and for Petatoes I wrote to Thornell and N Mcgan

28 I sent Jon wth ye bushills for beans to town wch he sould for 10:6 to broges for Nel 1:7½ needills 3d to steel 3d expences 6

29 I pd ye taylor for making Jack cloaths 1:1 to Tho: Jackson 1:1 to Sidr 8d then I went to Court where I dined and att 5 came hom wth Ned Mcgan

30th I went wth Ned to meet Thos. Studdert but did not till we were coming home he came wth me to expences 7s:5d to Patrick Hickey 6½

31 went wth Studdert to Arleman[7] but did no good so home expences 2:3

Aprill ye 1st I wrote to ye subsherriff and sent Jon to try to sell my barley I agreed wth thomas Harragan for ye west parte of Killmoreen for 12 years att 8s an acre and am to give him 2 acres of Medow in ye bargin earnest 2 8½

2d I sent T Harrogan and Jon to Studders to agree for sheep but they could not expences 5

3 I went wth Harragan to Killmoreen walkd about shewd him ye bounds so home

4 I gott a letter from Jon Rourk aboute Ms Chapman wch I answerd and wrote to Studers and Mr Thornell I sent Jon wth 2 barrills of Oats to Limerick wch he sould for 11:0½ to Costom 8 and Mr Hartstonge sent for me and I went to Courte where we had a disfactory night

5 to Church wth ye familly and back wth them

6 to Limerick wth ye Mises and gott 13 pistoles from Studders wch came 11:12:11 then I agred wth for 30 two year old wethers and 60 yr wethers for 22 then gave him a receipt for thirty three pound 12s:11 then I pd welch[8] for ye last year news 6:6 then I gave him 8:1½ for books for Mrs Hartstonge pd 7:7 for a hogshed for her then I pd Jon Bonfild in full 1:12:7 then I pd M fox ye half crown I borowd of him ye 17 of March last

Aprill ye 6th 1747 Continued then I pd 10d for peaper lent Ms Hartstonge

---

7 Ardlaman, Co. Limerick is 2 miles south of Askeaton.   8 Thomas Welch, printer, Limerick. Ferrar.

April 1747

17:11 gave to Andrew for her 1:1 then paid my bill wch came to 2:7 So to Courte without ye Lad

7th to Limerick gott ye sheep from Studdert and got to men to help Jon to drive them home to turnpike for them and us 3:8 so back to town aboute 12 ye Ladyes and I came away and got to Court by 2 and stayd to diner then gave Mrs Hartstonge 3 pistols to make good ye loss of sheep She had from me last sumer so home wth an promis not to lye a night at Court while D Bryen was on ye land expences 3:6

8th I sent 2 cows 2 two yr olds and 10 yrlings to Killmoreen and have 15 cows and 2 yrlings here I pd 4s hearth money and gave Jon 2s to pay his I pd Bry Boyle 2:6 in full

9 ye men and T Coagh diging for petatoes I gave Sypr. Pursill a barrill of Oats on acct

10 my men T Coagh M fitzGerald and C Slattery diging for Petatoes I gott Pheby shod I sent Jon to Limerick wth 3 barrill oats wch he sould for 18s to Costom &c 12d

11 diging I prepareing to go to ye County Cork

12 Jon and I left home att 6 baited at Charlevill to our bill 2:1 att Lisscaroll 3 we gott to Mr Chapmans by 2 I spoke aboute Ms and as we could not agree was going away but wod not be let to David Bryen on acct of his wayges 5:5

13 Spoke again aboute Ms and wth her consent we all agreed and I stay there yt day and night

14th I look my lave of them about 7 to ye serts 4:10$^{1}/_{2}$ att Liscaroll 3 Charleville 2:9$^{1}/_{2}$ so to Thos Welches and pd him 2:8$^{1}/_{2}$ I owed him so home

15th att home wrote to Ms Hartstonge and got an answer why I went from Courte so abruptly I gott a grate home

16 I had D Bryen C Slattery and my own men diging for Petatoes J fitzGerald fixing a grate and making up ye chimney for wch I gave him 13d and 4:4 to give Mick Williams for building ye walls of ye hen house and quarrying some stones for Killmoreen house I pd Honr Mcnemara 11:7 in full for her wayges I wrote to Ms Hartstonge an answer of ye letter she sent me yesterday

17 Ms wrote again to ye same porpose I brought two barrills of sidr from Thos Harragan T Coagh and my own men diging for Petatoes

18 went to Limerick to try if I co'd sell my oat but did not I took up att Bonfields a blew cloath for a coat 2 fostian weastcoasts a scharlet shag briches and a hatt then I pd 1:0:4 for a weding Ring 7s:7d for a lockett Ring 4:4 for a whip 1:3:0 for 3 dishes and 12 pleats and a sauce pan then I bespake a Side Sadle for wch I am to pay 3 ginies to Seeds 3d so away and fell att barnokile[9] expences snuff 1:1

19 att home unwell by ye fall

20th I sent 30 two yr and 40 yrling wethers to ye fare of Ballingarry but did not sell them to ye men 6

---

9 Barnokyle is 4 miles north-east of Adare, on the road to Limerick.

21 I wrote by post to Ms Chapman and to ye sub-Sherriff who desird I shod meet him at adear to morrow I sent ye sheep I had at ye fare to Killmoreen and had ye black cattle brought home I wrote to Mrs Hartstonge to allow Pattr Hickey 14s on my acct and to send me ye remaindr of ye butter money on what I laid out for her in Limerick I got but 5:5

22 I sent 1 cow 1 3 2 yr old bullucks 3 2 yr old heffers and 10 yrlings to ye fare of Croome but did not sell expences 6d to hoopes 3.

Aprill ye 23d 1747 I sent back ye Cattle to Killmoreen had D Gorman and anothe mending vessells I gave Ned Vallan 2 bushills Oats

24th I went beyand Rathkeal to meet ye Sub Sheriff on Studders acct but did no good expences 5d I went to Mick feigh for a barrill mault and gott itt

25th I went to Limerick and was dischargd by 1 but stayd for Tho Harrogan till 4 so home expences 15d pd for 2 pound hops 2:6

26 att home sleeping and reading I agreed wth Thos Harragan for 6 yrlings att 17s each and 3 two yr olds att 23s each wch comes to 8:14:0

27 I went to town and was kept on a jury from ten to 5 then I dined wth my brotr spoke to Henry Studdert I strave to have him taken till 8 so home

28 to town and by much ado gott Studdert taken and left him in ye Sherriffs hand then dined att Robt Peacocks aftr wch I mett Mr Thornell had some talk wth him pd for brandy 3:2 Rum 3 sope 1:11 Iron 2 Salte 11d expences during ye sizes 8 so home I gott 7 ginies from Thos Harragan

29 I sent Jon to town and he brought home my Cloaths and 6 knives and 6 forks

30 I borrow'd 3:0:8 from Martin fitzGerald and sent it and 6 ginies to Colonell Taylor and wrote to Robt. Peacock for 4:2:7 but did not get itt I tun'd my drink and spent ye day wth Will ffosbery I got ye gelding and black mare shod

May ye 1st 1747 I went to Killmoreen gave tho Harrogan Possession of ye west park and ye Lord make me thankfull and increas them I have this day 138 sheep 57 cows two bull 3 three yr olds 6 yearlings 16 lambs and 9 horses and mares I gott 5:5 from Mick Cosgary in full for ye grass of his yrling Willm fosbery and Jon Hartny satt wth me a while I gott a pr of Pumps from Shea so loiterd aboute

2d I spent settling out ye house I accted wth Charles Slattery and do owe him 2:11 1/2 I got 4 large sticks carried to Killmoreen

3 I went to Church dined at Courte so home

4th I orderd 30 wethers and 40 hogett and 4 yrling to ye fare and aboute 9 follow'd them did no good but gott 6l from Will ffosbery for ye gras of his Cattle gave david Bryen on acct of his wayges 2:2 so dined and made ye Servants drink so home expences 5:3

5th went early to Courte accted wth ye men and aftr diner came home pack'd my things wth an intent to go to be happy or miserable for ever and do rely on ye all might God for his blessing

6th went to Lisdogan but could not be maried till I gott a Lycence[10]

---

10 In order to get married immediately, Peacock needed a licence from the bishop of Cloyne, at that

June 1747

7th went to Cloyn[11] pd 18s:11d for a Lycence lay att Mr downss
8th came to Lisdogan all most Dr K [drunk]
9th I bless God I was maried and from thence take ye begining of ye date of my happyness I stayd att Lisdogan to ye **23d** then brought my comfort alonge wth me home ye allmighty God give us his blessings of love constancey and happyness expences whilst abroad 6:2:5
24th att home **25th** att home
26 we all went to Killmoreen rode aboute so home
May ye 27th 1747 I dined to Courte so back to ye grome 6½
28th I went wth Mr Hartstonge to bruff back att night expences 2:8½
29th I gott an acct from Jon yt he sould 1:0:16 of butter for 1:8:6½ he pd for board 9s:1½ for two quartrs Iron 9:8 a stone sope 3:10 nails 2:5 snuff 1:1 a small range 4 to ye blade of a knife 3 he gave me ye remaindr Mr ffosbery and family dined wth wth me
30 att home nothing materiall
31 dined att Castilltown Mr ffosbery gave me a barrill of sidr to ye man 1:1

June ye 1st 1747 I gott 3 ginies from Martin fitzGerald wch wth 1:16:0 he pd for Ceelers and 32 he laid out for his own use makes 5:7:5 and is ye Pryce of 4:1:5 of buttr for wch I gave him a receipt
2d I gave Pattr Hartney 1:2:0 in Parte of ye Pryce of ye side saddle to glew 1:1
3d I gott home my sidr
4th Catty and I were bleed
5th I gave Mick Cosgary by his brotr 2:5:6 in Parte of ye 3 gineys he borrowd for me from Ed Sheehy ye 2d of March I gave Sypr Pursill a barrill of Oats on acct and got moste of my Oats winowd **6th** and got a fish from Courte
7th Catty and I dined to Courte where she stayd but I came home
8th I sent 2 two yr olds 4 yrlings 1 bull 30 two yr old wethers and 40 hogetts to ye fare sould ye bull for 2:5:6 two yrlings for 1 8 0 one yrling for 13s and 20 of ye 2 yr old wethers for 6:5:0 gave Mick Cosgary 1:2:9 in full for Sheehy Money expences 6:7 so home
9th I went to Courte and in ye Avening brought home my wife Cate Cassow went away
10 I had 8 men and my own thrashing to nell on acct of her wayges 4½
11 I sent early for Cypr Pursill to draw my Cattys tooth he came aboute 8 and brok itt he then bled her and she continued very unwell I sent Jon to ye fare wth 20 hogetts and 2 two yr olds and 1 yrling ye yrling he sould for 14:6

time Dr George Berkeley. Before he issued a licence, the bishop required a bond to be entered into by both parties, to protect him from an action for damages if it was later found that there already existed an impediment to the marriage (as in *Jane Eyre*) or a legal object, when one party might have already contracted to marry another person. Bishops required two solvent persons, one probably being the bridegroom, to enter into a bond (the value of which was set proportionate to the status of the parties) that there were no such impediments. Preface, Gillman (1896). 11 Cloyne, Co. Cork. The seat of the bishop of Cloyne, 14 miles east of Cork.

12 I sent Jon wth 5:5:7½ to Tim Mahony for ye use of Mr Bryen in acct of May Rent and to Colonell Taylor wth 4:7:5½ in full to ye 1st of 9br last I had 7 men and my own trashing
13 I finishd trashing my oats and acct ed wth thos Harragan and alowd him for ye Rent of Caparow and ye Parte of Killmoreen he held till may and for all ye sidr I had from him and 1:1:9 on Jon Bryens acct then we gave a receipt in full to each other
14 att home Reading and walking wth my wife
15 finishd ye first winowing of my Oats had 5 barrills Maulte home
16 I had my sheep washed and fell to second winowing my Oats 17 att ye oats and finished and measurd 39½ barrills
18 dined to Courte
19 sent 5 barrills Oats to ye kill and Mill sould two 2 yr old for 2:5:6 pd 1 0 11 Countery Charges I had 12 barrills oats to send to town
June ye 20th 1747 I went to town to see ye Oats mesurd but it did not hold out I pd Pattr Hartney 13:3 on acct to gloves 12d ribon 1:3 to whistler in full 3:3 & 1:4 turnpike dined att Rob Peacocks so home to expences 1:0
21 dined at Court back att night
22 I gott home my Meal and pd 10d for drying it turnpike 1½
23d I got my sheep shorn and pd 3:4 I sould a barrl of oats for 6
24 I had my Meal sifted Hird Joan for 6:6 a quartr gave her 6½ earnest
25 sent Jon to town wth oats turnpike 9 I had ye Road from ballyassey laid out
26 I went to Limerick gott 9l for my Oats 8l for my wolle pd for Salte 1:10 snuff 2:2 Ed: Mcgans bill in full for me and ye me and 1:4 for letters 6:5 to Pitch tar and Reding 27 I lay att Mr Johnstons
28 I came home to ye Serts 13d to turnpike to day and yesterday 12
29th markd my sheep 30 nothing Material

July ye 1st 1747 nothing Materiall
ye **2d** gave Tim Herlihy for Richd Thornill Esq ye sum of 6l:17:0 to pay ye Quitt rent
3d nothing materiall ye black mare was shod
4 I sent 55 sheep 1 lamb to Killmoree and 2 yt strayd and 57 sheep and 13 lambs left here
5 dined att Court back in ye eavening
6th dined at Mr Downes back att night to ye Servts 13d
7th I gott 1l 17s 6½ from Jon wch wth 18:3½ he laid on for ye folloing Perticulers makes 2 15 10 he gott for two hondrd of butter to a cloak bag 10s:10d to Hary Evans 1:1 to hops stone blew powdr blew sterch and coperas 4:3 to cards 2s:0 turnpike and Costom 6½ then I gott from Mar: fitzGerald 10:1:3 wch he gott for 7:2:27 pound of butter then I pd Jon Rourk for Richd Thornill Esqr 19:15 5½ wch 6 17 0 I pd Tim Herlihy makes 26:12:7½ then having drank pritty hard and dined he went away then I pd Anst fitzGerald 2:8½ so to bed

August 1747

8th I pd Morto Henesy 3:3 on acct of his hire I went to see Mrs Hartstonge aboutte 11 back I had 22 men cutting turff

9 I had 14 men cutting turf I gave Elenor Harragan 6½ earnest and hir'd her att 5:6 a quartr I gott everything ready to go wth Gods blessing to ye County teperary in ye morning to tabaco for ye men to day and yesterday 9d

10 I baited att Bruff ye bill 2:8½ licorish and Cases 5 I gott to Greenfield[11a] by 4 where I stayd to ye **13th** to ye servants 3:3

13 I baited att Donarail[12] to ye bill 4:4 I gott to Lisdogan by 5

14 att Lisdogan **15** dined att Mr Bastabels

16 att Lisdogan **17** dined att Mr Thornills to ye Servts 5:5 I sould my gelding **18** att Lisdogan

19 dined att Mr Gores[13] to ye sert 13 **20th** I went to see ye Ld Persivalls Castle[14] to ye boy 13d

21 I left Lisdogan dined att Charlevill to ye bill 2s: att Ballingarry 3 so home by 10 att night to ye Servts at Lisdogin 2:2 my turf is drawn home

22 breakfasted to Courte back by 10 I gave Jon to give his son David 2:8½ and 2:8½ to give Tim McMahon

23 att home Rourk came to see me

24 I gott 6½ barrills of Mault home and pd 1:4:11 for Maulting all ye mault Marg: Edigan pd 12:6 and Jon Mane pd 12:5 wch I must allow them in their Medows

25 I went to Courte to seelibrate Mr Hartstongs birth aboute 7 I gott home

26 I spent reading and walking wth my wife

27 I sent Mick home wth mr Thornills mare

28 and **29th** nothing materiall

30 I had 3 mowers

31 3 mowers

Augst ye 1st 1747 nothing materiall

2d to Church dined att Court back in ye Eavening

3d I gott 3s from Jon Quain he owed me for Oats I pd ye women for spining frize 1:3 to B Hall 18d

4th Buckners Procters were here but did no good

5 my wife and I went to Killmoreen Rode aboute back aboute 12 then I gott 5:10:5 from M fitzger and he laid out for his own use 1:3:0 all wch comes to

---

11a Greenfield is in the South Riding of Co. Tipperary. 12 Doneraile, Co. Cork, a market town, 21 miles north-west of Cork on the river Awbeg. Lewis. 13 Possibly Francis Gore, vicar of Ballyclogh and Castlemagner, Co. Cork. The Gore family house was Assolas, near Kanturk. The house was acquired by the Revd Francis Gore in 1714. Brady. 14 Lohort Castle, near Castlemagner, Co. Cork, was built in the reign of King John. It was garrisoned by Sir Philip Perceval in the 1640s, was taken by the rebels and retaken in May 1650 when it was captured by Sir Hardress Waller. The castle was repaired in the early eighteenth century by Lord Egmont, and in 1740 was a tourist attraction. Lord Perceval was congratulated on its 'grandeur and beauty in it which are now apparent to every one'. Lewis. R. Brereton to Perceval, 11 November 1740. BL Add. MS 47009 f.66v.

6:13:5 wch he gott for 5 0 15 of butter then I gave him 7s to pay for Killmoreen tythes

6 I sould 65 sheep for 6:3 each and gott 1:4:11 in part

7th I sent for Jon wth a vale to town wch he sould for 18:3 he pd 2:8 for hops expences 6½ to a woman 1½

8 I gave nell 10s on acct of her wayges wch wth 2:1 I gave her att Severall times makes 12:1

9 att home gott 11:4½ from D Welch for medows

10 att ye Syzes pd 12 6 to Corns I owed for brandy &c:

11 att ye Syzes I gott from Jon for medows 2:10:0 from Jon Mane 1:5:0 from Jon Cartr 19s:6 from Patt Hanrahan 6:8 from Jon Newman

12 in Limerick pd Colonell taylor 13:1:7½ of wch I gott 4 gineys from Thos Harragan expen during ye Syzes 6:10 so home

13: 14 15th att home ye men reeping

16 att home gott 5:3 from Ed Vallan

17: 18: 19 att hom finishd reeping pd Jon for some of it men 5:5 to Jon Rohan for a hatt 5:8

20 we went to Jon Downss Christning to ye midwife 5:5 21 att Mr Downss

21 we came away Dined att Rob hewsons to home to ye Servants 3:3 from Thos collins 2:2

23d dined att Castletown

24th att home David Bryen carried away his cow and sould her

25th I borrowd 5l of thos Harragan and gott from Patt Boyle in acct 5:5

26 I began to draw oats and pd James Conr for Mr Bryen 5:5:5 and do owe him 8:9

27 went to Courte left my wife there lent Mrs Hartstonge 13d lent Ned Mcgan 13d so home

28th ye men reeping barley nothing materiall

29 finishd ye barley and gave tim Herlihy an ordr on Geo: Peacock for ye money of ye Road

30 dined to Courte back att night

31 drawing oats and finishd

7br ye 1st 1747 I brought Catty home to ye serts 2:2

2d finishd ye hay to pills and oyntment for me 2:8

3d pulling beans 4 finishd ye beans gott 11:4½ from Patt Nash on acct of Medow

5th I sent Jon wth 8:8½ to Tim Mahony wch wth ye money pd him in June and augt makes ye may gale I gott from M fitzGerald he gott for 5 0 17 of butter 6l:19s:0½d he laid out 1:10:2½ for his own use then I acct ed wth him and do owe him 1:10:1 then I pd tho Hanagan ye 5l I borrowd of him ye 25 of last month

6 dined att Castletown to ye boy 6½ so home

October 1747

7th drawing turff out of ye bog to nell 6½
8 a holy day nothing materiall to P Hickey 5:9
9th finishd drawing turf out of ye bog
10 I had my own 4 horses drawing home turf
11 12 att ye turff **13** Simon Henry and wife dined wth me Hen Downs sent for ye 2 pigs and gott them but did not pay for them
14th agreed wth Ricd Harnedy for his son Jon for seven year to comence from this day and am to give him an incalf heffer at six years end and cloaths ye men and horses att ye turff oyle 7d
15: 16 att ye turf and finishd
17 men diging Petatoes tim Calahan mending a truckle
18 2 men and horses draing hay for Calahan and one man and horse and Thos Harragans horse drawing hay for Den Bryen
7br ye 19th 1747 my 2 horses drawing turff for will hickey in full for ye men I had from him I sent Jon to town wth butter itt weighd 2c:0qr: 15lb for wch he gott 2:27:7½ he pd Patt Hartney in full for a saddle 1 12 6 to sale 1:1 to oyle 6 bandle cloath 5½ hops 2:8½ costom &c
20th Jon Geary and wife dined wth me I was taken vry ill att night wth a Pluresy
21 sent for things to Limerick wch cost 5:6
22 unwell still **23d** better **24th** still better
25th I just gott up sent 4 bus mault to ye mill and pd 1:6 for tucking my frize turnpike 18d
26 I bless God pritty well pd for a pr of basketts and brooms 13d
27 stalked aboute dined to Courte
28 went to Mick feighs
29 a holy day[15] pd 2:2 on acct of work
30 had tim Callahan a breaking out a window in my own Roome ye drink tun'd

8br ye 1st 1747 Callahan making bed sted
2d I spent fitt out my Roome **3d** att Dito
4th Judy and morse dined wth me I had a bed Coard and half a pd of snuff from Ned Mcgan wch I owe for to bread 2 to snuff to Jon Bryen
5 a removing my drawrs and desk
6 Catty and I to Courte I left her there and came hom
7th I went for her and brought her home to ye gleasier for leading a window 6½
8th I acctd wth david Bryen and pd him in broges and soles 2:4 in cash 1:1:1½ in Oats 6d and do owe him this day 1:5:6½ tim Callahan att ye bed
9 tim Callahan att ye bed ye boys drawing reed I diging
10 tim Callahgan att ye bed and finishd yt and enlarged ye roome ye men diging petatoes
11 dined att Court left Catty there

---

15 Feast of the Archangel Michael.

12 att Court and drinched ye Cattle ye men att ye hay
13 went for my wife and brought her home
14th I seett 400 of Plants ye men diging for them
15 ye men fencing ye garding and making beds in ye Stable I gave Bry: Boyles wife 1:1
16 I gott from Martin fitzGerald he gott for 5c:2qu:12lb of butter 6:8:11 he pd for hops 2:8 for cottning my frize 1:8 he laid out for his own use 2:1 pd Morton Henesy in full 7d I sent 3 bus mault to ye mill I pd Anstice fitzGerald 1:6d full
17th I had my bed and things removed down I pd May welch in full 2:8 I had a sheep killd
18 att home
19 I went to town pd ye taylor in full 19s:9$^{1}/_{2}$d to Geo Russill in full 1:1:0 to Mick on acct of his wayges one pound $^{1}/_{2}$ to yeamans in full 4:11 to bean in full for &c 13:6$^{1}/_{2}$ to Quittrent 14:4 to mending my gon 8 to a knife 8 to ticking 2:2 for trimings 9:1 gloves 1:1 Mcgan in full 4:9 tornpike &c 8 to a dram of shrub 4 so home
20 I borrow'd of Thos Harragan 12:6 an pd Nell fox her wayges 7:5$^{1}/_{2}$ Margett and Mary Hanall came to me
21 I gave 18d for 3 axelltrees and gave ye Haragans ye 2 axelltrees I borrow'd of them I sent 4$^{1}/_{2}$ barrills oats to ye mill
22 ye men diging Petatoes
23 ye taylor making my Coat
24 ye taylor finishd ye I pd him 2:8$^{1}/_{2}$ I had my Male home from ye mill turnpike &c 3
25 Harry downs and his wife M feigh and wife dined wth us I had 4 basketts home for wch I pd 6
26 I had a kill to dry flax made[16] and ye male sifted
27 I went to Courte and dined there aftr wch I came home
28 I borrowd 3:9 of Tho Harragan
29 I had an acct ye my sistr Mcgan died and sent Jon to town Catty unwell I sent for Syp Pursill to bleed her I spent ye day settling my Accts
30th 8br 1747 Syp: Pursill came and bleed Catty
31 Jon came from my sistrs funerall to steel 5 to thread 3d I sent him to Thos Harragan for ye rest of Patt nashs money and gott 4:4

9br ye 1st 1747 att home gave Jon Rohan 4d
2d I dined att Shanongrove calld to Courte where Catty stayd and att 1 I came home to ye Coachman for service of ye young black mare 3:3 to ye groom 13d to stone blew 4d

---

16 Here it is possible that Peacock had constructed a field-side kiln, which was a temporary structure, usually in the corner of a field and near a stream. It was a saucer-shaped depression, lined with stone and sunk into a bank. Hot air from a fire was drawn through the kiln until the sheaves were dry. Brunskill, *Traditional farm buildings*, p. 94.

December 1747

3 I went for Catty and brought her home to nails 5
4th James fitzGerald mending ye oven to Syprian Pursill 4 bushills mault
5 att home nothing materiall
6 and 7th att home
8 to Church back again 6 ye —— 6
9 att home nothing materiall
11 to Limerick pd for Iron 8:9 indigo 3 nails 10d Ned Mcgans bill for letters &c 3:10 snuff 1:1 a Sizers 1:1 a pan 2:2 ticking 2:2 other expences &c 2:9 so home
12 13 14 nothing materiall
15 Simon Henry wife &c dined wth us
16 17 nothing materiall Cyprian Pursill ill and boy spent 5 days making a coope to a rasp for him 6$^{1}/_{2}$
18th 19th sowing wheat and sow'd 9 pecks had 14 men trenching itt
20th and 21 nothing materiall brotr Billy came to see us 22 att home
23d to Court dined there and envited them to Diner Mr Hartstonge gone to Dublin
24 Billy and I dined att Jack Downs
25 I gott from Martin fitzGerald he gott for 5c:1qr:10lb of butter 6l 2s 9d he laid out for his own use 9d wayage 2 to 2 bottls Clarott 3s a bottle Lisburn 1:6 2 pd of Jemaca Sugar 1:4 1pd powdr sugar 10d pepper 5$^{1}/_{2}$ sand $^{1}/_{2}$ I pd 18s for ye wheat
26 Mrs Hartstonge and ye 2 Misses dined wth me I had a wether killd yesterday a pork ye day before I gave Catty 4:9
27th I sent 6 bus mault to ye mill
28 I had ye drink brew'd
29 I had 42 bandills of flaning home I had 18 pound of wolle to make itt and it carried 600 I gave ye weaver 3:4:26 of itt he is allowd for ye weaving I went to a Christning to Clampits
30 I borrowd 3:10 of Thos Harragan ye 10d he owed me in ye change of a giney then I pd Mick in full of his wayges 12:8$^{1}/_{2}$

xbr 1st 1747 I gave Jon 13d and sent him to mend ye hatchell I had ye drink tund
2: 3 a grate frost
4 Carried Catty to Courte and left her there
5 att home looking over accts
6 I went for Catty and brought her home to ye servts 13d
7th I agreed wth Tho Delane to dress my sheep for 5s a hondrd earnest to him 6$^{1}/_{2}$
8th sent 6 bus mault to ye mill pd for hops 3s handkercheifs 6s ribon 1:4 gloves 1:4 chequerd linin 6$^{1}/_{2}$ sope 2 10 snuff 2:2 to ye boy 2$^{1}/_{2}$d turnpike 1$^{1}/_{2}$ a mug
8 I had my pig killd
9 Geran Surveying I pd Tho Bouse 1:6$^{1}/_{2}$ in full for making ye boys cloaths

10 11 and 12 Geran surveying and finishd ye west parte of Killmoreen contains 10 0 32 I pd him 16:3 for surveing

xbr ye 13th 1747 att home

14th went to Courte back att night pd Jon Cloone for weaving flaning 4:10½

15th I gott 17l from Thos Harragan and acct ed wth hm and gave him a Recd for 9br Rent for ye west part of Killmoreen and Caparow then Martin fitzGerald gave me 4:17:6 for wch I gave him a Receipt

16th I sent Martin fitzGerald wth 10:19:9 to Tim Mahony and finishd ye Hackling of ten pound of flax

17th prepearing to go to ye County of Cork I gave Jon 4:4 to buy timber I pd him 9d for tabaco

19 we lay att Charleville ye bill 14:7 I pd 1:9 my boy had going wth mr Thornills mare

20 we calld att Buttyvan[17] bill 13 gott to Lisdogan by 3 wre we stayd till ye **9th of Janr 1747/8** to shoes for Catty 2:8 to cambrick 4:4 a tare thred &c: 13 to ye Serts 2:8½ bastables Servt 13d Cards &c 2:4d

---

17 Buttevant, Co. Cork on the river Awbeg between Charleville and Mallow.

1748

Janr ye 8th 1747/8 we came away calld att Buttyvan to our bill 11d so to Charlevill dined att Mr Egans[1] aboute 8 to our lodgings at 9 to bed

9th to ye bill 9:8 to Sidr att andersons 6 so home ye all might God be praise

10 to Courte met ye family on ye Road so turnd

11 to Courte and back att night pd Quittrent for ye lott of Killdeemo 6:15:8½ to Will breen in full for last years tythes 5:0:0 to Pattr Hickey on acct of his wayges 5:5 to snuff 13d to Jon 8 to ye miller for drying my Oats 8 in my absence Jon sould a cow for 2:11:1 a sheep for 8 a calfs skin for 1:9½

12 I went to Jon Gearys Christning to ye midwife for my being Godfather 2:8½

13 I had my 2 pigs salted to coambs 2 needill 1

14 I gave 1:9½ for Sticks to make a plow soll

15 I had a cow killd and sould ye hide for 5:4½ I gave Patt Pursill 5 5 on acct of his wayge

16 att home they took 40 wether for Mrs Hartstone

17 to Killdeemo and back again

18 att home

19th sent Jon to town wth a cask of butter wch weighd 2qr 10l gross he gave it to Ned Mcgan but did not get ye pryce I pd to frank yeamons 17:9 for ye following perticulars viz: Lisburn 3s rum 6:5 whiskey 1:6 powdr sugar 3:1½ oringes 6 rabitts 6 stone blew 4 sope 1 10 to soles from Jon Rohan 6

20 I went to Courte to envite ye Ladies

21 they and Geo: fosberys familly dined wth me

22 I went wth Catty to Courte

23 I acct wth ye men before Ball and gave him ye acct

24 came home to ye maid 1:1

25 I accounted wth Martin fitzGerald and gott from him in full for last years acct 1:9:5 and got on acct of ye ballance of this years acct 5 0 3½ and allowd him 8:11 he laid out then gave him his dinner and sent him home ahordd [?]

26 unwell wth a pain in my year hird Thos Coagh att ye rate of 12s ye quartr earnest 6½ hird David Bryan att ye same but am to give him a pr of broges

27 Ned Mcgan sent for his mare and fole to a teapott and slop bowl 12d

28th to Killmoreen vew'd ye Cows back by dinr

29 I gave mick a note on Ned Mcgan for 8s

30th I wrote an agreement between me and Mar fitzGerald pd Mick Harting

---

1 Maurice Egan of Charleville, will proved, 1781. Phillimore, vol. 3.

6:2¹/₂ in full for Countery chargs to sugar &c. 9 10 to Corks &c 2:2
31 dined att Castletown back att night

febr ye 1st 1747/8 att home
2 dined att Killtinan back att night
3 I had hemp dressers att work I pd John willson for Patt Hartney 1:9:2 in full for a pillyon and all acts
4th I went to Killmoren and sould Patt Daniell and M fitzGerald 22 sheep for 4:12:6 and sett them ye north east Park for 17l:10s:0d ye first year and 18l:5s:0 a year for every year aftr am to give them straw thatch one cabin and help to build ye walls of 2 cabins they gave me earnest 13d
5th att home putt in a stock beans
6th I began to Cros cutt my fallow sent to Charleville for milly to ye boy and horse 1:7¹/₂
7 att home nothing materiall
8 att home ye Plow going ye hemp dressing
9 ye Plow and hemp going
10th att Dito I gott a pd of tea from D Bryen to ye maid 6¹/₂
11 finishd ye hemp and pd 6s:3d for dressing it I gave Jon 13 to Cate Coffow 3:3 dined at Court
12 att home ye Plow going I gott a pig from Court
13 att home ye frost hinderd ye Plow
14 I agreed wth thos. Harragan for 6 cows att two pound 9s:6 each wch come to 14l:17s:0 am to pay for them any time between this and Novembr next earnest for them 13d
15th I gott ye cows and all most lost one of my own
16 I had beans trashd
17 att Dito I pd Mr Hunt 3s:1¹/₂d for ye church rates of Killmoreen I had oats put in
18 ye Plow going oats trashing
19 I sent ye men to dig Petatoes ye Plow going 20th I sent Jon to town wth butter ye casg weighd half a hondrd and odd he pd for rum 1s:7d whiskey 9 snuff 13d two pd jemeca sugar 1:4 hops 3:0 oyle and bladder 7 ballsom 3 expences 4 mendeing ye dish 4
21 I gave willm breen 10 10 Procters fees Catty and I went to Court and lay there
22d I came home and sett ye men to trash I got some beans winow'd
23 I went to Court and stayd there I began to sow beans
24th I came home to ye maid 13d
25 I gave Mauris Geran 4:1 in full for two sticks for schafolds and stooles I gave Thos Coagh 2:8¹/₂ for tallow and in parte of his wayges
26 I gave Joan Daly 2:8¹/₂ on acct of her wayges
27 I gott 2 salmonds from Courte

April 1748

28th to Church and back again Thos Harragan and his wife dined wth me
29d I wrote to Frank yeamons for Iron he sent me 4l4 1/2 ~~wchI pd him for~~ 7s:4 1/2d to a baskett 4d turnpike 1 1/2 Had a hogett drownded

March ye 3 1st 1747/8 I had a new Plow iron made 2 men and a boy from M fitzGerald
2d noth materiall
3d I had a cow skind and sould ye hide for 9s
4 two boys from M fitzGerald
5 ye men trashing 1 boy from Martin
6 att home reading to ye spiners 10 1/2d
March ye 7th 1747–8 Martin and a boy I sent 4 bus beans to Court and a barrill of Mault to ye mill turnpike 1 they finishd spining
8 two boys from Martin and my own diging Petatoes
9 3 men two boys and my own diging Peatatoes
10 3 men 2 boys diging for petatoes
11 ye men trashing 2 boys diging Jon Donevan making a break
12 3 men 2 boys trashing and fencing ye new garding dined att Jack downs
13 to Courte lay there
14 5 men diging Petatoes I began to sow Oats
15 ye men att ye oats Rum 1:7 whiske 9 sugar 8 fustian 11 Mader 2:9 my Sistr in law came to see us
17 Hary Downs his wife and J crow dined wth us to a man from court 6 1/2
18th att ye Petatoes
19 we dined att Jack downs expences 6 1/2
20 rob Hewson and wife dined wth us
21 my sistr went away to a cow hide 4:11 1/2
22 I gott some eelefry ye boys 8 1/2
23 went to Courte for money but got none I agreed wth a man for sleating my house at 13s a squear he to find everything to him earnest 13d to thread 6 1/2
24 att home ye men diging I gott 2:0:8 from T H
25th I pd 4s hearth money Cook lay wth me
26 27 28 att home ye plow and petatoes going
29 went to court back att night
30 went to ye Syzes gave up James fitzGerald pd for pins 8 sweath [?] 6 1/2 thred 3 Iron I had from frank yeamans ye 29th of last month 7:4 1/2 to nuttmegs 8d sinamon 9 to snuff 13d M Can 6 1/2 seeds 2:9 expences
31 att home very unwell

Aprill ye 1st 1748 att home sowing seeds
2d still unwell 15 dow candills in ye drawr
3d att home nothing materiall
4th I gave Margett 2:2 on acct of her wayges

5 ye men trashing duch barley I gave Patt Daniel 6 bushills
6: 7: 8 nothing materiall
9 I sent Jon to town wth buttr and Oats he gott for oc:2qr:16lb of butter 16:6½ for a barrl oats 5 he pd for two shovells 2s:6d steele 2d bread 6d to a portr 1d to him in town 1:1 to Jon att home 5:5 to him to pay Welch for work and tallow 4:0½
10 to Church from thence to Courte and lay there
11 home to ye Maid 13d brotr Cook was att home as before
12 I gave Mary hanan 2:8½ on acct of her wayges
13 I gave Jon 2:8½ and sent him to look for hemp seed he gave me 3:15:10 he gott for 13 sheep he gave thos. Coagh 6½ his expences att ye fare 12d
14th Ned Mcgan lay wth me
15th I went wth him to Mr Odels expences 8 2 10 so home very unwell
16 I gave Jon 2:10 to pay for 5 barrills of hemp seed wrote to frank yeamons for sugar rum whiskey powdr shott wch he sent and I owe for sent my watch to be mended but did not come back 17 unwell and took a Puke
Aprill ye 18 1748 still unwell tho better
19 I sent 5 barrill of Oats to ye kill
20th gott a message from Colonell Taylor
21 I wrote to Mrs Hartstonge for money but did not gett itt I gave mick 14:2 in full for his wayges and 4d I owed Jams Nash
22d I wrote to Colonell taylor and sent ye men and women to ye mill to drying ye Oats 10 to cottening 2½ bandils of frize
23d ye male was sifted I pd ye women 3:4 in full for spining
24 I send some yearlings and 2 cows to Coroheen to greas sent for ye midwife to M expences 6½
25 I wrote to Frank Yeamons for a quartr of a Stone powdr Sugar ye same of jemeco and a pound of rice wch he sent I wrote to Ned Mcgan for a cheese and half a stone of sope wch he sent
26 Catty unwell
27 Catty ill I wrote to Mrs Hartstonge to come to her sent Mick to town for lemons oringes and spermacitty pd 2:2 for them and I bless God Catty was deliverd att a quartr aftr 6² Mrs Hartstonge went home and Sistr Wallis and Brotr willm came he gave me 17 ginies wch mes 19:6:9
28th brotr Willm went away
29 I sent mick and ye black mare wth Motr Handy³ to Balingary back att night
30th sowing hemp seed

---

2 Price Peacock became a solicitor, and acted for his cousin, Lord Glentworth for whom he collected rents and fines. He married Mary Brereton of Stradbally, Queen's County, on 3 June 1783. He was an executor of Mary Pery's will. His son, also Price Peacock, was born in Limerick and entered TCD 2 November 1812 aged 16. He took BA in 1817 and became archdeacon of Limerick. IGI; Peacock to Glentworth 2 April 1797; Thomas Knox to Peacock, 13 March 1806; Limerick papers, NLI PC 875/2/15/58/31 and /28; King's Inns memorial; *Alumni Dubliniensis*. 3 Mother Handy: the midwife.

May 1748

May 1st 1748 Harry downs and wife dined wth us I pd Simon kent for Jon Wallis 8l:17s in full to snuff 13d turnpike &c 8d

2d I bought a cow from Martin for 2l:11s am to allow for her when we acct to spring thread 7d

3d I gave Jon 2:8 $^1/_2$ he pd 1:4 for a gallon and 1:4 $^1/_2$ for his own use

4th I went to addear to envite Mr Quin to stand for my child and for ye minstr to Chrsten him I sent Jon to town for ye following things for wch he pd viz: Clarott 3 Lisbon 3 shrub 3:3 whiskey 3 2 bottles 5d sugar 5:3 punch ladles peaper and 2 $^1/_2$ to 5l8 of beefe 9:8 wheat 4:2 credle 1:7 $^1/_2$ saucepan 4 other things 1:5 $^1/_2$ 5 I had a sheep killd and fitted ye cradle

6 I gave David Bryen on acct of he wayges 1:2:9 and sent him to town he pd for 3 hondrd of sparrowgras 4s 2 lobsters 1:2

7th I had ye child christend to ye 1s:6 $^1/_2$ to Mick for his trouble 2:8 $^1/_2$

8th to ye midwife 11:4 $^1/_2$ and sent her home

9th My brotr Chapman and I dined att Court

10th att home a well day

11 att home marget on acct 2:10 I gave James fitzGerald a barrill of Oats on acct

12 and 13 at ye hemp and petatoes

14 Mrs Wallis and I dined att Court **15th** att home

16 17 putting out dung for Petatoes I got 40 bandill of Blanketting from ye mill wch cost ye milling 1:4 and I owe for

May ye 18th 1748 I went home wth Sistr Wallis **19th** I came back expences every 5:5 to John bryen on acct of his wayges I gott 15 ~~18~~ 4 from thos Haragan on acct of Rent and 13:4:4 from Mrs Hartstonge in full for Sheep and beans

20 I pd Jon Rourk for ye use of Mrs Thornell ye sum of 37 14 3 $^1/_2$ in cash and ye Qutt rent receipt for 6 15 8 $^1/_2$ in all 44:10:0 bought a Shovell for 1:8 wch I pd for in barley

21 I acct ed wth thos Haragan for last may gale and gott 15:18:4 and 2:0:8 he gave ye 15 of march last I allowd him 1:12:0 for a hogshed of sider for 4 barrills of hempseed 2:0:0 and allow'd him on Jon Bryen acct 14:6 in all 22l:5s:6d and tho: he owes me I gave him a receipt in full

22d att home nothing material

23d I sould a bushill of barley for 1:7 $^1/_2$ gott a busl of hempseed from Jon Geery wch ~~I owe for~~

24th I sent 3 busls mault to ye mill and pd ye miller for tuckng ye blanketts 1:4 to two pd of hopps 3:4 turnpike 1

25 I had ye mault brew'd and finishd ye hemp and ye trinching ye Petatoes

26 all ye men att ye Peas

27 I began to cutt turff to salte 2d Betty kill my [?]

28 att ye turf I had a sheep killd

29 att home **30th** ye wettness of ye day hinderd my sending cattle to ye fare

30th I went to Mr Hartstonge to hunt to ye huntsman 13d to sidr 3

31 Martin fitzGerald gave me he gott for 4c:2qr:26lb of butter 6l:9s:1d and he laid out for his own use 4s:5¹/₂d then I gave him a receipt for ye butter

June ye 1st 1748 I borrowd 6:0:1 from Martin fitzGerald and made up 12:10:1 and sent it by him to Hary Page and got his Receipt
2d Catt and I dined att Court to snuff 13d
3 I borrowd 7l:1s:2¹/₂d from Thos Harragan and sent it by John to Tim Mahony
4th brotr Chapman came he and I went to Court back in ye eavening
5 he went home Catty and I went to Jon Geery to try if they wod keep ye child till we came from ye county of Cork so back
6 att home
7 we sett of att 5 and gott to Lisdogan by 4 where we stay'd to ye 10 and got home by 4 Catty very ill expences every way 8:10¹/₂
11 att home Catty very unwell tho better
12 I went to Court to Margett on acct of her wayges 3:3
13 I sent Jon to ye fare of Croom⁴ wth 2 cows and 4 bullshins 3 of ye bullshins he sould for 2:16
14th I gave thos Harragan 2:15:11¹/₂ on acct of ye Money I borrowd of him ye 3d inst and sent my sheep to be washd
15 I loyterd aboute 16 nothing materiall
17 I sould my wolle att 2s a fleice and 2 wether for 1:4:0 and had them shorn earnest 6¹/₂
18 I went to see ned Mccan calld att Mr Johnston so home expences 8 salte 6
19th Catty and I dined att Court
20th David and Tom went away I gott half a stone of Iron from fishier ~~wch I paid for~~ then I sent 5 wethers and 19 ewes to Corroheen and have 10 ewes 1 Ram and a lamb att hom
June ye 21st 1748 I had 4 sticks from ballygeasy wch cost 2s1 1 men futting turf
22d 4 men att ye turff
23 I sent Cattle to ye fare but did not sell
24 Catty and I dined att Bally England
25 I gott from Martin fitzGerald 7:0:7¹/₂ he gott for 1d:3qr:22lb my own butter and 2c:2qr:27lb his butter then I pd him ye 6:0:1 I borrowd of him ye 1ˢᵗ inst to sale 1:10 sope 2s blew 4 snuff 13d I sould him 11 lambs for 1:7:6 of wch he pd me 18s:1¹/₂d owes 9:4¹/₂ then I paid Thos Harragan ye ye 7:1:2 I borrowd of him ye 3d inst and he pd me 5:10 he owed on ye ballance of his Rent then I pd Moll ye 5:5 I borrowd goeing to ye County Cork then I gave Catty 2:8¹/₂ to pay Margett in full then I paid Moll 2:8¹/₂ on acct of her wayges then I pd Jon Ryan in full for Countery Charges 13s:9¹/₂d

---

4 Croom is 9 miles south west of Limerick on the river Maigue, which is crossed at this point by a six-arched bridge. A castle was built to protect the ford of the river by Dermot O'Donovan in the thirteenth century, but by the eighteenth century it had fallen into ruins. Lewis.

July 1748

26 I pd fishier for ye Iron I had from him ye 20th then pd Patt Pursill 5:3 in full then was bleed
27 walkd to Killmoreen rov'd about pd Tige Varih in full for hire 2:9 so home
28 to Killmoreen and back again
29th to Hary Downs thence to ye fare Agreed for two pr of truckel wheels and 6:6½ for them to back to downss dined there so home to tape 4 expences every way 2:4 I gott 7:7 from Martin on acct of ye lambs
30 att home finishd ye 2d footing in my turff

July ye 1st 1748 ye two Jons thatching ye Cabin att Killmoreen
2d I sent Jon Rohan to Limerick he pd for Cottening 28 bandles of frize 2:4 for Iron 3:9½ snuff 13 &c 3½
3d Jon and Hary Downs and their wives dined wth me ye young couple went away I hired Ned welch for 1l:16s:0d a year to him earnest 13d
4th walkd aboute
5th gott from Martin fitzGerald on acct of his arrears 2:10:7
6 I went to Askeaton and pd Hary Page in full for 9br gale[5] 2:12:2½ and gott a Receipt in full and gave him up a Receipt dated Aprill ye 30 1747 for 9:16:9 another dated June ye 12th 1747 for 4:7:5½ another dated augst ye 12th 1747 for 13:1:7 and another dated June ye 1st 1748 for 12:10:1 then he and I had in sidr and dram 12d then I pd for moulds buttons and mohair 1s:7d so home rid to Killmoreen to ye Childern 9½ so home
7th Jack downs and wife went away I pd for a quartr of Motton I had a Sonday last 2:1 I had my Oats winowd and gave Jon McTamos a barrill in full for his hemp seed and Jon Bryen half a barill on acct of his wayges and orderd ye Best to be dryed
8th I sent ye oats to ye mill turnpike 1½ Catty and I dined att Court
9th I sent for hopps and pd for a pd and half 2
10th I sent Jon Rohan to Lisdougan to him 13d
11th I went to Limerick gave my note for [illegible] and took up two tun and a half of balk to Portars 1:7 a permitt 6 snuff 13 bread 5 Manna 3 sidr 6 peapr 10 fery 4 I borrowd from Syp Pursill 1:7 from Robt Peacock [illegible] E Mccan [illegible] and owe my reconing
July ye 12th 1748 I sent Jon to borrow 8:1½ from Thos Harragan and to give to ye boat for bringing ye timber
13 Sent by Martin for a pound of Sugar Hary downs and his wife lay wth me
14 Catty and I dined att Martins
15 prepareing to draw turff
16 my own 3 horses drawing ye little mare and Martins puting out ye turff
17 to church dined to Courte back att night
18 19 20 21 22 23d att ye turff and finished

---

5 November gale day, half-yearly rent due.

24th att home

25th I borrowd 5:5 of Thos Harragan Catty and I went to Court to selebrate Mr Hartstong birth to ye Servt 13d

26th Jon had my 3 horses drawing turff to ye sayers 2:8½ to nell 6½ I borrowd of her to bread

27th: 28th loytering aboute

29th ye Ms Hartstong and Perrys came to us and went wth them to Court and back att nigth I sould a cow to Mr Melsop of 2:10:0

30th I had 5 mowers Mowing

31 Catty and I went to Ned flanigans in ye aftr noone

Augus ye 1st 1748 I had David Bryen Conr Harragan and my own boys at ye hay

2d I had 8 and my own people at ye hemp I went to Killmoreen and got Syp: Pursill to wright ye agreement between me and ye Sleater

3d I had 11 and my own at ye hemp

4th I had 14 and my own att ye hemp

5th I had 14 and my own at dito I went to Limerick sould my own butter and Martins and got ye money I pd for 6½ yards Cotten 10:3 a yard Lining 10 stone blew 4 pins 8 cadis 4 bread 6 snuff 13 Ribon 3s Sugar 12 leas 2 then I pd 5:6:9 to Jon Kelly and took up my note to me 3:6 so a way and got home before night to Jon Geary 3:8½

6 I had [blank] And my own att ye hemp and hay I went to Killmoreen and pd 2:0:9 for 4: 10 0½ of sleats

7 att home pd Patt Hickey 6s on acct of his wayges I got from thos Harragan 6:0:0 and ye money Jon brought me from him ye 12 and 25 of July last

8 I went to Limerick pd 6:14:8½ quittrent to ye drivers fees 4:6 then I pd welch 6s for ye news then I pd Ned Mccan 15:4½ and gott his Receipt to Mader and Allom 1:2 to gloves 13d to me 2:8

9th I gott a giney from Jon Quain and went to Killmoreen pd 1:17:9 for for 3 10 0½ of sleats so to Court dined there and back in ye eavening

10th att ye hemp 11 att Dito gott 1:2:0 from Den Bryan for hay I pd ye soyers 15:6½

11 12 att ye hemp and bog'd ye most of itt

13 I pd for 900 of sleats 19:8½ I finished ye pulling ye meadew hemp I pd ye taylor for making ye boys cloths 4:4

14 to church and back again

15 att home to Mary Hannan on acct 6½

16 I agreed wth Bally Craheen men for 2 acres medow att 2s6 an acre earnest for ye next year 6½ an acre wth Jon Quain for 1a:1r:0p att Dito earnest 6½ allowd tim Callahan for 3 acres against next year for 2s6 ye acre Willm welch Jams Nash David Bryen reeping barley

17 will welch David Bryen and Jon Rohan reeping wheat to ye weaving 42 ban-

September 1748

dles of cloath 3s pd I weight 16½ pound of woolen yarn and 12½ pound of linin yard for Cortains

18th to will Breen for Procters fees 11:4½ D Bryen will Welch Jams Nash and Jon Rohan Reeping

19th D Bryen W Welch J Nash and J Rohan a making a pleace to bog ye hemp I gott from Jon Quain on acct of medows 8:8 Dan Shea gave me a pair of slippers I gave James ffitzGerald 11:4½ on acct of Measons work

Augst ye 20th I pd for 5800 of sleats 3:0:7 and gott Smiths Receipt for 14000 of sleats I had D Bryen J Nash and W Welch att ye Dike J Rohan ye 2 boys and 3 horses drawing turff for Conr Harragan for 6 days labouring work

21 att home to ye sawyers 2:8½

22 to pulling hemp 5s D Bryen J nas and Jon Rohan at ye hay I sent 4 bus of mault to ye mill to turnpike and nails 2½

23d D Bryen J Nash and Jon Rohan att ye hay 2 men from Martin Honr Coagh my own maids spreading hemp att ye bog hops and serus 2:2

24 att ye bog 2 men reeping ye drink brewed

25 att ye bog 2 men Reeping

26 two men Reeping my own maids binding I pd ye weaver for ye bandle cloath 3s

27th Some of ye hemp drawn from ye bog

28 I gott 5:5 from Jon and went to Lisdogan

29 I came home expences every way 4:4½

30th ye rest of ye hemp brought home ye reeping and binding going on

31 att Dito and stackd ye wheat

[illegible] ye 1st 1748 att ye Oats

2d I wrote to Ned Mccan for Nail &c: for wch I owe him 5:6 to a letter 4 turnpike &c: 2½ pd

3 I and Catty dined att Court

4th I had 3 women 4 men and my own and finishd Reeping and pulling beans

4 [sic] att home Jon Geary dined wth us

5 I began to draw hay 5 men and my own 6 women pulling hemp

6 I gott 1:0:11 from Jon Mellsop on acct of ye Cow ye men and horses att ye hay and finisht and drew a little oats and duch barley ye women att ye hemp

7th I hird Mick Collins att 3l a year earnest 13d 6 women 1 boy 4 men att ye hemp I gott [illegible] from Jon in full for James Bryens and Medow I went to Killmoreen walkd about and sott wth Pursill a freaming ye Roof of ye house

8 Catty and I dined att Killtinane back at night to Jon Rohan to buy broges 2:2 to David on acct of his wayges 5:5 att he hemp to sidr 3d

9th ye hemp pulling to nails 3d

10 ye hemp pulling I sent Tim Callahan to buy timber to make a hemp break pd for itt 1:5

11 to Church back to dine Sent Milly to Limerick for a Spark to snuff 1:1 to bees wax 4 expences 4

12 to att ye hemp to Cate and Neere 4 to ye Meason 13
13 to Syp Pursill by Alick Burk 6 6 I gott from Martin he gott for 4 0 15 of butter 7:19:1 1/2 to tornpike 2 lent him 2:8 1/2 and ye black mare to go look for his mare ye men drawing oats
14 3 men trashing Oats
15 4 men trashing Oats
16 3 men att ye Oats and finishd
17 ye men att ye place to bog ye hemp I gott 4s:5d from Thos Collins wch wth 2:4 he pd in work make 6:9 he owes 6 I pd ye sayer 5:6 in full
18 I gave for a pound of hops 13d to old Marget on acct of her hire 1:4 lent Mick 2:8 1/2 pd 4d for a snuff box
19 I had my Oats 5 barrills sent to ye mill ye men cutting Reed
20 att ye mill ye men drawing wheat and beans
21 I gave Den Lynes[6] my note for 3:6:8 Mr Quins parte of ye tythes to drying ye Oats 15d
22 I gott 5:5 from Den Bryen wch more than pas for his medow wth 2 days mowing to coperas 2
7br ye 23d I gott 13s from David Welch in full for 1/2 an acre of medow I pd 1:1:5 in full for pulling hemp to this day I gott my hay made into a large stack and a small one to salte and coperus 6 I gott ye 11:4 1/2 I gave Breen as Procters fees
24 I finished pulling all my hemp
25 to Church and dined att Mr Hunts to 7 1/2
26 I pd Moll Hannar 1 6 and 8 1/2 in full of her wayges and turned her away I sent 3 peck of wheat to ye mill
27th I sent Jon wth 3:18:6 1/2 to Tim Mahony wch wth seven pound I and 2 1/2 pd him ye of June makes 10:19:9 ye may gale Catty and I dined att Court back att night
28 I gott 13s from Jon he got for 1/2 an acre of Medow from Matthew Hanrahan I sent Jon wth a cask of butter to town wch weigh'd neat 1:1:9 att 31s a hondrd come to 2l:1s:2 1/2d he pd frank yeamons for 2 pd of powdr Sugar 1:1 one pd of jemeco 8d 2 quarters whiskey shrub 2:4 a glas tumbler 3 to 30 1/2 Iris Iron 4:11 to ye ballance of 6 teaspoons five shilling 1d to nails 4:2 salte 1:10 to snuff 2:2 a stone sope 4:0 cards 1:3 expences 1:1 in all 1:8:6 he gave me 12:2 to fisher for a dowsin sidr 3s:0d
29 I had a letter from Syp Pursill for 14s I sent him a four pound peice to take it out and desird he wod send me ye remaindr I gave Jon on acct of his wayges 8:8 Mr Jon Hunt and Jon Geery dined wth us
30 Syp Pursill sent me ye chaing of ye 4l peice

8br ye 1st 1748 att home I sent for Cattys Saddle I gott from Martin ffitzgerald 3:15:9 he gott from 2:1:22 of butter and he laid out 10:8 wch I must charge to his acct

---

6 Dennis Lyne[s] a farmer, leased the glebe at Croom for 31 years from May 1753. His will was proved in 1777. PRONI, Kenmare Estate papers D/4151/B/1/1 f.51; Clare.

October 1748

2 Catty and I dined att Hary Downs

3 I sent a bulluck and Cow to ye fare ye cow was sould for 2:14:6 he pd Cleary 4 for bringing a letter from Tim Mahony I pd me for a quartr of an acre of medow yt Torly Mc Mahon had 6:6 Jons expences att ye fare 8

4th Robt Peacock lay wth me

5 I borrowd 3l of Robt Peacock and he went away I went to Ballynort and pd 13:19:1 1/2 in full for May Gale and gott a Receipt in full and 2 wheels and a Reel so home to me 6d

6 went to [?]nong rove aboute ye sleater to little Porpose calld att new markett[7] aboute my Cortain and gon to me 6d to ye weavers 6 1/2 meet Mr Hartstonge honting dined wth him so home

7 and **8th** att home ye frise finishd to ye women 1s:3d

9th to Church dined att Court so back Ms Bork very ill I sent 2 men and 2 horses to Ed Mccan **10th** sent to see Ms Catty and I att Court

11 Sent to Ms who is better Sally Perfected her Indenturs and I stroled aboute

12 I wrote to Geo: Hewson[8] for Mault but did not get itt turnpike 1 1/2

13th sent Jon to town for mault he borrowd 9:6 from Ed Mccan and pd for half a barrill he took up a pd and half of hops and v a pd of snuff turnpike 1 1/2 ye men and horses came home

14 15 16 att home

17 ye men and horses boging hemp 18 att Dito

19 I gott 6:2:7 1/2 from Martin he gott for 3:3:23 of butter he laid but 6 1/2 I pd ye sheperd 2 2 I went to Croagh[9] spent 3 mogs sidr wch I owe for pd Janr th

20th I pd James Cusick 12 for a thousand of sleats I sent Mick to town and he pd Robt Peacock 3:3:3 I borrowd of him he pd Ned Mccan 9:6 Jon gott ye 13 pd Frank yeamons for a pound and half of hops 1:9 a quart whiske shrub 1:2 to 2 quarts whiskey 6 a nuttmeg 1 1/2 half a hondrd of nails 1d he pd 4d to Ned McCan for a letter and broak ye shrub ye fox killd 6 geese Cate hird att 4s a quartr

21 I pd gillbreath 14:2 for weaving 35 yards of Cortins

22 nothing Materiall

8br ye 23d 1748 I sent to Jon Rohan to Limerick for Judy and he brought her turnpike &c 10d

24th Mr Hartstonge came to hunt ye fox

25 I began to sow wheat to Roberts for mending my gon and in part for another 1:1

---

7 Newmarket, an early name for Pallaskenry where there was a bleaching green. Peacock may have been checking there on the bleaching of linen cloth for his curtains. See 17 August 1748. Lewis. 8 George Hewson of Attyflin died on 28 October 1749 aged 42 and was buried in Kilceedy. His wife Jane died on 26 February 1781. The Hewson seats were Attyflin, Patrickswell and Castle Hewson, Askeaton. *Memorials of the Dead*, vol. 7 (3), (1909) p. 6. 9 Croagh is 3 miles north of Rathkeale, on the road to Adare. It was of importance in the 11th and 12th century when it was incorporated. It had an abbey and two castles. Lewis.

26 Sent John Rohan wth Judy he pd for nails 1:4 to chequerd Linin 4:4 to [illegible]
27th att ye wheat I pd Mauris Welch 1:4 in full
28 att ye wheat
29 att ye wheat and finishd and sowed 10 bus
30 I pd Ed welch 2:8½ in part of his wayges Morse Bryen and David dined wth us
31 we made 36 dowsin Candills Mr Bastable calls to see us

9br 1st Catty and I att Court 2 att home
3 4th att Court 5 6 att home ye Child unwell
7 Continued unwell
8th Sent for Syp Pursill who wrote for things for him for wch I owe 2d /pd/ and 1:1½ for more to ye boy 6d he was bled and gott ease
9 10 11 12 13 14 15 growing better and pritty harty
16 I gave Martin a bill Pursill drue one me for 1:12:0
17 Smith ye sleater drue a bill on me for 1:2:0 he is to pay Martin for Supplying him wth lime and sand
18 Jon and Harragan Measurd ye ground he has from me it contains 1a:3r:9p
19 nothing materiall
20 me father in law came and pd me 7 ginies and a 40s peice wch makes 9:17:7 I pd ye Clark 10 for Christning
21 we all dined att Court back att night to rum 1:6
22 Tho Welldon and I went to Killtinan about Betty Courtship I gott from Martin he gott for 2c:2qr:13lb his own butter 3:13:2 for qr qe my own butter 16s he laid out for his own a/c 2s:2d he pd Davis[10] for ye things ye Child had ye 8th 2 Catty pd Margett on acct of her hire for spinning 2:8½ to Rum 2s Sugar 11
23 my father in law went away Cattle pd ye Pedlar on nels acct 1:4 wch wth 3½ she had befored makes 1:7½ I sent Jon to town he pd 6:15:8½ Quitt rent for ye lott of Killdeemo and 14:4½ Ye Quitt rent of Port and drivers ffes 4:6 to nails 2:6 Iron for his son David 1:2 oyle 3 turnpike 1½ I had 2 dowsin of Sidr from Thos Harragan wch I must allow him for in acct ing he wod not charge
24th I charg'd Thos Harragan for ye Rent of Caparow 3:5 0 ye Rint of Killmoreen 19:4d of Rent of his holding in Portacaca 0 15 1½ and for 3:1:0 medow 4:4:0 his creddit is Augt yr 5th to Cash pd 6:13:6½ 8br ye 5th 1:2:9 Reep hooks 1s:4d to 25d pd of hemp 7:4½ to James Harragan for medow 2:5:6 allowd Conr Harragan for 73 days work 1:10:5 allowd Dito for Jon Bryen for

---

10 Samuel Davis, apothecary, Shamble-Lane, Main-Street, Limerick. He advertised that he imported a 'large sortment' of garden seeds, with a great variety of the earliest Pease and Beans, some Flower Roots and Flower Seeds'. His wife was Elizabeth and they had four children at his death. Anne married John Anderson and Frances married another apothecary, George Henehy. He was the brother-in-law of Christopher Carr and Francis Sargent of Limerick. His will of 31 October 1776 was proved on 11 March 1778. His wife's will was proved on 23 June 1793. *Munster Journal*, 23 January 1755; NLI, GO 139; Wallis, p. 156.

frize 9s:0d to Cash pd 15:1:2½ I gave him a Receipt in full and Conner lent me a giney for wch I gave a Receipt or Cash note to Cottening my frize 7½
25 I pd Jon Rourk for Mr Thornell 1l:10:9 and in two Quittrent Receipts – 13:11:4 in all 29:2:1 to David Bryen on acct 6½ to pay for he making a Spead I wrote to Geo ffosbery for a barrill of mault wch I owe for
26 sent ye mault to ye mill Rourk went away I went to Killmoreen to see ye Roofe of ye house I pd James Harragan in Cash 5:6:7 in medows and oats 2:7:6 in all 7:14:1

9br ye 27th 1748 att home

28th to nell in full of a quarters wayges 3:10½ to old Margett in full 3:9½ to hanan in full for weaving canvis 1:6 60 Cloone in full for weaving frize 1:3 to ye spiner 1:1 for knitting 6½

29th I gott from Patt Daniel on acct of ye Rent of his holdings in Killmoreen 3:17:4 I gott 3 pound of hops from fisheier wch I owe for
[line deleted]
30 sent John to ye fare wth a bulluck wch he sould to ned ffitzGerald for 2:10:0 and only gott 7d pence

xbr ye 1st 1748 2d 3d 4th 5th nothing materiall
6 I gott from martin fitzGerald he gott for 2 2qr:1lb of butter 3:6:5½ he laid out for his own use 6½ ye I gott 3:13:0 on acct of his arrears then gave him a receipt for yt and 3:0:7 he gave me ye 5th of July last on acct of his arrear then I sent Jon wth 6 18 10 to Tim Mahony on acct of ye 9br gale
7th nothing materiall
8th I went to Limerick pd frank yeamons ye old acct I owed wch was 14:1d then I pd him for 2lb Reasins 1:8 1 oz nutmeg 9d ½ of pepper ½ stone blew 4 powdr blew 4 starch 4 Rice 4 Salte peter 1 in all 9:11 then I pd him for 4 yards Sarge 3:4 ½ a yard Buckram 3½ canvis 2d thread 4 buttons 6 mohair 2½ a pd powdr 1:6 shott 1:5 in all 17:8 then I pd him for wine 4:4 then we ballanced accts and he owes me 15:5 then I pd whistler 1:4 I owed then I pd for Manna oyle and Serces 7 then I pd Mick O Daniel 4:6 for Mick then I pd Mrs whiddon for tartpans and potting pans 4s to Rug 4 sarge 2:2 bottons 1d to Hary Evan for gyrths 3:9½ then I pd for gloves for Mick and Jon 2:2 Ribon 6 to ye cottening my frize 6s so to Ned Mcgans and drank some sidr wch I did not pay for so home
9 10 11 12 nothing materiall
13 I sett ye taylor to make my cloaths 13 att them
14 att them **15** finishd but I did not pay him
16 I gott my Sow killd
17 went to Court and dined there
18th Catty and I dined att Court and lay there
19 home and gott my wheat winowd
20 gott my Cow Killd

21 Sent Jon to town wth ye hide wch he sould for 8:7 and he pd Puttney 2:2 onions 5 Salte 1:11 ballsom 3 I wrote to frank yeamons for 2 gallons of whiskey 2 quarters of whiskey shrub two quarts Rum a cheese yt weighd 14lb att 2½ a pound and a stone Iron and a gross of corks all wch he sent but did not send ye acct

22d att home

23d att home

24th I pd Joan Lowry 1s and Anstice 1s in full I gave John Bryen two bus oats on acct of his wayges I sent 8 bus of wheat to Court wch wth 2 bus they had before makes 10 bushills Syp Pursill made ye lafe of a table to nails 2d

25th att home no Company 26 I gott 2:18:1½ from Martin he gott for 2c:9:22lb of butter I sent Mick to town for Robt Peacock turnpike 3d

27 28 29 Rob and wife here and Hary downs and wife here

30th they all went away I to Courte gave ye Servants as a Chrismas box 8:1½ so home

1749

Janr ye 1st 1748/9 Robt Peacock sent a corroll and bells[1] to Pryce to Rum 2:8 to Cotten 3:0 for ye horses 1s turnpike 3d

2d I acct wth Martin fitzGerald his debt as by acct was 36:13:11 his Creditt was 36:14:10½ then I accted wth James fitzGerald for building ye walls of ye house my debt was 18:13:6 and my Creditt was 16:9:8½ then I gave them a dinner and sent them away Brother [illegible] and dick came

3d I pd Will breen for Mr Buckner 1:9:2

4 we dined att Hary Downss back att night

5th Jon borrowd 8:1½ for me I went to Lisdougan and stayd there to ye **9th** so home expences 5 7 I pd for ye sidr I had att Croagh and a mug now 1:3

10 11 12 att home treated wth Martin about ye dear 9

13 Catty and I dined att Court back att night

14th att home agreed wth Martin [erased] on ye same lay only yt I am to allow him 1:13:0 on acct of his having his freedom amoungst ye cows

15 I swapt a strapper wth Patt daniel for an in calf heffer and 12:6 to allow him in his May Rent then I went to Harry Suppels wifes funerall aboute 3 back

16 I wrote to Mr Jackson for my Martingall and cavison yt Peter Hogan [?] but did not gett this to nails 4d turnpike 1½ I borrowd 2:8½ of Mick and pd ye taylor in full 3:3

17th strold aboute

18 to Courte mett all ye Quins back att night

19th att home

20th Brother Will came to bid for Lisdongan

21 Bill and I dallying

22d went wth him to Croagh but did no good he went home I came back

23d Sould ye big Sow to Martin for 16s he took her a way I bespoke a barrill of barley

24th Sent Jon for ye barley and to carry itt to Mick feighs and pd him one halfe barrill I owd him and to gett mault for ye other ye mault was not ready Jack Downs lay wth me

25 he went away I sent ye big pott to be mended wch was mended but do not know what ye mending will cost sett 600 of plants

26 27 28 nothing materiall

29 to Church dined att Court back att night

---

[1] A stick of coral mounted on silver with bells attached. Dating from the seventeenth century, it was given to teething babies for them to hold and chew.

30th I sent Jon for ye mault to feighs wch he gott and sent itt and a busl of wheat to ye mill I borrowd 5:5 of thos Harragan Hary Downs lay wth me
31 he went away

febr ye 1st 1748–9 I gave Thos Harragan a minitt for his holdings att Kilmoreen for 10 years from May next I had 2 sheep drounded I gott a pound and half of hops a sheet of pins and half a quair of peaper from Frank yeamons but did not pay for them to snuff 1:1 sould 2 sheep skins for 3s
2 Catty and I went to Court and lay there
3 came home gott water drawn
4 we brewd sent Mick to see Court family
5 6 I pd Pursill 1:6 in full of Christning money
7th I set some garlick
8th I gott 4 ginies from thos Harragan and gave him my note for yt and money he pd for laths and ye 81½ I got att Christmas and ye 55 I gott ye 1st inst in all 6:7:2
febr ye 8th [sic] 1748–9 I sent Jon wth ye remaindr of ye 7br gale to tim Mahony and got his receipt for 4:0:11 then I gave John 3s to pay for schallops and 2:8½ to buy Iron for Spades to Cate in full of her wayges 4:4
9 nothing materiall
10th I pd tim Herlihy 2:6 for ye Church rates of Killmoreen and bid him call to Thos Harragan for ye remaindr
11 I went to see Mr Hartstonge back att night I pd 2d for ye snuff I had from Pursill
12 Catty and I went to Church dined att court and lay there 13 we came home Jon went to town wth ye cow hide and some butter ye butter he did not sell he gott 5:5 for ye hide he pd 1:9 for Iron 1:1 for snuff 4d expences 14th I had a peice of beefe from Court to ye man 6d
15 16 17 nothing materiall
18 Jon went to town aboute ye butter but he did not sell itt he pd for sope 2s stone blew 4d tape 2d salte 2 sugar 1:8 a puke 2d allmanack 3d spur rowells 3d nails 9½d herrings 3
19 20 21 att home ye Plow going
22 to Courte back again
23 24 Catty and I att Court to ye 1:7½
25 we came home and I wrote by Smith for 6000 of 4d nails Jon gave a pd of snuff he took up in town yesterday
5 att home nothing materiall dined at home
6 7th nothing materiall
29th att home Syp: Pursill came to me and tould me Smith Run away

March ye 1st 1748/9 I sent John to town to enquire for Smith and he brought me word yt Smith was gone and took up ye nails wch comes to 12s he gave me

April 1749

an acct of ye butter it weighd 0:2:6 and sold it att ye rate of 1:4:0 it came to thirteen shillings and 11d he pd ye for ye snuff he took up and pd 1d for spickets and fossels turnpike 3d then I gave him 5:5 I gott from thos Harragan

2d Catty and I went to ye County cork where we stayd 6 then came home expences 6:2d

7th to court and back ye diner got ye 11d yt Caromady owed me and pd 1:0½ Countery Charges

8 9 10 att home ye Plow going

11 Jon went to Town wth a letter to Colonell Taylor pd 1:1 for Snuff 1½ turnpike and pd 1:1 to Harragan for mending ye pott wch I must charge to nells acct

12 att home Brotr Cooke dined wth us to nell on acct of her wayges 1:5

13 I went to Killmoreen gott a cow yt was hipshott killd and sould ye hide for 6:9 so home

14 15 sowing oats in ye Petato garding

16 sowing oats in ye lay corcas to hops 1

17 to Patricks Pott 4:9

18 Sowing oats and finishd ye lay Corcas

19 att home 20 sowing oats

21 to Shanongrove to ye Servants 3:9

22 23 24 Plowing and sowing Oats and finished all ye Oats aboute 8 acres and I took up a barrill of Oats from Thos Haragan

March ye 25th 1749 att home

26 Catty and I went to Church but were nott lett in so back well wet

27th Catty and I dined att Court back att night

28 to Killdeemo Vestory by disputes did no good but came home and dined on shouldr of vale I got from Court

29 ye Plow a plowing for flax seed

30 Sowed a Bushill pd for part of itt 1:1

31 nothing Materiall

Aprill ye first 1749 I had a vale killd

2d 3 4 5 6 a diging for Petatoes in ye orchard and set aboute 3 barrills

7 8 att ye Petatoes and finishd

9 to Church and to court and lay there

10 came home one of ye Cow att Killmoreen died

11 I sould ye hide for 8:8

12 ye men fencing and setting beans

13 I gott 6:1 from Thos Harragan and sent yt and ye 8:8 I got for ye hide by Jon to town and pd 12 for ye nails Smith took up and he brought me 5000 nails home and gave me 2:6 turnpike 1½

14th nothing materiall wheat to ye mills ½d

15th I sent Jon to Town for Plants he gave 8d for three hondrd and I wrote to Robt Peacock for 6:0:0 wch he sent to snuff 1:1 turnpike 1½

16 att home gave ye sleater 2000 nails
17 to Courte and back att night
18th I had a barrill and ½ mault home and some beans sett in ye flax gardin
19 Court stayd to ye 21 so home to ye maid 1:1
22 att home ye men diging gave ye sleater 3000 nails
23 wrote to court for 2 pantiles and got them
24 25 ye men Second diging Petatoes
26 lent Martin fitzGerald 1:2:9 he used for to make a Coffin for his mother 35½ feet of eight in board and some time ago ye same to make a Coffin for his father
27 ye men diging for Petatoes
28 I wrote for 3000 of sleats to Court and got them to nails 6s
29 ye men and horses drawing ye Sleats I gott but 2100
30th Hary Downs and his wife dined wth me to brandy 2s sugar 5d

May ye 1st 1749 I went to Court and pd Ball on acct of greasing 3 gineys a moydr and 13: and was allowd for wheat 2:3:4 in all 7:13:8 and 20s yt Law Daniel gave him then I went to Corroheen and 1 cow 2 two year old 2 mares and left them att Killmoreen and sent six yearlings there **2d** nothing materiall
3 I sent 2 cows to ye fare butt did not sell
4th I sent 2 cows and some yearlings to ye fare and sould a cow for 2:4:0 a yrling for 1:5:0
5th [erased] and got 4:11 more from Thos. Harragan
6th I made up 13:2:6 and caried it to Ballynort and pd it to Mrs Taylor so to ballyengland and Greags so home expences 6
7th I had 3 cows from Killmoreen to Church din'd att Court back att night
8th sent 2 men to tend ye Sleaters and Jon to look for laths he sould 2 yrlings for 2l and gave me ye money
9 May 1749 I accted wth thos Harragan for his holdings and took up all my notes and gave him a receipt in full sowed a barrill of duch barley and got my sheep washd to Joan for tabaco 6½
10th I finishd ye duch barley and gave ye sleaters 5 5 gott my sheep shorn pd 6½ to ye sheperd
11th my brother came to see me I had 400 and odd of laths from ye wood wch I owe for
12 my brotr went away I sent Jon to town he pd Puttney 1:1 I owed him and 1:1 for snuff pd Elenor fox 14:1 in full for her years wayges and Hired Joan Caromady
13 Catty and I went to Lisdougan where we stayd to ye **18th** in wch time Jon sould 1 two yr old heffer for 1:9:0 1 yrling heffer for 19s:9d and a bullshin yr for 1 6 3 he pd Thos Harragan two of ye 4 ginies he brrowd ye 5th and I gave him Creditt for ye other 2 in his acct expences whilst abroad 13:2
19 I pd George ffosbery for ye mault I had from him some time agoe 19s and pd ye widow covard 1s:7d I owed on one of ye boys acct

June 1749

20th I wrote to Geo: ffosbery for 5 barrills Oats and sent it to ye mill
21 to Church dined to Courte back att night 8½d
22 I began to break my hemp ye sick cow died I gott 6s for ye hide
23d att ye hemp christy brought home 52 yards I pd him 3:4 in part for weaving it
24 25 att ye hemp I got home ye male
26 had it sifted ye hemp breaking I got a Salmon from Ned flannagan to ye boy 6½
27: 28: 29 nothing materiall
30th I sould 3 sheep for 1:9:0 earnest 6½ I gott 5:15:8 from Martin he got for 3c:2qr:26lb of butter to Salte 1:9 snuff 1:1
31 nothing materiall

June ye 1st 1749 I went to Killmoreen walked aboute Sypryean Pursill measurd ye house, ye rooffe contains 19 square and half
2d I gott 19s:4 on acct of ye sheep pd Mick Nash in full 5:1 dischargd Mick Collins and orderd Jon to pay him his wayges in town he sould 1c:1qr:23lb of butter for 2l:5s:1d and pd Mick 1:12:9 to snuff 1:1 indego 5d coperas 2 an all 1d nails 2 turnpike 5 to ye Coopers 4d and gave me 10d I got 19:8 for 2 stone of woll
4th Church dined to Courte home att night pd Cloone 1:8 in full for weaving ye Canvis to Joan Caromady 6½ I pd Tom Coagh on acct 2:2
5 Mr and Mrs Henry Hartstonge breakfasted wth us so home I began to Cutt turff
6 att ye turff I gave Jon Bryen on acct of his wayges a bushill of beans
7th att ye turff to will welch a bus of beans
8 to Iron to make a Slane 10½d
9th I pd Geo ffosbery 2:10 for ye 5 barrills of oats I had ye 20th of last month I pd cristy 11s:19d in full for weaving I pd Den Bryen 6:6 in full for Countery Charges and gave Tho Harragan he gave me to pay his share of ye Parrish charges Patt Days heirs clamd 1:2½
10th to tin things 11d ye hemp and turff going on
11 att home
12 I pd James Nagle on acct of sleating 17s:4d I sent Jon to ye fare wth cattle and sould 2 cows for 4l:6s:0d one bull for 2l:2s 0 and pd Patt Daniell ye 4 ginies I brrowd of him ye 5th of may last
13th nothing materiall
1749 June ye 14th 1749 Catty and I went to court pd Hen: Ball 5 7 11 wch wth 8:13:8 pd ye 1st of May last make 13:1:7 and [?] in full for ye gras of ye Cattle I had att Corrohan last year I came home Catty stayd expences 1:7½
15 I went for Catty and brought her home to ye md 1:1 my own and 9 men from Ballylongford att ye turff
16 my own men at ye turff I brewd 2 bushills mault
17 att ye turff and finishd **18** att home killd a lamb

19 to Jon Bryen on acct of his wayges 5:5
20 to Court carried Mrs Hartston to Croagh from thence to Killinan[2] where we dined back to Courte and came home aboute 9
20th my men cutting turff for Jon Bryen to sope 4s:2d
21 my men cutting turff for David Bryen
22 ye men weeding
23 I sent Jon to town wth a Cask of butter it weighd 1:1:6 att 32s a hondrd comes to 2:1:7 he pd for an ax 2 6 snuff 1:1 streaner 3$^{1}/_{2}$ he took up from frank yeamons an ounce of nutmegs 9 pins 5$^{1}/_{2}$ Pepper 5$^{1}/_{2}$ stone blew 8 sugar 1:4 nails 4d Iron 5s:2d all wch comes to 9:2 of wch he pd him 1s 1$^{1}/_{2}$d due to yeamons 7:2$^{1}/_{2}$ Costom and turnpikes 3$^{1}/_{2}$
24 went to Killmoreen got from Martin he gott for 5:0:7 of butter att 32s a hondrd 8 2 0 so home
25th I pd James Michaell [erased] s1:10:6 wch wth 17:4 I pd James Neagle makes 2:7:10 and is in full for sleating 20$^{1}/_{2}$ squair of work
26 nothing materiall I got ye young black mare shod
27 my brother went away and I gave him ye gray mare and her fole and 2s
28th nothing materiall finishd ye futting my turff
29th prepareing to goe to ye County Cork
30th I gott to Lisdonogan by 10

July ye 1 and **2d** att Lisdongan
3 to Cahiramee[3] fare did no good expences and to Patt Daniell 17s to back to Lisdongan
4th came home expences every way 19s 0d
5 6 7 8 att home to buttens 3 knee buckles 8d snuff box 6 to ye pedlar Catty owed him for ye hankercheifs 1:4
9 Catty and I dined att Jack Downss to me 3d
10 11 12 and 13 att ye turff
14th Jon Rourk came and lay here
15 I pd him on Mrs Thornells letter 14:1:3 wch wth 2:8 he gott from my father in law make 34:1:3
16 he went away
17th I gott from Martin he gott for 4c:3qr:4lb butter 7:16:11 att 38s a hondrd to Iron 0:18:6
18th I gott from Thos Harragan 3:12:8 from James Harragan 1:2:9 Den Bryen Clonogollen 2 16 10 from Den Bryen Killbreedy 1:2:9 Patt Hanrahan 1 0 1 all wch is in acct of Medows
19 I sould a 2 yr old bulluck for 1:8:0

---

2 Killinan/Killelonehan/Killonahan is 8 miles from Limerick off the road to Croom. Lewis. 3 The great horse fair at Cahirmee, Co. Cork was held for over 200 years (or in one record since 'the memory of man runneth not [to the contrary]') in a field in Cahermee townland, two miles west of Doneraile. Originally it was held on 1 July every year, increasing to five after 1771. Cronin, pp 124–8.

August 1749

20th I made up 10:19:9 and sent it by John to Tim mahony and gott his receipt
21 I went to greages aboute ye cassanet and pd 6 ginies for itt and went wth it to Courte
22 23 and 24 att home Catty very unwell wth tooth ake and in rack all night
25th borrowd 1:16:4 of James Harragan and sent it by Jon to Lewellin on acct of my note to Mr Quin for ye tythes of ye year 1747 sent for to Court but Catty illness hinderd me
26 27 Catty still in racking pain to Hongary water $6^{1/2}$
July ye 28th 1749 Sent Jon to town wth butter he did not sell he took up from Yeamons sands tack $4^{1/2}$ nails 5 indigo $4^{1/2}$ tape 5 sugars 1:4 lisborn wine 1:4 all comes 50 4:4
29 att ye forge getting truckles shod to Ned Mccan 1:1
30 and 31 att ye forge and fins ye truckles

Augt ye 1st 1749 I gott $6^{1/2}$ and borrowd 1:9:1 and pd Lewellin ye Remainder of my note wch is 1:10:4 and pd fisheier in full 3:9 tim Callahan at ye truckles
2d I began to draw my turff
3 and 4th att ye turff 5 a wett day
6 John Geary and wife dined wth me
7 8 9 att ye turff
10 I agreed for my tythes for 3:14:4 and 5s for Davids gardin make in all 4:2:3
11 I sent Jon to town he sould ye butter for 2:1:3 itt weighd 1c:1qr:1rlb att 30s a hondrd he pd for snuff 2:2 to Brummell ye nailer[4] in full 10s nails 4 to other things 8 my Brother cooke came to see us
12 I pd Clampitt in full 3s to a quarter of Motton 1s:$4^{1/2}$d
13 brother Cooke went away I sent my note to Mr Buckner for 1:7:5 and a note to Lynes for 2:14:19 for this years tythes and pd 10 10 in full for last years Procters fees and pd 7:$6^{1/2}$ in part of this years Procters fees
14th and 15 Catty and I putting up ye bed I borrowd 2s:2d and pd 2s for 3 reep hookes one of them is for will welch
16 I dined to Courte back att night
17 aboute 6 I sent for Mrs Hartstonge and Mother Handy to come Catty who was taken ill aboute 2 and continued so all day aboute 7 Mrs Hartstonge went home ye men Reeping wheat and Oats
18 Catty ill all night and aboute 6 she was deliverd of a boy I gott from martin he got for 5c:1qr:22lb butter 8:3:4 he sould it att 30s a hondrd I gave Syp Pursill on acct 18:11$^{1/2}$
19 I sent Jon to town he pd John Bonfield in full of all accts 3:19:1 to tea 4s:9 sugar 4s:9 pins 5 Ric 3 4d Rum 1:4 peaper 10d fustian 2:2 snuff 1:1 hops 2:3 he pd ye 2:2 ye borrowd ye 15 I pd 7:7 for bleaching 52 yards of cloath I pd 1:7:1 in Part of ye 1:9:1 I borrowd ye 1st inst

---

4 In 1766, Benjamin Bromwell was a smith at John's Gate, Limerick. Ferrar.

20 I sent thos Coagh to Lisdowgan to him 1:1
21 he came back Ned Mc lay wth me to him 1:1 ye men Reeping wheat to ye taylor for making filims cloaths
22 ye men Reeping wheat and finishd and began to bind Oats 2 women wth them
23 ye men reeping oats
24 drawing turff 25 finishd ye turff
26 Jon Bryen thos Coagh gott 9:5$^{1/2}$ from thos. Gibons in full for hay to Coagh in change for manna and Serus 9$^{1/2}$
27 my Sistr Ireland came to us and stayd to ye 29 I gott 4:4 from will Hickey on acct of medows ye men and horses wth David Bryen
30 and 31 ye men att ye hay and reeping

Sept ye 1st 1749 ye men att ye hay
2 a wett day to Serus 3 Manna 8 tar 5$^{1/2}$ I sould Thos Coagh 4 stone toe for 4:4 to be chargd to his acct David Bryen 4 stone pd in part to Joan Lowry 2 stone anst fitzGerald 2 stone will welch 1 stone Will Hickey 2 stone Dennis Bryen 2 stone thos Collins 2 stone
3 att home
4 to Court and sould Mr Hartstonge ye young black mare to 6 ginies he lent her to George Quin to go to ye Corroh[5]
1749 7br ye 5 I went to Frank Cookes
6 I came home expences 5s to will welch 1:1
7 att home 8 making up accts for Mrs Harts:
8 and 9 att home 10 att home
11 to court in ye aftrnoon back att night
12 to ye county Cork 13 there 14 I came home expences 3:9$^{1/2}$
15 Jon gave me acct yt he sould 1 1 4 of butter for 1:17:3 he pd for stuff 8:9 exchange of ye soope dish 1:2 two dale boards 2:11 for 3 yards chequerd linin 3:7$^{1/2}$ snuff 11 chalk and thred 1s 100 nails 3$^{1/2}$ 2 quarts whiskey 1:6 a quart Lisburn 1:4 a quart clarott 1:4 w quarts shrub 3:3 powdr sugar 2:4$^{1/2}$ jameco 1:4 a pd of resins 6 tape 2 Catty gave allice lowry 2:2 Jon sould a cow for 2:0:0 he recd in part 19:7$^{1/2}$ he recd from Jon Quain 2:8$^{1/2}$ from Jon Carter 1 1 he gave Joan on acct of her wayges 1:8 for two milk gallons 1:7$^{1/2}$
16 a wett day I gott 1:2:9 from Jon quain pd for weaving 54 bandills flaning 2 3 I gave Jon 4:4 to pay for making ye mault ye hemp was finishd
17th aboute ye house I sent to town aboute ye hemp but did no good to turnpike 1$^{1/2}$ I pd hearth money for me and Jon 6s I gott 1:2:9 and 5 5 from James Harragon wch wth 9:7 due for Petatoes and 5s for frize makes 2:2:9 and ye money he lent me wch was 1:16:4 ye money I gott ye 25 of July last make in all 3:19:1 wch was for his and Jon Newmans medows I got 1:98:11 from thos. Harragan

---

5 Possibly the races at the Curragh, Co. Kildare.

wch wth 3 12 8 I got ye 18th of July last makes 5:1:7 wch more than pays for his medows by 5s d I gott 1:2:9 from Den Bryen Killbreedy ye men att ye hay

18th ye men att ye hay and oats 19 att Dito

20th att Dito Catty unwell

21 sent for Syp Purrsill to draw her tooth she wod not let it be drawn but her gum lanced and bled in ye arm I borrowd 4 2 8 of thos Harragan and gave him a note for yt and ye 5s 7d he owr pd of ye medows I borrowd 1l 18s 10d from James Harragan ye men and women att ye hay and Corn

22 Sent Jon to town wth ye 6l I borrowd of Robt Peacock ye 15 of Aprill last wch he pd and he d for snuff 1:1 to a creadle 2s Silk 1d turnpike 1½ I gott 14:4½ from Jon Carter in full for his medows

22 I gott 3:3 from Den Bryen Killbreed in full for his medows

23d and 24th att ye hay gott 11:11 from Mat Hanrohan

25 drawing corn from Den Bryen Chanagillen[5a] in full for medows 1:11:1½

26 att ye Corn and finishd ye drawing

27th att ye hay 2 men and my own I gott 15s:9d from thos collins on acct of medows

28 3 horses 2 men and my own drawing in hay to ned Mccan 6½ to pay a man for carrying [illegible] to dick fitzGeralds

29th I gott 9 3 from Jon Nash and allows 1:2:9 for a rode to ye bog wch makes 12 0 yt price of an acre of medow

30 2 men and my own at ye hay

8br ye 1st 1749 a reading and walking

2 I got what hemp had trashed and bog'd

3 I went to Lisdowgan 4 I came home to expences 4:1

5 I had George ned to beefe 2:8½ Shrub 1:6 whisky 1:6 a tumbler 4 to ye boy &c 1:4d to bees 8:1½ I gott from Martin 7 17 :7½ he gott for 5:1:20 to ye midwife 11:4½

8br ye 6th 1749 I went wth my father in Law to Court back att night to my Servts 2:8½

7th I gott 1:7½ for 1½ stone of toe [erased illegible]

8th to Lisdowgan wth Catty and mett Jon Bryen comeing from cork he sould ye hemp for 28s a hondrd it weighd neat 11:0:22 ye money came to 15:13:6 he pd ye carrier 1:2:6 his expences 3:1 to ye hacklors 8:1½ in my absence Jon pd Colonell taylor in full for May rent 14:15:9 I pd Jon Longan in full of my note for Lynes parte of last years tythes 3l:9s:8d I stayd att Lisdowgan to ye 15th so home expences 8:1½ to Catty 5s:2d I gave him ye 1 18 10 I borrowd of James Harragan ye 21 of last month

17th I gott 2:8½ from Den Maddin for toe and gave him another Stone

---

5a Chanagillen, probably Shanagolden, Co. Limerick.

18 Catty and I to Courte

19th home to ye Servants 2:2 to ye tinker: 9d I pd Charles Slattery 3:6:0 in full for dressing my hemp

21 ye men diging Petatoes

22 att home

23 ye men putting in Oats

24 I pd Den Heas 1:8 for 2 days mowing

25 26 27 a trashing oats and send 4 bus to mill

28 29 30 att home 31 nothing materiall

9br 1st 1749 I pd for 21 pound of flax 6:s to snuff 6½ to oyle 5½

2 3 att home

4th Catty and I went to Court I came back att night I sent Jon to town wth butter flaxseed and bandlecloath he sould 1:1:7 of butter att 26s:6 a hondrd Comes to 1:14:9 flax seed 3:8½ ye bandlecloath he lost he pd a stone stope 4:4 salte 1:7 snuff 1:1 stone and powdr blew 1 podr 1 6 oyle 1:½ mending my rasiers [illegible] nails 4 thread ½ expences 5d I gave thos. Coagh on acct of his wayges 2:8½ and gave Jon 2:10 and orderd him to go in ye morning to look for ye bandle cloath I pd Margrett Ceasy 10½ for spining frize and then finished it to Court and all ye family went to Dublin Catty and I dined att Mr Melsops so home to ye servts att Court 1:7½ Jon did not get ye bandle cloath expences aboute it

6th I sent to ye mill to have 3 barrills Oats ground and had my frize warped ye oats came home and was sifted

7th I went to Killtinane to see George Quin back att night dung wett

8 9 10 att home ye Plow going wheat trashing I gott 6 ginies 3:4 and 2:8½ and 1½ ye pryce of 5:1:16 butter he sould ye 4th inst

11 sowing wheat and prepearing to go wth Catty to Lisdowgan in ye morning I carry wth me 8s:8d 3s:4d gras 6½ and 11:4½ of wch I gave ye spiner women 1:1½

12 wee left home att 6 and got there by 4 were I stayd to ye 15 so home and left Catty there I gave her 4:5 pd brotr 2:2 expences 5s

16 I gott from thos Harragan by his wife 17l:0:0 and from Martin on acct of horn money 6:9:4 and made up 33:6:1 and pd it to Jon Rourk on acct Law Danel put a cow to Killmoreen to greas

17 18 nothing materiall ye frize came home

19 pd ye weaver 5:5 in full for ye bandle Cloath and frize I finishd sowing wheat

20 21 nothing materiall

22 I pd ye quitt rent drivers fees 2:8½ and had ye men thatching ye corn

23 ye men att ye hay

24th 9br 1749 ye men thatching ye hay I pd Joan Caromady 6:7½ in full of her wayges

25 I went to Lisdowgan to see Catty where I stayd to ye 27th so home expences 2:2

28 ye men att ye Petatoes to needills and an allmanack 3½

December 1749

29 ye men att ye Petatoes to snuff 1d ye hackler att work and finishd I pd him 1:3 in full for hackling a stone and half of flax
30th ye men att ye Petatoes

xbr 1 1749 ye men att ye Petatoes
2 I went to Lisdougan 3d there ye 4th Catty and I came home expences 5:8
5 to Margett fitzGerald as earnest [illegible] to Jon Rourk for &c 1:1 ye men att ye Petatoes all ye while
6 I sent Jon to Limerick wth my frize to be cotten'd to snuff 1:1 turnpike and Jon 4 att ye Petatoes
7th att ye Petatoes
8th att a pertition a wett day I gott 1 0 5½ from welch ye butcher in full for a cow sould him ye 15th of 6br last I gott from Martin he gott for 2:3:2 of butter 3:14:8
9 I pd Geo ffosbery 9s for half a barrill mault ye men at ye Petatoes
10 att home reading to Marget 1:6 as wayges
11 I sent Jon to ye fare wth a cow but he did not sell I borrowd 4 ginies from Patt Daniel I pd Nick Sheas boy for ye use of James Hewson and by his letter 11:11 in full for ye Service of my two mares
12 ye men att ye Petatoes I borrowd 1:18:10 from James Harragan
13 I wrote to Jon Bonfield to town but did not get them Jon sould 0 3:3 my own butter for 1:0:11 I pd for Iron 2 10 hops 3:3 ye cottening my frize 2:5 a letter 4 snuff 1:1 costom and turnpike 4½ he sould 10 torkeys for 8:4
14th I sent Jon to tim Mahony wth 10:2:4 on acct of 9br Rent Robt Peacock pd ~~14:8~~ for ye years Quitt rent 9br 1750 pd
15th I sent Jon to town for brming wch I bought of will Burk and came to 1:0:4½ of wch he pd 16:6 turnpike &c. 6d
16 Nothing materiall
17 att home reading
18 ye taylor att work ye men att ye turff
19 ye taylor finishd a Coat and weast Coat for me but did not pay him
20 I pd Margett Harrolld 10 in full for her son and gave 6½ earnest to a spiner girle I gott 19:6 from Thos. Harragan on acct of wch I gave Jon Bryen on acct of his wayges half a barrill oatts and 1:1 and gott ready to goe to Lisdougan in ye morning

1750

21st [December] we gott to Lisdougan where we stayd to ye 5th of Janr 1749/50 expences every way whilst abroad 12s so home

6 and 7 nothing materiall

8th I had my cow and pig killd and went to town to settle wth Studdert wch in some measure did and sent Martin to treat wth ye Quittrent

Janr. Ye 8th 1749/50 Continued drivers wch cost me 11:5^{1}/$_{2}$ I stayd in town yt night

9th Jon came to town wth ye hide wch he sould for 12 10 he pd for rum 2:8 whiske 1:6 sugar 1:3 vinager 2^{1}/$_{2}$ saltepeter 6 [illegible] bushill salte 3:3^{1}/$_{2}$ costom and turnpike 3 I pd for a handkercheif for Horse 3:3 to will Burk in full of acct 3:10^{1}/$_{2}$ lent Ned Mcgan 1:8:2 to me 5:3 so home

10 I had my cow and pigg cutt up and salted and paid James Cloone 4:4 in full for weaving Will Harding lay wth me

11 he went away I pd Patt Caromady 3 4 in full for his hire Ed Mcgan and wife lay wth me to him 1:1

12 I had a stack of wheat putt in Ned and wife went away

13th sent Jon Rohan to Lisdowgan to see how my Mother to him as expences 1:1

14 he came home and brought word yt she departed this life aboute 8 a clock last night

15th I went to ye funerall got there by 10 she was buried aboute 3 I stayd att Lisdowgan till ye **18th** so home and found George in ye Measells expences whilst abroad

19 ye Plow going for barley ye wth Plowing for Petatoes

21 att home and Pd Patt Day 2:8^{1}/$_{2}$ for making my Coat and weast coat

22 ye men trashing Catty wrote to Ms Hartstonge I accted Martin fitzGerald and he owed me on ballance 1:19:8 to ye Clark as ning money 10

23 Sistr Ireland went home

24th I borrowd 5:0:6 from thos Harragan wch 19s6 I gott ye 20th of xbr makes 6l and gave him my note for itt to schollops 12d

25th ye men trashing wheat

26 I borrowd 4 ginies of Patt Daniel and gott ye 1:19:8 Martin owed on ballance

27 I gave John 11:12:11^{1}/$_{2}$ and wrote to ffox for to give him 2:0:0 wch he gott and pd ye years Quittrent for ye Lott of Killdeemo to snuff 1:1 to thred and Jon 6 Salte 2d to ye Clark 10d

28 Mrs Mellsop and Ball dined wth us
29 30 31 ye men trashing wheat

1749/50 febr 1st nothing materiall
2d I had a letter from my sistr to go to Killtenan
3 I went to Killtinan had some talk wth Mrs Quin aboute Betty Ireland and Studders so home to sugar 10 to me 6
4th Mrs Mellsop went home
5 I wrote to rourk and sent Jon Rohan to Lisdowgan for ye things Catty mother left yer to him 1:1 serverall Peopple a taking Killmoreen
6 thos. Harragan agreed for Killmoreen for 21 years att 11s an Acre ye first year and 11:6 an acre every year aftr he to keep up ye house and improvments Robt and Hew Cox dined wth us
7th Jon Rohan came home no letter from Jon to Countery Charges 3:3
8th I gott my barley first winow'd and gave fishier two bushills wheat wch he owes for
9 I had ye barley piled and got ye bay mare shod
10 I went to Limerick and Recd for o 3:2 of butter my own 19:10$^{1/2}$ for 20 Stone of wheat att 11d a Stone 18:4 then pd Matt fox 1:2:9 of ye 40s I borrowd of him ye 27 of last month
1749/50 febr ye 10th Continued then I spoke aboute a husband for Betty Ireland then I pd for snuff 11 black thread 3 large 3 silk 5d 13$^{1/2}$ yard Poplin 16:10 Cadis 3 linin 1:1 Ribon 1 8 mode and love[1] 2 10 for Salte 1:8 hops 2:2 to ye Childern 1:1 to me 2:8$^{1/2}$ so home and gave Margett Ceasy 2:8$^{1/2}$ in full for her quarters hire
11 att home unwell **12** I began to make a dich to yet new Petato gardin still unwell **13th** unwell
14th I wrote to Ned Mccan Sistr Ireland and Studdert to pepper 5 Ginger 2 licorish 2 reasins 1$^{1/2}$
15 to Thos Coagh 2:2 I went to Killtinan aboute betty Irelands affair to me 3
16 ye men slitting ye Petatoe ground I had a letter from Rowrk
17 I went to Limerick aboute Betty Irelands mach and gave my note in favour of Mrs Quin for 20l payable in ten days I did not sell my wheat expences to me and men 1:10 so home
18 I sould Paddy Donrel a strapper for 2l in parte of ye money I owe him to eale fry 3d
19 I had a taylor making my Suttoote I had my Pig killd sent ye men to Killmoreen aboute ye house Jon sould my filly for 6:14 recd earnest 11:4$^{1/2}$
20th I pd hearth mony for me 4s for Jon 2l pd ye taylor in full for making my surtoote 2:2
21 I sent Jon to town to try to sell ye wheat he sould itt for 3:3:1$^{1/2}$ he pd Matty fox in full of all accts 17:3 to Plants 4$^{1/2}$ a crock 5 sarge 3 expences 2:4

1 Possibly *Love à la mode; or the Amours of Florella and Phillis* (London 1732).

March 1750

22 to a bushill beans 4:2
23 to thos Coagh in full of his wayges 14s:2d I went to Killtinan and gott 20l to give Betty Ireland alias Attkins I dined there so home and in my way home pd ye Millor of Cloonshere[1a] 1:1 in full of all acts
24th I sent John Rohan wth ye 20l to Limerick and gott Betty Attkins and my sistrs receipt for itt and my note I gave for itt to turnpik 1½ tar 7d I pd Pursill in full for Christning mony for George 1:6 6 to snuff 1:1
25 Catty and I went to Greags where we lay
26 home to ye maid 6½ I gott 4:12:9 from Martin on acct of ye filly to eelefry 3 I had 2 cow skined and sould ye hide for 14s expences 6
27th I gave Patt Daniel 4:12:7 on acct and Paid Mr ffosbery 9s for half a barrill mault I sould Patt Daniell 6 hogetts for 1:14:6
28 29 nothing materiall

March ye 1st 1749/50 I gott some oats putt in
2 3 4 trashing 5th I began to Plow 6 att ye Plow
7th 8 att ye Plow and sowed ye west garding
9th I gave Rourkes man 6½ to will welch on acct of hire 5:5 to John Bryen on acct 1:1 and 1½ I borrowd hackles from Tim Mahony and began to hackle
10 11 nothing materiall
12 to a vestory att Killdeemo did no good but I spent 2s so home
13 hacklling and finishd
14 I went wth Ball to vew Corrohan wood
15 I went to Killmoreen and sould 26 cows and a bull for 75l and only recd 1:17:8 of wch I gave Sypr. Pursill 1:2:9 to Patt Daniell 2:8½ and spent on ye byers 7s½ so home late
March ye 16th 1748/9 I gave Jon on acct of his wayges and for to buy flax seed 2:2 and wrote to Mr Hartstong I sould Martin fitzGerald 6 cows for 19l earnest 1:1
17 to Mauris Geran 4s in full for laths to ye Servants 2:8½ to drink Catty and I alone
18 att home Recd from Mau Gibon for a cow 3:4:0
19 I sent Jon rohan to town he pd for 3½ powdr sugar 2:4½ tea 4:7 nutmegs 9d pins 7 seeds 1:10 more sugar 1:9 Iron 4:4 schrees 8 snuff 11 thred 3 he pd Jon Griffiths in full for church rates 6 1½ to him and turnpike 4½
20 to greags and back att night
21 to Mick Nash in full for his hire of 13 days att 5d a day 5:5 I gott 3:4 from welch on acct of ye cow hide I sould him and 1d earnest for a cow I sould him for 2:15:0
22 I bought a cow from Welch for 3l:3s and pd him by ye 2 15 0 he was to give me and allowd him 8s of ye Pryce of ye hides I got my 2 mares shod
23d I wrote to Lisdowgan and to rourk

[1a] Clonshire, Co. Limerick.

24th I sent Jon Rohan to Lisdowgan and gave him half a giney to pay for ye barley and 1:1 for his expences pd for 2 sives 10d
25 March 1750 att home alone
26 ye men trashing
27th I went to killmoreen and gave him ye Possession and sould him a cow for 2:14:0 so home and swapd a cow wth Martin and am to allow him 20s as boote he bought 24 ceelers for 1:8:0 I pd Jon Griffin 1s yt remaindr of ye church rates
28 I finishd ye sowing my oats and gott from Jon Harney 1l1:10s :0 in full for 5 cow sould him ye 15th inst I pd 2:8$^{1/2}$ to a sleater for mending ye house att Killmoreen and I gave Harry downs 7 bushills Oats
29 my Plow and horses wth Jon Bryen and I gave Pattr Daniel 2:5:6 wch more than pays him by 1:10:7
30 and 31 nothing materiall

Aprill 1st 1750 att home reading
2 3 4 5 att ye Plow
6th I sent Jon Bryen to town and gave him 2 ginies one to pay for Caleminco ye other to give frank Yeamons he brought 8 yards Caleminco wch cost 10s and 20 yds mohair 2:4 wch I owe for as ye ginie wod not Pass ye other ginie he gave yeamons out of wch he gott snuff 1:1 a hondrd of nails 10d 200 of nails 5d collyflower Seeds 2 hops 2:2 vinager 3 shrub 1:8$^{1/2}$ fish 1:7 in all 8:2$^{1/2}$ out of 1:2:9 ye remnaindr is 14:2 wch I am to get credit for so home ye men and horses putting out dung
7th ye men puting out dung
8 I pd 4:4 for maulting 2 barrills of barley
9 att ye dung **10th** att ye dito I pd 1:1 for Quilting a coat for Catty **11 12 13** att Dito **14th** att ye barley and fins
15th att home **16** att home and gott 1l1:11 from Martin fitzGerald on acct
17 Catty and I dined att Killtinan to ye boy 1:1
18 I gott 1:14:1$^{1/2}$ from Thos Gibon on acct of his note for a cow
19 went to Limerick took up Cloath and triming for a coat to a wigg 6:6 to a hatt 6 to Sundrys to Frank Yeamons 2:10 Sope 4:0 a grater 4 cowcumber seeds 2 a file 3 spurs 10d Scheals 6:8 to me 8s:3 then I pd Alderman Mansell[1b] 13:10:1$^{1/2}$ for ye Rent of Porte ye 1s^t of 9br so home
20th aprill 1750 I went to Jon Hartnies to try if he wod take back ye light ginies he gave me wch he refused[2] I returned Martin fitzGerald one of ye ginies he gave me ye 9th inst. And returned Thos. Gibon 11:4$^{1/2}$ he gave me y 18th
21 I gott 12:3:5 from James Harragan on acct I paid Will Bork in full for Caleminco 12:9

1b Probably Richard Mansell, 4th son of Col. Thomas Maunsell and Anne Eaton. He was mayor of Limerick in 1734 and MP for Limerick City from 1740 to 1761. He married Margaret Twigg and they had two children. Maunsell, p. 46; Limerick papers NLI PC 875/2/15/58. 2 Light guineas – money where the weight of the gold coin did not match its face value – were liable to be seized and held by collectors of customs, who reimbursed the value of the gold. Minutes of the Irish Board of Customs and Excise, 26 September 1750, 10 October 1750, CUST 1/49/53 and 62.

May 1750 213

22d att home
23 I went to town paid for timber &c 3:10:7$^{1}/_{2}$ for a grind Stone 2:2 for a spoke sheave 1:8 for a knife 6$^{1}/_{2}$ for a bras inkhorn 1:1 snuff 2:2 then I lent Ned Mccan five ginies to me 7:9 so home
24 I pd ye taylor in full for making my Coat 3:9$^{1}/_{2}$
25th I sent y bushill barley to ye mill
26 27 and 28 att ye Petatoes and trinched ye new ones to Will welch on acct 2:8$^{1}/_{2}$
29 I sould Thos. Harragan 4 lambs for 10 10 wch he pd I gave Jon Cristy 9 pound of 4 dowsin yarn
30 I delivrd Harragan his lambs and sent six of my own lambs to Caparow to be weaned

May 1st 1750 nothing materiall
2 no men att work gott ye bay mare and bay garron shod to ye smith for oars 2:8$^{1}/_{2}$
3 I sould 2 cows for 4:11 I pd for hoops 3:3 for Pigins 1 expences 1:3 so home
4 I gott my sheep shorn pd ye man 10d I gave Henry Downs 3 barrills and a bus oats wch wth 7 bus he had before makes 4 barrills
8 Catty and I went to Lisdougan where we stayed to ye 10 so home to ye maid and boy 1:7$^{1}/_{2}$ expences 3:8$^{1}/_{2}$ otherwise 9:9 bread 4 tabaco 3
11 att home sent 4 bus mault to ye mill and 3 pecks wheat
12 I sent Jon Bryen to town he pd for hops 12:2 starch 2 rum 1:4 shrub 1:6 Salte 1:8 pypes 3 300$^{1}/_{2}$ of brads 2:11 1000 nails 1:8 60 nails 5 tar and blader 5 hob nails 3 caleminco 7$^{1}/_{2}$ tapes serus 1 he brought change 1 5:1$^{1}/_{2}$ to him and turnpike 3$^{1}/_{2}$
13 to Pegin full for her wayges 6 6
14 to Pat Hannan in full 1 1 wm Rourk came
15 I pd Hew Russill 11:4$^{1}/_{2}$ yt Jon Bryen borrowd of him ye 3d of May for his own use to be charg'd to his acct I got 11 12 10 from Martin 40:0:3 from Thos Harragan and 9:8:11 from James Harragan
16 I made up 63 11 8 and gave it and a Quitt rent receipt to John Rourk for Mrs Thornell ye whole makes up 77:2:0$^{1}/_{2}$ then bought 2 heffers from Rourk for 4:11:0 and gave hm 4 light ginies if he can pas them he is pd if not he is to return them I gave Catty 5:5 to pay Moll
17th nothing materiall
18 I went to Ballyengland for hair but got none from thence to Jack Morpleys where I got itt from thence to Rathkeal and got from fisheer 6:11 wch wth 6$^{1}/_{2}$ I owed him and 1:2 reconing make up ye Pryce of 2 bushill of wheat he owed me expences 1:6$^{1}/_{2}$ so home
19 I sould 4 sheep for 2:4:0 got 5:5 earnest
20 to Church and back to me wth Kitt 1:1 I gave old david 1:1 to buy broges
21 Syp Pursill his man and ye meason att work Catty hyred Joan Dondon att 6s a quarter to her as earnest 1:1 I gott 200 of bricks from Geo ffosbery

22 ye joyners and meason att work

23 lowry took one of ye sheep he pd 5:7 wch wth 5:5 he pd earliest make 1:1 ye Pryce of one of ye sheep

May 23 1750 continued Syp Pursill and men went away ye man has 8 days he two and ye meason 2 days I pd Ivers[3] 5 5 for a new light and repairing ye 2 old ones I gave John Bryen and his son 7d and sent him to town wth butter I gave ye meason a stone wolle and will welch a stone of wolle Jon Paid for joney 1 $1/2$ wax $1/2$ 6d ram vardigrace 3 turpintine [illegible] comondale boards and sawing 9:6 to 8 tyles 8d to pepper 5 $1/2$ to a quart vinegar 6 shrub 1:6 sugar 1:4 Iron 5:10 $1/2$ indigo 5 1000 nails 20s snuff 1:1 expences &c 8 $1/2$ ye butter weighd 1:1:10 att 30s a hondrd 2:2:6

24 and 25 nothing materiall

26 ye stones breaking I gave Syp Pursill half a barrill oats I gott 11s from Lowney for a sheep

27 I agreed for cutting 200 turff for 1 16 0 to them as earnest 6 $1/2$

28 I gave Thos. Harragan ye 4 bushills oats I borrowd last Spring ye meason mending ye kill he had 3 days

29 I went to Court to me 6

30 I gave Jon Bryen a barrill oats on acct of his wayges I had turff drawn and ye kill set on fier I returned David Welchs cow and am to allow him 6:6 for her milk

31 I gave Patt Hannon 4 bus Oats and to will welch 4 bus more ye turff drawing and ye kill burning

June ye 1st 1750 ye kill filld and finishd ye turff I gave Martin a stone wolle

2 from Tim Lowry 11:11

3 from Dito a quartr Motton weighed 18l att 1 $1/2$ a pd ye pluck 4d in all 2:7

4 to Thos Coagh an acct 6 6 I Recd from Thos Gibon in full 1:8:7 Recd for a bull 2:8:10 to Jon Rohan on acct 1:1

5 to mill 4 barrills Oats to David Bryen 4 bus oats

7th ye meason at work ye male came home

8 to Syp Pursill 4 bus: oats I gott a salmon from Ed: flannagan to ye boy 6 $1/2$ I gott from Tim Lowry in full 7:2 ye men drawing sand

9 ye men breaking stones to Syp Pursill 4 bus Oats

10 ye meason att work to Cha 1:1 Jon pd for 8 dale board 12 4 glew 6d nails 1:1

12 ye meason att work 13 ye meason att work

14 ye joyner att work

15 2 joyners at work I gave David Bryen four barrills oats for Mr Hartstonge

16 3 joyners att work

17 to Buckners Procter 1 13 4 and took up my note

---

3 A member of the Ievers family of Castle Ivers, Co. Limerick.

July 1750

18 ye joyner and Meason at work to ye later in acct 9:9 **19th** ye joyner and Meason at work Syp Pursill and I went to town and pd for boards 12:1 to nails 3:9½ 7 pr of Hinges 4 8 to a plain Iron 5 a trowell 1:1 a crock 6½ a coamb 1½ to ye widow Gardiner for dyeing threads 1:6 60 a skin 2:8½ to Cambrick 8:0 to Poplin 15:4 sugar 3:7 tapes 6 Rum 2:6 glew 6 [illegible] to me every way 4:4 I borrowd 2:8 of Ned McCan so home

20 Syp and meason at work

21 both at work 22 ye meason att work Jack Pursill brought 8 Melon to him 1:1

22 ye meason at work ye 2 joyners came aboute 12 to nails 4

23 2 joyners att work 25 1 joyner

26 will came aboute 3 I got 1:2:0 from T Harr

27 2 joyners att work

28 2 joyners at work to nails 3:5 locks 1:10½

29 1 joyner to hops 11d

30 1 joyner I gott 1:11:1½ from David welch and allow'd him for a calf 4s and for James fitzGerald 6

July 1st 1750 att Mick Hartneys funerall

2d 2 joyner **3d** 2 joyner **4** 1 joyner Catty and I att Thos Coagh weding expences 3:9½

5 two joyners I pd them 2:0:7 wch wth 18s for a barrill and half of Oats makes 2 18 7 and is in full for theyr hire for 39 days

6 I brought home a hogshed of Sidr for wch I owe 1:10:10

7th I gett 2:16:10½ from Martin in acct July 3d I allowd Thos Harragan 1:3:4½ on John Bryens acct and allowed James Harragan on John Bryens acct 1 8 11½ allowd Tom for bread and wine 1s:0 for hinges 2:5 wch wth 6:19:1½ I Recd in cash makes 10 4 10½ and is in full of all accts to ye 1s of May last then I gott 14:11½ from James Harragan in full

8 to Harry Downs and back

9 I Paid Thos Welch for Mr Bryen 10 19 9 of wch I gott 10:10 from Thos Harragan I agreed for this years tythes and gave my note to Mr Quin for 2 14 0 and to Buckner for 1:1:0 ye meason att work 10 att work

11 12 13 14 ye men att ye gardin

15 I gott 11 2 from Ed Nash in acct

16 Ned flannag and wife dined wth us I Paid will frost 4:1 in full for Countery Charges 2 mowers att work

17 2 mowers

18 I sent John to town wth wheat it weighd 14 stone att 11d a stone wch come to 12:10 he pd for snuff 1:1 sand 1 Cadis 1:6 hinges 1:6 a lock 8 a plain iron 5 a board 1:4½ Costom and turnpike 4½

19 2 mowers I pd them 5s in full I gave Jon 1:1 and orderd him to go to town wth butter wch he sould for 5:0:7 itt weighed 2:2:2 he pd frank Yeamons 5:3

for ye things thos Coagh took up for me he pd Ned McCan 3:4 yt I borrowd of him and he pd for Nails yt Coagh brought for ye exchange of six spoones 4 and he pd Catty 2:12:9

20 I went to Lisdowgan 21 there

22d home expences 4:10½

23d nothing materiall 24 nothing materiall

25 Syp Pursill here

26 he and man att work I pd Mr ffosbery for a barrill mault 1:0:0 to nails 4d

27 Syp att work to Norse 2:8½

28 Syp finishd

29 to Church dined att Mr Hunts

30 I gott 1:16:4 from James Harragan on acct of Medows

31 tim Callaghan and 2 measons at work

August 1st 1750 I went to Limerick and paid for 4 chairs 16 sope 1 snuff 1:1 Cadis and tape 11½ Lase 1½ hops 3:8 glasses 2:1 Saltes 7 Cinnamon 8 shrub 3 rum 1:2 sugar Candy 2 mug 2 metheglin 1:3 to ye men 1s 0½ these I pd 3:4 earnest for briches and lent Ned Harragan 5 05 turnpike 4d a looking glass 8½ left in yeamons hand 16:2

2d Augt 1750 I began to Reep wheat and aboute 12 sent for ye midwife She came att 8 and God blessd me wth a boy[4] about 12 att night and aboute 4 sent ye midwife back

3 Christy brought home my Cloath and I gave him a note on frank yeomond for 11:11 wch in full for weaveing 26 yards

4 my brotr Chapman and tom welldone I got a quartr motton from Lowry wch ~~I owe~~ for it weighd 17 pd Augt 19 pd

5 Some of ye nighbour came to see Catty

6 ye same I got 17:11 from Jon Quain and from Thos Harragan 18½/21

7 my Brotr went away I sent Jon to town he pd frank yeamon 2:4 wch wth 4:3 ye owd me is ye Price of ye things he took up this day then I gave Sally 1:1 Jon Pd Mcdaniel[5] in full for ye briches 1:1:8 turnpike &c 4½ snuff 1:1

8 I gave Joan Daly 2 10

9 nothing materiall

10 11 nothing materiall

12 Mrs Mellsop dined wth us I gave Syp Pursill ye red briches

13 I went to Pallas and left 26 yards of Cloath att ye bleach and pd Christy 9:8½ for weaving 33 yards of cloath now in ye loome att 3½a yard and pd him 1:4 for a pd of powdr then I came to Russills and spent 1:6 in Company wth ball so home

14 I gave ye Quittrent drivers 6½ and got my wheat stacked

---

4 William Peacock.   5 Michael McDaniel was a warden of the Fraternity or Company of Skinners, Saddlers, and Tanners of the City of Limerick. Certification of admission to the company to Edmond Pery, 7 April 1741. Limerick papers, NLI PC877/3/8.

September 1750

15 a violent storm but I bless God little damage done me by itt
16 nothing materiall
17 I gott 2:1:11 from Den Bryen Killbreedy from Patt Hogan 1:2:9 from Jon Moleheene 15s 9d Matt Hanrahan 10s:11d from Patt Daniel 1:2:9 and wth what money I had made up ye Quitt rent
18th I sent John to town he pd 6:15:8½ Quitt rent and 4:6 fees he pd for a srline beefe 3:1½ Salte 1:6½ turnpike expences 5½ I wrote to frank yeamons for ye following things wch I did not pay for 3½ pd English flower 3 quarts Vinager 292 Cloves ½ a pd pepper half a stone jemaco sugr ½ a stone powdr sugar a pd of Resins a gallon Shrub 2 quarts whiskey
19th att home pd 2:2 to Lowry for ye motton I had from him some time agoe and sent for another wch I owe for
20 Prapareing for ye Christning I got a Pale from Haly gave ye boy 1:1
21 I had Willm christianed Mrs Hewson his Godmother Jon Pordon and Jon Hunt his Godfather to ye Poor 8 to Mrs hewsons boy 1:1
22 Settling ye house
23 I went to ye bog and measur'd 90 kishes of turff
24 I sent for Pad Daniel to see ye black mare yt got a swelling in her belly I gave Will welch ten bandills of bandle cloath wch come to 2:11 to be chargd to his acct
25th of Augt 1750 I went to see Court family yt came home last night back in ye eavening
26 nothing materiall
27 I had 5 men 3 horses puting out turff
28 I had I had 2 horses puting out 6 drawing home
29 I had 2 horses puting out 8 drawing home
30 I had 2 horses puting out 7 drawing home

7br 1st 1750 I had 8 horses drawing home and finisd ye drawing 280 kishes wch stands me 4d a kish and 2:8½ over every way Catty lay att Court and came home in ye Eavening to ye maid 1:1
2d I had Conr Harragan and Thos Coagh and my own men Reeping
3d ye above men a man from Martin Jon Pursill 2 women binding allmost finishd
5 finishd ye Reeping began to draw hay
6 finishd ye hay 7th turned ye Oats
7th Jon went to town wth Plums wch he sould for 5:8 he pd for Snuff 1:1 blew 1s Serus 2 Costom and turnpike 4d
8th to Mrs Williams 11:4½
9 to Church dined to Courte back at night
10 Cutting Reed I bought a mare from Dan Shea for 4:15:0 am to pay at 9br
11 I got her and ye bay mare shod
12 ye People drawing hay

13 I pd Ballycraheen tenants 11:4½ in full for ye liberty of a Passage
14th I allowd Patt Boyle 1:8 for mowing he claimd since last year and gave Jon Bryen on acct of his wayges 11:4½
15 to ye County of Cork and County tiperary and came home ye 25 expences 1:5:7
26 went to Courte gave Mr Hartstonge yt Mr Bastable sent her 30l so home
27th I paid tim Lowry 2:2 in full of all accts and got my hay made into a Reek
28th I had my Corn made up
29 30th Thos Wallis wth me

8br 1st 1750 thos Wallis att Cashletown
2d I joyned him in a note to Geo ffosbery for 22 13 10 in Pledge of wch I left 21:17:10 then I lent Jon Molcheen 1:11 to pay Mr Wallis then I went home
3 and 4 ye men at ye Corn
5 I hired Nell fox gave her 1:1 earnest
6 ye men thatching Corn
7 Catty and I lay att Court and borrowd a Ram and sent it home
8th came home and gott 1:3:6½ from David Welsh on acct of Medows
9 10 ye men at ye Corn
11 I pd 3:3 for 300 apples puting out dung
12 13 puting out dung
15 16 Catty at Courte to ye maids 2:2 snuff 1:1
19 finished sowing and trenching wheat Conner Harragan has in all 9 days Jon Pursill 10 days ye young sow brimed
20 21 att home Jon Bryen att his house
22 from Den Bryen on acct of Medows 6 10 I pd ye man at adear 1:4 I owed him for Rum and 11d for hops 23 we brewed
24 25 26 at home 27 I had a bottle of Rum and a quartr of Motton wch I owe for Jon Rourk came 8br 28th 1750 Rourk wth me
29 and 30 wth me
31 I got from James Harragan 2:2:4 wch wth 1:16:4 I gott ye 30 of July makes 3 16 0 in full of 3 acres medow then I allowd him 10 10 for Jon Bryen then I gott 6:8 from thos Harragan and allowed him four pound for ye black colt wch wth 18:1 I got ye 6 of Augt last makes 4 18 9 in full for his medows I gott 11:10 in full for medow from Patt Hogan 5:3

9br 1 I sent Jon Bryen to ye fare of Nantenan[6] he gott from Henry Downs 1:2:9 in acct then I gott 3 4 12 5 from Thos and James Harragan
2 I got 14 1 from Jon Quain and allowd him 7:9 on Will Welshs acct
3 I pd Jon Rourk for Mrs thornell 37:18:9 wch wth 6:15:8 by a quittrent receipt makes 44:14:5 then he went away

---

6 Nantinan is 3 miles south east of Askeaton, on the river Deel.

December 1750

4 att home began to dig
5 began to dig Petatoes
6 Cristy brought home my Cloath from ye bleech I pd him for weaving and bleaching 16s9d and bought 30 yards of Cloath at 1:5 1/2 a yard and gave him my note for ye money payable 25 xbr
7th 8 9 att ye Petatoes [illegible] I got 9 11 0 from thos Harragan and allowd him 1 10 0 for Mick Molcheen I owed him for Sider 11:11 for horse hire and 2l 2s 0 for cuting turff and 6so for Jon Bryen and 10s 5d for 25 mens hire in all 14 11 11 1/2
10 I sent Jon to town he pd Mansill for Colonell Taylor 13:19:1 1/2 he pd frank yeamons Sugar 1 4 rice 4 nutmegs 9 in all 2:5 Snuff 11 Sallys flannin 1d Salte 1:6 expences 6 1/2 ye I gave him 3:4
11 I went to Church back to diner
12 13 14 ye men at ye Petatoes and all most finishd
16 I sent Jon Bryen to town he pd frank yeamons for clarott 1:2 Lisborn 1:2 rum Shrub 3s whisky 5 3 teacups 1:8 whiteing 1:2 snuff 1:1 surfitt water 6 expences 6 1/2 he brought home a cap to Catty to Joan Dondon in full of her wayges 4:2 to Hon Coagh 1:3
17 I had a Sheep killd and sould ye skin to Will Welsh for 2:6 I pd Robt Peacock by my Aunt in full 14:4 to ye man att Adear in full 2:8 in swap for a spead 1:1 to Will Hickey 2 11 in full Den Bryen has 7 days and Thos. Coagh 13
18 I gott 4 10 0 from Andrew tomson in full for a filly
19 thrashing oats I put in my Pigs
20 thrashing
21 to Dan Shea in acct 1:6:1 Mrs Hartstone and Ms Nance Dined wth us
22 23 ye Plow going
24 Catty and I att Court 25 att home
26 ye Plow going
27 my aunt went away to ye horse 2 1/2
28 I see some beans ye Plow going
29 ye Plow going one of ye Cows calved
30 a holy day[7] Catty Cut out my shirts

xbr 1st 1750 I sent Jon to town he pd Quitt rent for Killdeemo 6 15 8 1/2 for Port 14:4 1/2 he pd yeamons for starch 4 Iron 2:4 snuff 1 0 pins 1/2
2d Catty and I to Church and back to diner ball Pegy and Dolly dined wth us a violent Storm att night but bless God no grate dammage to Tom McMahon 1:1
3 ye man att ye Corcas bank 4 att Dito
5 6 nothing materiall
7 sould 2 cows to Patt Daniell for 2:10:0 recd in Parte 1:2:9

7 Feast of St Andrew.

8 went to Askeaton wth Martin to pins Cotton and expences 3 6 when I came home I pd Matt Harragan in full of his bill 4:4

9th Catty and I to Courte 10 home to ye maid 1:1 I sent Jon to Mick feighs for a barill Mault wch he pd 18s:6 he pd 3:4 for 2lb hops

xbr 11 1750 I sent Jon rohan to Greag wth 2½ stone toe and 2:8½ toward spining

12 I gott drink brewd to knitting 4d

13 ye men trashing

14 brewd again ye white mare sick

15 sent Jon Rohan wth my watch to Rob Peacock sugar Candy 3 peper 10 brimstone 2 tar 3 whip 3 soles 6 snuff 9 bread 3 thread 1 turnpike 1½

16 Paddy Daniel wth ye white mare

17 some Peas from Ed flanagan to ye boy 3

18 to Jon Rohan for Sweeping ye Chimney 6½

19 I sent Jon to town wth butter wch weighd near 0:3:4 for wch he got 17 3 he pd for Salte 1 6 a saucepan 1:5 expences 8d

20 I killd my cow sould Tim Lowry 2 sheep for 19 6 earnest 1:1 I pd him ye 1:6 I owed him ye 27th of 8br last

21 I sent Jon Rohan to town wth ye hide wch he sould for 1:1:1 he pd for oysters 7d expences 2 I went to Adear pd Gleesan for a gallon Brandy 5 6 Dito Rum 5:2 sugar 1:9 resins 1:4 an Almanack 3 steele 3 Sugr Candy 6 expences 9

22d nothing materiall

23 I sent Jon to Adear he pd Gleesan for whiskey 3s Sugar 2:4 4 bottles 8 brimstone 1½

24 sent Jon to town to Snuff 1 1 gloves 2 2

25 Catty and I to church back to dinner

26 we had Company dined wth us to ye P 6

27 to Lisdougan there ye **28th** home ye **29th** to me whilst abroad 3:2 I gott 1:5:10 from Thos Wallis to pay Mr ffosbery to Thos Welsh for Mr Bryen on acct of Rent 4:1:10

30 Catty and I to Courte home att night

1751

Janr ye 1st Thos Harragan his wife James and wife dine wth us I borrowd 3:8:3 from Martin

2 I sent 5 13 0 by Jon to Thos Welsh for Mr Bryen pd in all 9 15 8 do owe 1:4:1

3 I gott 15:1 from Tim Lowry on acct of ye Sheep

4th I put in some wheat 5 att ye gardin dich

6 att home

7th att ye wheat I sent Jon Rohan to ye forge I gott from James Harragan 1:2:9 of wch he owed me in ballance of Medows 17 3$^{1}/_{2}$

8 I sent Jon to town he pd 18d for tracle 7 honey 8 brimstone 10 tar surfitt watr 6 vinager 5 tred and tape 3 snuff 1:1 turnpike 1$^{1}/_{2}$

9 I had ye men att ye wheat

10 brotr Chapman and Cooke came to see us

11 12 13 both here

14th Cooke went away Chapman and I to court

15 Chapman went away

16 17 18 ye men att ye dich I gott 2:1 from Tim Lowry wch wth 15 I pd Jon 1:1 earnest 1:3 for Motton make in all 19:6 in full for 2 Sheep

19 nothing materiall

20 att home 21 Catty and I went to Court

22 home to ye maid 1:1

23 sent Jon to town wth wheat he sould 10 Stone 2lb for 19:6$^{1}/_{2}$ he pd for Salte 1:6 to Sally 1:1 snuff 1 1 ballsome 3 Sand 1 Costom turnpik 6 to Jon 3 he had for his own use on acct 2s

24th I paid Will Couse for handkircheifs 8:6 I pd Tho Welsh for Mr Bryen in full for 9br Rent 1:4:1

25th Catty and I att Castletown I pd Mr fosbery 1:6:0 for Thos Wallis so home a giney of wch I borrod of J Rohan

Janr 26 1750/51 Meloge was killd I sent for Thos Harragan and Jon Quain to see her who awarded me 20s ye hide and welsh to Rare ye calf or pay me five shillings in lew of ye Calf

27th I sent 2:2 to Adear to gleesan he sent $^{1}/_{2}$ a pd of hops and kept ye change in acct

28 I gott 7:7 for ye cow hide ye men trashing

29 I had my Pig Salted men trashing

30 I sent Jon to town to sell wheat wch he sould for 16:2 he pd for bottons 3 expences 9

31 I gave Jon Bryen on acct of his wayges 2 2

Febr 1st 1750/51 I gave Dan Shea on acct of his Mare 2:1:9$^{1}/_{2}$ wch wth 1:6:1 I gave him 17 9br last makes 3:7:10$^{1}/_{2}$ do owe him 1:7:1$^{1}/_{2}$
2 Catty and I to Bally England and back to 6d
3 att home
4 ye men diging for Plants
5 ye men diging for Peas beans and Parsnips I sett some of each
6 ye men diging for Petatoes I had 15lb of Iron from Thos Gleesan and pd him in full of all accts 5s:9d
7th I borrowd a giney from Andr Tomson for Mrs Williams and sent a sheep to Courte
8 9 nothing Materiall
10 to church and back to diner
11 Jon Borrows 11:4$^{1}/_{2}$ for me I gave Thos Coagh a barrill Oats and David Bryen a barrill Oats
12 13 14 ye men trashing and at ye Sally bed
15 I wrote to gleesa for a quartr a stone of Powdr Sugar and a quartr a stone Jemeco
16 I sent Jon wth 2 barrills Oats to town wch he soald for 18s he pd for bread 4 Mana 8 Rice 4 expences 6$^{1}/_{2}$
17 Catty Pryce and I went to Courte and lay there
18 we came home and pd Cristy in acct of linin 1:8:2 19 ye men at ye Sally bed
20 ye men trashing
21 I wrote to Geo ffosbery for a barrill mault wch I gott
22 I sent some of ye mault to ye mill and wrote to Gleesan for a pd hops I gave tige Vary a barrill of Oats on acct
23 I finishd my Sally bed and set 3700 stocks and left Standers I gave David bryen a barrill Oats and barrill to willm Welsh I put in a reek Oats Dan Shea gave a pr of Pumps to Jon Bryen
24 Catty and I to Court 25 home Jon Sould Gleesan 6 barrills Oats for 8:6 a barrill he gott 1:1 earnest and spent ye
26 to Adear dined att Ned flanigans back att night to ye boy 6$^{1}/_{2}$
27th a violent storm stript ye houses
28 mending ye houses

March ye 1st 1750/51 I sett some shilats Cabig seed and Onion seed
2d I went to killtinane and back
3 Catty and I dined att Harry Downs calld att Robin Hewsons sould ye white mare for 6 16 6 and pd Martin ye 3 gines I borowd of him ye 1st of Janr last
4 I deliverd Gleesan ye 6 barrills Oats and gott 2 11 0 to snuff 3
March ye 5th 1750 I gave Dan Shea 1:0:1$^{1}/_{2}$ wch wth 7s allowd his son make 1:7:1$^{1}/_{2}$ and is in full for his Mare I sent Ball 1:2:9

April 1751

6 sent Jon Rohan to town he pd for stone blew 4 licorish 1 sugr candy 3 pins 4 snuff 2:2 sope 4

7th I pd for sidr 1:6 rum 1:4 dudly joynt[1] and Syp Pursill dined wth us

8th I sowd a barrill Oats Brotr Chapman lay wth

9th he went away I gave Jon 13d and sent him to look for Schollops he pd 14d for 700

10 att home

11 I spent drawing up Harry Evans acct

12 I went to Croagh to advise wth mension aboute it but he was not att home I pd Jon Cristy in full 16:7 I had some beans sett

13 Catty and I went to Geo: Hewsons Cant but did no good expences 1:9 I had 10lb Iron Gleesan

14th on ye bed ye men trashing

15 I gave Thos Coagh 13d and wrote to Gleesan for Shrub but got only a quart of Rum

16 no men att work

17 H Downs and wife dined wth us I gave Jon Rohan to drink 1:1

18 I was over flowd but thank God no gap

19 I pd Gleesan 8:1 in full to ye Sherriff baliff[2] for not returning me 1:1 to frost in full for Countery charges 8½d

20 I went to Croagh gave Counsiller Wollcutt[3] 1:2:9 for his opinion in Harry Evans affair he advised not to make up ye affair to ye Servt 1:1 to me 3d

21 I putt up 77 stocks

22 ye Plow going 23 ye Plow going I sent Jon Rohan to town wth Brotr Chapmans letter but got no answer to it turnpike 1½

23 ye men att ye Plow

24 Ball and Dolly dined wth us

25 Pryce very ill 26 still ill and was bled

27 a little better I went to court for an answer of Brotr Chapmans letter so back

28th I sent Jon Rohan to Lisdougin wth an answer of ye letter to him 1:1 ye men att ye bank

29 no man att work

30 Jon Rohan came back my father in law ill ye men att ye Plow to Harry Downs boy 4d

Aprill ye 1st 1751 ye men att ye Petatoes

2d I got my Oats winowd

---

1 Dudley Joynte died c.1760. Described as 'gent', his stock, including wool, sheep and horses, was auctioned on 27 and 28 April 1761 and fetched £461 19s. 11½d. National Archives Chancery pleadings, unidentified material, box 52. 2 Sheriff bailiffs carried out much of the work of their sheriffs, to whom they were bound by articles of agreement. Their duties included serving warrants, collecting fines and, as in this case, seizing goods in cases of debt. Garnham, p. 95. 3 John Minchin Walcott, of Croagh, Co. Limerick, was MP for Askeaton 1747–53. His will was proved in 1753. Vicars; Lenihan.

3d I sowed 1 1/2 barrills oats I sent Jon Bryen wth a message to Harry Evans I pd for seeds 1:2 tea 1:6 thread and tape 4d snuff 1:1 sugr Candy 3 turnpike 1 1/2
4th ye men Plowing for flax and at ye petato
5th att ye Plow and Petatoes **6th** Dito
7th to church dined att Court so home to bed
8th att home gott 13s from Thos Carra and pd 4s hearth money
9th Catty and I went to Lisdowgin where we stayd to ye 13 so home expences 5:3
14th att home to Jon on acct 1 1/2
15 to Courte pd Mrs Hartstone 49:13:2 yt arther Bastable sent by me to her so home ye men att ye plow
16 had a pound hops from Gleesan and brewd
17 18 sowing Oats and digging in ye gardin
1750 April 19 Plowing for barley
20 att Dito and second diging Petatoes
21 to church dined there and lay there
22d home to ye maid 1:1 ye men at Dito
23 att Dito to Snuff 0:4 to Jon Bryen 3d
24th I sowed 2 bus barley ye men att Dito
25 I sowed some Peas in ye yallow gardin to a samon 1:1 I sent a jole and my clock to Mr Hunt
26 Plowing for Barley
27 att Dito and Petatoes
28 I got my Clock home mended
29 I had ye last of ye toe thread home Mr Hunt dined wth me and set ye Clock agoing I sowed 4 bus Duch barley and by much ado finished
30 ye men att ye Petatoes I acct ed wth David bryen and owe him 16s to snuff 6

May 1st 1750 I borrowd 2:8 1/2 from James Harragan and went to Sheas daughters weding to ye Priest and Poor 2:9
2d and **3** att Court **4th** att home and finishd 2d diging my Petatoes
5 I got two 4 os peices from thos Harragan
6 I sent for a barrill of Mault to George fosbery and pd him 18s for it and 18s for ye last barrill I had in all 1:16 0 and I sent some of ye mault to ye mill and wrote to gleesan for hops
7 I gave Jon Rohan 2:1:6 on acct of ye ginie I was to pay him on thos Coaghs act and ye ginie I borrowd of him
8 ye men trenching Petatoes
9 I got 3:3 from Thos Harragan on acct
10 Catty and I went to Court
11 att home got my big sow killd
13 went to Courte got some hemp seed and lave to send my filly to ye Park
14th sent her to Jon Bryen half a bar Petatoes

June 1751

15 Catty and I dined att Thos Harragans I borrowd 6½ and gave it ye Pyper
16 17 18 att home to ye men fencing
19 to Church dined at Court back at night
20 Cristy broug home 34 yards of 3 quartr Cloath for weaving of wch I pd him 6:0½ he ows four pence ½d change ye men diging for hemp
21 ye men att dito ye litle mare foald I marked 25 lambs I bought from Thos Harragan for 3 15 0 wch I am to allow him for Robert brought home my gon wch I owe for
22 I sowd about 3 pecks hemp seed
23: 24: 25 nothing materiall
26 att home in expectation of my bro in law
27 I gott 12s from Jams Harragan and 5s from Thos Harragan I gave Patt Coagh 1:1 in full I hired Nell Pordon and gave her 6½ earnest 28 Catty and I dined at greags to ye Child 1:1 pd hearth money for Jon Bryen 2s:0 expences every way
29 I got my sheep shorn to ye man 1:3 in acct I gave Jon Bryen on acct of his wayges one and 20 pound ½ of wolle I began my turff
May 30th 1751 Cutting turff
31 Cutting turff Sistr and brotr Cooke came

June 1st 1751 att ye turff I got a lamb killd
2d att home
3d att ye turff I gott from James Harragan 11:4½
4 Sistr and brotr cooke went away I gave Nell fox 5:5 and 2:8½ my wife gave her at other times make in all 14:2 expences 1:3 I finish ye cutting my turff and spent 83 men in all wch comes to 1:14:2
5 I had some drink brewd to Chickins 8d bread 2
6 I gave Jon Bryen 1:1 on acct of wayges I gave Jon Rohan 1½ wch 4:10½ he got from Conr Harragan makes 5 in particas ye money I got from him to pay Cristy
7th nothing materiall I lent my mare to Jon rourk to go to Limeric
8 I got 6:17:7 from Thos Harragan and sent Jon to town to Pay ye Quittrent in ye eavening I gott from him 24:6:1½ and from James Harragan I gott 10:11:11 to snuff 1:1
9 I pd Jon Rourk for Mrs Thornell 44:10:4 and I paid him 13:6½ yt was dificient of ye 4 ginies I pd him for ye two heffers I bought from him ye 15th of May 1750 [illegible] then he went away and paid Jon Rohan 5s 0½ in full of ye ginie I borrowd of him ye ginie I was to pay him for thos Coagh and ye money I borrowd of him to pay Cristy
10 Catty and I went to court and pd Andrew Tomson ye ginie I borrowd of him to Lend Mr Williams
11 I went to Bruff aboute business of Mrs Hartstongs expences 1:7 calld at ye fare sould a yearling for 19s to me ye men 1:5 so to Court and in ye eavening brought home Catty to Jon in acct 1½

12 I had my turff futted
13:14:15 I spent a trying for a quarry
16 to church dined at Courte back att night to expences 1:1
17 I gott a Salmon from Ned Flanigan to ye boy 1:1 I settled accts wth ye two Harragans and gott 6:2:3 from Thos Harragan wch wth 9:4 James overpaid makes 6 11 7 and is in full for their holdings to ye 25 of March and 1s of May last to Church rates 8:8
18th I wrote to Mrs Hartstonge for ye giney I lent Ball and she sent it and 13s she owed me for a sheep in all 1:15:9 then I borrowd 2:10:11 from James Harragan and 1:8:2 from Tomas Harragan and made up 13:19:1½
19 I sent Jon to town he pd Mansill for Colonell Taylor 13:19:1½ turnpike 1½
20 att ye quarry
21 in ye gardin sould 5 sheep for 11s each earnst [illegible]
22 sent Jon to town wth butter and wolle ye wolle he sould for 12s he did not sell ye butter he pd for snuff 1:1 peaper 10 blew 8 starch 4 pitch 7 expences 10d
23 att home 24 to Jon for Thos Coagh 1:1 Catty and I to Courte back att night
25 I was served wth a supena by Hen Evans I wrote to Wallis and inclosd ye supena and wrote to counsiller Wallcott aboute it
June ye 25 1751 Continued I sent Jon to town wth ye letters and gave him 3:6 60 pay Harry Evans for work done me some time agoe Catty and I went to Ned Flanigans back att night to ye boy 6½ Jons expences 6
26 att home sent ye Side sadle to Mrs Roice⁴ but did not sell it
27 ye men diging in ye gardin
28 att ye mill and forge to chickins 9s
29 I sent Jon to town aboute ye butter he sould it for 12 9½ he pd for sope 0:1:0 expences 6½ I got 1:0:7 from ye butcher
30 I gave Jon B 1:2:9 to give Tho Harragan and 10s to give James in Parte of ye money I borrowd ye 18th inst

July ye 1st 1751 I went to ye Vestory dined to Courte home allmoste drunk
2 to Courte and back again I got 0 19 6½ from Jon fitzGerald on acct of medows
3 att home nothing materiall
4 will Quinlan came to make a Rack and Manger in ye Stable I gott 1:12:9 from James Covard on acct of medows and from Dan Shea 1:2:9 from his son on dito acct 8:1½ Matt Hanrahan 8:8 Dan Gormon 5:5 from Jon Bryen 1:2:9 and from Thos Harragan 1:11:5 all on acct of medows I gott ye giney I lent Mr Williams last Spring
5 I sent Jon to Thos Welsh wth 8 13 0 on acct of Rent and got his receipt I gave Jon Bryen a bill on Patt Nash for 15s to ye Sheariffs bealiff 6½

---

4 Thomas son of the Revd Henry Royse, who was born at Kilcornane, married Annabella Russell, daughter of Col. Henry Russell, governor of Galway. Thomas Royse was prebendary of Dysert 1739–47. *Alumni Dubliniensis*; Leslie, 'Diocese of Limerick'.

August 1751

6 ye men Second futting turff I gott 17s:0 from ye butcher on acct Will Quinlan 3d att ye rack

7 Catty and I dined at Castletown expences 1:3 home at night

8th I gott six bottles of Sidr a bottle of Rum and a pr of Garters from gleeson I pd Jon Longane 17s on acct ye Priest lay here

9 he went away

10 11 nothing materiall

12 I gott 8 $1^{1/2}$ from Jon Quain and gave Thos Coagh $2:8^{1/2}$ in acct

13 Catty and I att Court home at night

14 att home

15 att court setting medows to countery Charges 2:2 I called at Ryans so home

16 I gott 4:11:0 from Jon Nash

17 I sent Jon to town he pd for boards 8:6 nails 2:9 a letter from Wallis 4 Snuff 1:1 Salte 1:6 6 then I gave Jon 3 ginies and 5:5 to give John Longane for ye tythes 1750 and fees for 1751

18 went to Lisdougin 19 there 20 home expences 2:0 ye 21 att home

22 I hired thos Melcheen for 12:6 a quarter to him earnest 3d ye men at ye turff

23 I gott 4 ginies from Thos Harragan and sent it by Jon to pay to pay Nash ye 4 ginies I got from him ye 17 inst

July ye 23 1751 continued we were sent to Court we had ye assurd acct of Sr Standish death[5] so back att night

24 I gave Will Quinlan $2:8^{1/2}$ on acct and I got a SrLoyn of beefe from Court to ye boy 6

25 Catty and I went to Court and lay there

26 home Jon gave me 1:1:0 he got yesterday for 2 yearlings then I sent him to Adear and he pd Gleesan in full 18:5

27 ye farrier came to see ye black mare I gave him 2:2 to buy Phissick for her I gott $11:4^{1/2}$ from Thos Harragan on acct

28th to church and back Jon Hunt dined wth us to ye 2

29 to Courte aboute Madams business

30 wrote to Geo: ffosber for a barrill mault I hired owen Nele for a year att 5s a quarter or as good Cloaths as he brings

31 Jon Rourk came for Parte of ye arrears I gott 30lb from James and thos Harragan and gave my note for it and am to pay them interest for itt Rourk went to town

Augt ye 1st 1751 Rourk and Jack Downs lay wth me I had a pd of hops from Gleeson and ye drink brewd

2 I pd Rourk for Mrs Thornell 30l he and downs went away to snuff $6^{1/2}$

3 ye men claining ye hay yard I dined att Jack hunts

---

5 Sir Standish Hartstonge of Bruff, Co. Limerick, was MP for Co. Limerick. His will was proved in 1751. Vicars.

5 att home 6 7 att home began to mow
8: 9: 10 at Court aboute Balls affair
11 att home wrote to Mrs Thornell
12 att Court and canted Balls Cattle for [?] hired labourers
13th Catty went wth Mrs Hartstong to Hally Park Jon took my tythes for 3 10 and fees earnest 6¹/2d
14 I sent a cow to ye fare Snuff 3d
15 I went for Catty but she did not come on Porpose to wean billy
16 ye mower att work ye men at ye hay
17 I brough Catty home
18th att home I lent Hary downs ye side sadle
19 wth ball to Courte snuff 2
20 att ye hay and Reeping Oats
21 att ye hay I got 15:2 from Jon Quain and 6$^{1}/_2$ as earnest from Matt Hanrahan
22 I went to Limerick and gott a Coppy of Hary Evans bill from Jon Wallis so home to snuff 1:1 reasins 3 N McCan 5$^{1}/_2$ expences 9d I sent Jon Bryen and Pattr daniel to ye fare of Killdorery wth ye bay mare and colt and gave them 10 10
23 I bound some Oats
24 Patt Daniel came home they sould ye mare for 9:10:0 expences 8:8
25 1751 to Church Jon Ger [illegible] and wife dined wth us to Jon Longane in full for this years Procters fees 7:7
26 to Catty to pay ye weaver 9
27 I sent Jon to town wth buttr it weighd near 1:1qr:21lb att 2s6 a hondrd comes to 1:1$^{1}/_2$ he pd for vinager 5 tar 6 syze 4 other expences 7d
28 ye men att ye hay 29 att Dito
30th a very wett day I gave Pattr Coagh 5s on acct of Monding
31 Reeping and binding english barley and made one Stack Oats

INDEX

accounts 163, 181; NP balances 70, 79, 140, 128, 164, 165, 180
Adare, Co. Limerick: NP visits 61, 62, 79, 82, 87, 91, 100, 110, 112, 118, 128, 140, 144, 145, 148, 161, 187, 220, 222; visits with Mrs Hartstonge 120
Aghakillare, Co. Limerick 53
Aherly, —, paid for Mr Badham's saddle 112
alum 190
Anderson, —183
Anstice, knitter 149
appraisers: value NP's stock 112; fees 118
apprenticeship: indentures perfected 193; terms of 179
Ardlaman, Co. Limerick 172
Art, Tim 82
Ashton 126
Askeaton, Co, Limerick: barley sold at 100; oats sold at 101, 146; rent paid at 165, 185, 189; fair at 99, 100
assizes: Grand jury called 157; NP at 25, 157, 158, 162, 178
Attkins (*née* Ireland) Elizabeth: courtship of 194; marriage settlement paid by NP 210, 211
attorney: not at court 115
Austrian Succession, news of war of 111

Badham, Brettridge: NP leases land from 87; NP meets 87; death of 118; estate of 149; his bond 80; letter from 112; visits NP 111
Ball, Henry 210, 211, 216, 222; and wife dine with NP 223; his affair 228; his cattle canted 228; pays for grazing 200
Ballinemona wood 95
Ballingarry, Co. Limerick 104, 173, 177
Ballyashea/Ballygeasy, Co. Limerick 79, 188; road laid out 176
Ballycraheen, Co. Limerick 115; liberty of a road through 149, 218
Ballyengland, Co. Limerick 71, 101, 111, 128, 133, 149, 188, 213, 222
Ballygeasy curragh 109
Ballygran [Ballygreana] Co. Limerick 127
Ballylongford, Co. Limerick 49, 57, 74, 128, 147, 152; land leased at 77
Ballynegolla, Co. Limerick 147, 160
Ballynorte, Co. Limerick 57, 69, 87, 93, 193, 200; NP visits 71, 101, 133

Barker, William: grazes two horses with NP 81, 137
Barker, Mary 81
Barnokyle, Co. Limerick 173
Barrott, — 55
Bastable, Arthur 154, 177, 194; NP writes to 155; not at home 153; his account 146; sends money to Mrs Hartstonge 218, 224
Bastable family 147
Bastable, Mrs 125
Bay Bush, Co. Limerick 91, 113
beagles 110
Bealan, —, NP bets with 96
bees: bought 205; swarm lost 63; bees, swarm of 51; bee hive 50, 63; bees wax 191
Belvedere, Co. Limerick 112
Bendon, —: his men 156
Berkeley, Dr: mare given to 136
bet, on when cuckoo sings 96
Birr, King's County 113
blacksmith: paid 155; horse shoes 50; oars 213; forge 203, 221, 226
Bleach, Jon 59
Blunt, John 52
Blunt, Ms, marriage negotiations 52
bog 64; road to paid for 205; drawing out the 84; measured 217
Bonfield, John, of Limerick 55, 69, 81, 110, 122, 150, 207; NP pays 101, 126, 138, 142, 144, 172, 203; makes coat 136
books and reading 52, 55, 58, 60, 63, 67, 69, 73, 76, 75, 76, 85, 88, 93, 95, 97, 98, 103, 117, 120, 122, 123, 125, 128, 129, 132, 133, 135, 136, 137, 138, 139, 149, 150, 151, 152, 157, 158, 160, 162, 163, 165, 166, 170, 174, 176, 177, 185, 205, 207, 212; almanac 89, 107, 198, 206, 220, 169; bible 149; books for Mrs Hartstonge paid for 172; brought from Portacacha 126; on Christmas day 129; *Elegy on the Potatoes* 48; François Petis de la Croix, *The thousand and one days: Persian tales* 135; *Love à la Mode* ... 210
Bork, Bridget 60, 61, 62
Bork, David: bill on 158; pays rent 159; legal affairs 154; to be driven 151; replevied distress 153; writes letter to NP 132; NP writes to 137, 150; threatens replevin 152; stock driven 153

229

Bork, Jon 48, 86
Bork, Tim 51, 55, 56, 59, 84, 137, 140; Hartstonge furniture sent to 109; NP sells barley to 99, 100; NP lends money to 139; pays for hay 125
Bork, Tom, tailor 96, 101, 126, 127, 132, 143, 145, 146, 149, 160
Bork, Will 51, 53, 57, 126, 181, 212; dead 59
Bork, Law 109
Bork, Richard 131
Bork, Miss 193
Borris, Co. Tipperary 140
Boyle, Patt 64, 68, 69, 71, 94, 96, 97, 98, 99, 100, 117, 128, 148, 178, 218
Boyle, Will 54, 65, 83, 128
Boyle, Bryen weaver: canvas 134; blanketting 156; wife paid 180
Breen, Will, tithe proctor 183, 184, 191, 192, 197
Brettridge, Mary 12
brewing, coopers paid 201
Brodeen, John, hackler 89, 90
Brown, Jon 65, 82
Bruen, David 92
Bruff, Co. Limerick 147, 150, 153, 166, 175, 177, 225
Brummell [Bromwell] Benjamin, nailer 203
Bryen, David 108, 109, 110, 117, 121, 124, 125, 126, 128, 131, 132, 133, 135, 137, 138, 139, 140, 142, 144, 146, 148, 152, 155, 157, 158, 159, 164, 166, 169, 171, 173, 174, 177, 178, 179, 183, 187, 190, 191, 202, 204, 222, 224; serves subpoena 158; accuses Tom Hanrahan and Patt Daniel 134; NP fears he has forged his name 119; not trusted 173
Bryen, Den, Kilbreedy 52, 68, 129, 135, 172, 218, 192, 201, 202, 205, 217
Bryen, James 191
Bryen, John 49, 50, 51, 52, 53, 54, 57, 58, 60, 61, 6265, 69, 70, 76, 77, 82, 98, 107, 109, 114, 118, 119, 121, 122, 123, 125, 127, 128, 131, 135, 139, 142, 143, 159, 161, 164.165, 172, 202, 204, 211, 212, 214, 218, 224; accounts for butter money 85, given pumps 222; goes to Cork 205; goes to Limerick 64, 66, 89, 90; his wethers 66; lends Mrs Hartstonge money 93; paid on account 211; paid wages 68, 69, 87, 96, 99, 105, 131, 169, 187, 189, 218, 222; NP pays rent to 88, 111; pays Joan Daly wages 89; pays binders 67; pays for pot hangers and scales 69; pays rent for NP 92; refuses to let land to NP 79; sends for fat sheep 73; sent to fair 104; sent to Miss Lister 129; settles accounts to NP 73, 79, 98, 141; son 55
Bryen, M. 135
Bryan, Terence of Limerick 84, 85, 86; sent rent 93

Bucknor, Revd. William, rector of Croom, Adare and Dunamon 102; NP pays tithes to 197, 203, 214, 215
building: barn built 64, 65; barn roof 65; barn roofed 69; barn door, wood for 68; foundation stone money under 64; barn thatched 69; stable 180; henhouse built 173; masons building 64; new foundations 84; sand 65, 86, 181, 194, 203, 214, 215, 221; henhouse 165, 169; couples 53; pig stye 56; stones 86, 165, 173; manger 90; slates 85, 141, 191, 193; pointing 85; wattles 87, 90; plastering 86; slating 121; turpentine 214; cobberd 91; cowl 90; quins 88; hinges 90, 215; boards 90, 118, 170, 175; thatching 62, 88, 127, 189; gadds 89, 96; sticks 87; prop [scaffolding?] 89; scallops 89, 164, 170, 198, 209, 223; stable 86, 89, 90; window glazing 124, 134, 155; pitch 99, 141, 176, 226; tar 83, 115, 141, 176, 204, 213, 220, 228; grate 173; chimney 173; lime 194; roof 191, 195; bricks 213; laths 198, 200, 211; serus 191, 195, 204, 213, 217; carpenter 216; masons 214, 215, 216; slater 155, 190, 200; tiles 214; trowel 215; chalk and thread 204; joiners 214, 215; repairs after storm 222; walls of hay yard 62; ceelors 161, 212; ceelors 212
bullrushes 53
Burk, — 166
Burk, Alick 192
Burk, —, his horse 64
Burk, Will, paid 209
Burn, Frank, paid 99, 100, 103, 125, 164
Bury, Thomas 47, 114, 156, 162
Bury, John of Shannongrove 85, 102, 108, 114, 115, 116, 121, 123, 125, 138,
Bury family, dine with Hartstonges 162
Bury, Thomas 136, 142, 147, 145, 146, 150; NP asks for warrant from 170; his groom 143; his gardener 151, 158
Bury family, dine with NP 124
Bury, Mrs 127
Buttevant, Co. Cork 182, 183
Buttler, W. 157

Caffow, Cate 129, 175, 184
Cahirmee, Co. Cork 202
Calahan, Tim, carpenter 53, 64, 65, 69, 70, 71, 73, 74, 82, 83, 86, 87, 89, 92, 128, 148, 152, 160, 190, 191, 203, 216; makes doorcases for new house 85; makes rail for tester 75; makes gauges 83; mends truckle 88, 116, 179; makes beams for new house 90; makes bedstead 179; makes stable door 90; putting in posts 107; throws out new window 179
candles, made 92, 131, 154, 167, 185, 194

Index

cant, at Rathkeale 99
Cantilon, David, his horse 139
canvas: bought from Sister Ireland 95; woven 195, 201; sowing sheet made 78; NP makes bags of 95
Caparoe, Co. Limerick 48, 51, 53, 56, 134, 136, 194; ploughing 60; fenced 62, 82; stock to 62, 63, 65, 82, 213; rent of 176, 182; harvesting 66; stock brought from 131
Caramedy, Luke 86, 88
cards: bought 125, 176, 141, 182, 192; NP plays 152; NP loses money at 89, 90, 94, 96, 97, 98, 102, 124, 154
Carigoginall Castle, NP and Hartstonges visit 171
Cark, John 81
Caromady, Joan, paid 200, 201, 204, 206
Caromady, Luke 84
Caromady, Mick 100, 102, 199
Caromady, Patt 209
carpentry: augers 95, 97; boards 128, 204, 214, 215, 227; brads 213; carrier 205; chisels 59, 95; cross-cut saw 93; file 98, 212; gate mended 56; glue 128, 175, 214, 215; laths 140; locks 215; nails 70, 73, 82, 98, 102, 110, 135, 140, 151, 155, 175, 181, 191, 192, 193, 194, 197, 198, 199, 201, 203, 204, 212, 213, 214, 215, 216, 227; rasp 181; saw 97; sawing 214; sawyers 190, 191, 192; screws 149, 169, 211; size 228; tevanes and rafters 117; tools 114; timber 64, 160, 162, 163, 182, 189 213,
Carra, Thos 224; takes cider to Dublin 111; bring Mrs Hartstonges goods 161
Carroll, Den, 97, 98, 162, 164, 165, 166,
Carter, Jon, pays for grazing 139, 178, 204, 205
Carty, Cate spinner, given bandles of flannel 145
Carty, Cate spinner 87, 89, 90, 93, 144, 151, 157, 159, 169, 192, 193, 198
Carty, Charles 112, 120, 148, 151
Casey, Margaret, spinner 158, 206, 210
Casgary, James 86
Casgary, Ned 86
Castlemagner, Co. Cork 147, 153
Castletown, Co. Limerick 125, 131, 218; NP and Catherine at 175, 178, 184, 221, 227
Cenedy, — 56
cereals: barley 172; barley bought 197, 211; barley bound 53; barley dried 165; barley to be malted 79, 212; barley sold 69, 98, 99, 104, 135, 187; barley milled 63, 165, 213; Dutch barley threshed 99, 185; English barley threshed 74; milling barley 63, 66, 67; mowing and staking bear barley 65; ploughing for barley 224; pot for barley 59; reaping barley 65, 66, 84, 178, 190; reaping Dutch barley 145; reeping English barley 52, 228; sowing barley 54, 68, 224; sowing Dutch barley 49, 61, 200, 224; sowing English barley 49, 59, 60; stacking barley 85; threshing barley 58, 59, 80, 91, 98; threshing bear barley 76; winnowing barley 58, 65, 69, 80, 91, 151, 170, 171, 210
corn 205; broken corn threshed 88; corn cut 51, 218; corn milled 74; corn winnowed 79, 117; rick of corn thatched 206, 218
meal dried 176; meal ground 180; meal sifted 74, 176, 186
oats binding 204; oats drawn 163, 178, 192; oats dried 48, 59, 61, 89, 115, 186, 192; oats milled 50, 71, 74, 102, 166, 176, 180, 183, 189, 192, 201, 206; oats not sold 173; oats riddled 135; oats sent to kiln 79, 82, 176; oats sold 156, 176, 222, 186, 161, 170, 172; oats sold 90, 156; oats sown 134, 171, 184, 199, 206, 211, 212, 223, 224; oats sown 56, 57, 58, 62, 74, 75, 78, 79, 155, 185; oats swapped for tow 58; oat stack 228; price of 134; ploughing for 171; ploughing for oats 90, 155; reaping oats 52, 66, 123, 146, 228; stacked 53, 66, 87; threshing oats 47 56, 58, 59, 67, 73, 74, 77, 90, 98, 132, 135, 176, 184, 192, 206, 219; winnowing oats 56, 70, 74, 77, 78, 93, 115, 117, 133, 134, 136, 155, 165, 175, 176, 189, 223
quern 61, 74
wheat: binding stooks 65; corcas wheat 54; Dutch barley, wheat and bear barley harvested 87; reaping wheat 65, 190, 191, 203, 209, 216; rick of wheat 68; sowing wheat 55, 68, 103, 181, 193, 194, 206, 209, 218, 221; reaping and stooking red and white wheat 86; wheat sent to mill 192, 198, 199; wheat sold 80, 81, 84, 90, 101, 102, 121, 133, 151, 210, 215, 221; wheat stacked 191, 216; wheat threshed 50, 54, 67, 68, 83; winnowing wheat 50, 55, 195
Chanagillen [Shanagolden], Co. Limerick 205
Chapman, Catherine, see Peacock, Catherine 172
Chapman family 147, 153
Chapman, —, brother, with NP 221, 223; and NP dine at Court 187, 188
Chapman, Nicholas 202, 205, 216; cannot endow his daughter 154; cannot agree terms of settlement 173; refuses to let NP leave 173; very ill 223; visits NP 194, 223
Chapman, Mrs, death of 209
charity: to priest at wedding 224; to poor 217
Charleville, Co. Cork 92, 118, 129, 147, 153, 173, 177, 182, 183, 184

childbirth: Catherine in labour 186, 203; midwife paid 187, 205; midwife sent away 216; midwife sent for 186, 203, 216; spermaceti oil 186;
children, paid by NP 210
chimney sweeper 123
christening: NP goes to at Clampitts 181; NP pays clerk 194, 209; NP pays midwife 178, 183; preparations for 217; invitation 79, 164
Christmas Day, celebrated 152, 196
Christy, John weaver 210, 213, 216, 219, 222, 223, 225
church attendance 49, 52, 56, 61, 64, 66, 69, 74, 84, 87, 89, 96, 98, 99, 101, 104, 114, 117, 120, 123, 124, 127, 129, 132, 134, 136, 137, 142, 144, 145, 146, 147, 148, 154, 157, 159, 161, 162, 164, 164, 165, 167, 169, 170, 174, 177, 181, 185, 186, 189, 190, 191, 192, 193, 197, 198, 199, 200, 201, 213, 216, 217, 219, 220, 224, 225, 226, 227, 228; attended, no clergyman 140; attended, no sermon 91
church, visitation 54
church, visitation fees 54
church, rates 56, 60, 62, 63, 64, 65, 86, 115, 121, 139, 149, 161, 184, 201, 211, 212, 226; on Court, Kilcollum and Coroheen 139; on deer park 133; on Kilmoreen 165
church, vestry: churchwardens chosen 157; disputes at 199; meeting 61, 84, 115, 121, 159, 211, 226; vestry accounts 61, 159; vestry book bought 64
church: glazing of 156; wall mended 160
Clampitt, — 147, 149, 150
Clampitt, A. 152, 203
Clampitt, Franklin: dead 57; funeral 58; injured 57
Cleary, Patt 82, 84, 193
Clery, — 118
clock 105, 116; attempt to mend 116; cleaned 123; mended 104, 224; pendulum broken 105; set going by Mr Hunt 224
Clonans, Co. Limerick 86
Cloone, Jon weaver 182, 195, 201, 209
Cloonshire/Cloonshere, Co. Limerick 91, 92, 211
Clorane, Co. Limerick 58, 62, 89, 138
Cloriphest, Co. Limerick 113
Clotakey 162
cloth: buckram 75, 195; caleminco 212, 213; cambric 74, 182, 215; cassanette [cassinette] 203; cotton 190, 197, 220; cost of 89; flannel 58, 71, 85, 182, 219, dyed and pressed 64; frieze 59, 69, 70, 74, 80, 121, 148, 156, 165, cottened 57, 73, 180, 186, 189, 195, 206, 207, finished 193, tucked 179; fustian 203; linen 144, 190, chequered 181, 194, 204; mohair 189, 195, 212; poplin 210, 215; rug 195; serge 195, 210, black 110, blue 103; silk 205; worsted 86, for skirts 142
clothes: brought back from Limerick 81, 174; mourning 88; NP counts his 108, 129, 164; wedding, made 173
breeches 82, 83, 117, 135, 138, for servants 118, leather, deposit on 216
buckles 64, 82, 114, 158, knee 202, mourning 89
caps: NP making cap 165, for Catherine 219, nightcaps, NP makes two 165
cloak, for Matty 150
coats 81, 136, 138, 157, 207, 212; for David Bryen 150; for Matty 123, 148; tailor cutting out 122; riding 122; tailor mends 110; trimming for 122, 212; quilted for Catherine 212
garters 49, 99, 138, 165, 227
gloves 103, 121, 137, 138, 176, 180, 181, 190, 220; for servants 195; gift of 84; glover 144; mittens 90
handkerchiefs 55, 69, 122, 126, 181, 209, 210, 221; NP makes 144; lost 124
hat 81, 103, 126, 157, 178, 212; funeral hatband 145; servants made 151
shirts 105, 122, 143, 165; bought 154; Catherine cuts out 219; does not fit 104; NP comes home for clean 146-7; NP makes 103; made by Sister Ireland 103; stockings: knitted 54, 92, 143 145, 149, 220; worsted 88, 96; yarn 96
stocks 79, 116; NP makes 131, 132
studs 64; surtout 210; tailor 59, 84, 150
waistcoats 144; calf skin 133; fustian 81; old friezecoat made into 68; rug 150; tailor mends 110; trimming for 122
wig 81, 103, 105, 118, 212
Cloyne, Co. Cork, NP gets licence from 175
Co. Tipperary 218
Co. Cork, NP and Catherine go to 188, 199; NP goes to 182, 202, 204, 218; NP goes to marry in 173, 174
coach, to Limerick 126
coachman, given tip for service of mare 180
Coagh, Honor 164, 191, 219
Coagh, Joan 61; dismissed 64
Coagh, Mary 61
Coagh, Patt 225, 225
Coagh, Thos 171, 172, 173, 183, 186, 201, 206, 211, 214, 216, 217, 222, 223, 226, 227
Coagh, Thos, marriage of 215
Coagh, Thos, sent to Lisdogan 204
coal, measured 88
Collins, Mick, hired 191; discharged 201
Collins, Thos 79, 86, 100, 128, 178, 204, 192, 205
Collins, Tho, his wife 165

Index

combs 138
Connill, — 137, 138
Connily, Jon 78
Connor, James 178
Conway, James 52, 133
Conway, D. 57
Cooke, brother 186, 199, 203
Cooke, Sister and brother visit NP 225
Cooke, Frank 204
Cooke, — 221
coperas salt 64, 75, 92, 176, 192, 201
coral and bells 197
Corbitt, widow 74
Corcamore, Co. Limerick 141
corcas: bank 89, 219; banks, fear of overflow 121; bank, repaired 223; big 78; covered with water 170; farming of 19, 58, 66, 78, 104; lay 199; rush, NP sows wheat in 88
Cormock, Den 79
Coroheen/Curraheen, Co. Limerick 55, 102, 103, 124, 131 188, 200, 201, 211; deer park 101, 105; list of cattle at 139; rent of grazing at 138, 150
Corroh [Curragh], Co. Kildare 204
Cosgary, Mick 69, 70, 104, 113, 118, 123, 126, 127, 135, 140, 142, 145, 175; his horses 149; laths and hoops 121; pays for grazing 165, 174; sent for Hartstonge pictures 111; sent to Limerick 160
country charges 50, 51, 65, 132, 162, 184, 188, 199, 210, 215, 227; for deer park 137; for Court, Killcollum, Corroheen 156; NP pays High Constable 109
court: hearing 115; fees 115
Court, Co. Limerick 47, 57, 58; all in confusion 114; barley sent to 80; barley delivered 59; beans sowed 184; breakfasts at 177; Catherine Peacock stays at 178, 179, 181, 217; cellars in 103, 151; dines at 145, 178, 180, 181, 174, 190, 226; filly sent to Park 224; glazier puts panes in new window sashes 134; goods come to 160; grazing at 134, 135, 145; keys given up 144; NP accounts with men there 103, 174; NP and Catherine visit 198, 199, 200, 218, 219, 220, 221, 225; NP and Catherine stay at 222, 227; NP and Chapman visit 221; NP, Catherine and Pryce stay at 222; NP and Catherine dine at 175, 176, 183, 184, 186, 188, 189, 190, 191, 192, 194, 198, 199, 201, 206, 226, 227; NP mends locks at 146; NP on watch at 86; NP visits 59, 60, 61, 62, 63, 64, 65, 66, 67, 68, 69, 74, 76, 77, 78, 79, 80, 81, 82, 83, 84, 85, 86, 87, 88, 89, 90, 92, 93, 95, 96, 97, 98, 99, 100, 101, 102, 103, 105, 107, 116, 117, 126, 129, 131, 132, 133, 134, 141, 142, 142, 146, 147, 151, 152, 152, 153, 154, 156, 162, 163, 165, 166, 169, 170, 175, 182, 190, 205, 214, 218, 224; ploughing at 148; potatoes planted at 97; room cleaned at 142; sheep sent to 222; stock brought from 131; surveying land at 138; visits for Christmas 167; wheat sent to 196
Couse, Will, paid 221
Covard, Connor 48, 56, 68, 128, 146, 148
Covard, Widow, paid 200
Covard, James, pays for grazing 226
cowboy, NP with the 98
Cox, Hugh, dines with NP 210
Cox, Robert 116, 153; and his wife 169; asks NP to stand godfather 124; dines with NP 210
Cox, Tob, visits NP 75
Crataloe, Co. Limerick 92, 94
Croagh, Co. Limerick 193, 197, 223; bowls 50; crockery; dishes bought at cant changed 121; earthen dish, 91; list of china 104; mended dish 184; mug 181, 216; plates and dishes 124, 140, 173; slop bowl 183; soup dish 204; teacups 219; teapot 183
Croom, Co. Limerick 117, 174, 188
Crotty, Mr 152
Crow, J., dines with NP 185
Cunraty, —, sells slates 85
Curzon, Thomasina 12
Cusick, J. 193
Cusick, M. 163
customs, turnpike and wayage fees 64, 70, 75, 76, 82, 82, 83, 85, 92, 93, 96, 98, 99, 100, 101, 102, 107, 115, 116, 121, 127, 129, 143, 149, 164, 165, 167, 170, 172, 173, 176, 179, 181, 202, 207, 209, 215, 217, 221
cutlery: knife 180; knives and forks 75, 125, 126, 153; pewter exchanged for new 90; pewter spoons 49; spoons exchanged 216

dairy: butter 'above his complement' 154; butter sold 64, 74, 83, 84, 96, 98, 99, 101, 115, 116, 117, 118, 120, 121, 129, 133, 141, 143, 145, 146, 148, 151, 153, 154, 162, 167, 174, 175, 178, 179, 183, 186, 188, 190, 192, 198, 203, 204, 205 6, 207, 210, 214, 215, 220, 226, 228; butter sold to Sister McGan 129; farming, agreement for scheme 165; milk 68, 204
Daly, Joan 49, 54, 58, 62, 70, at mill 82; given money to buy tobacco 119; moves in with NP 117; paid 87, 96, 132, 136, 142, 148, 157, 162, 167, 184, 216
Dan, —, at the shooting butts 93
Dane, Mr 169
Daniel, Patt 51, 57, 59, 60, 61, 70, 78, 83, 86, 91, 94, 96, 98, 109, 116, 123, 135, 142, 143, 147, 153, 157, 159, 160, 165,

197, 202, 217, 219; buys big mare 78; buys sheep 132; examines black mare 217; given barley 186; goes to fair 121; lent a strapper 170; lent horse 83; makes gridiron 90; NP borrows from 209; NP repays debt 201; pays for grazing 195; pays NP 86, 91; paid on account 211, 212; rides Dublin mare 140; sold sheep 184
Daniel, E. 55, 60, 61, 68, 90, 116, 117, 148
Daniel, John 64, 86, 95, 100
Daniel, Law, blacksmith, 84, 128, 133, 149, 200, 206
Daniel, Ma, dresses buckskin 132
Daniel, Mick 108, 116, 151
Daniel, Peter lent money 135
Daniel, Simon 120, 160, 165
Daniel, Tho. 132
Dohig, Arthur 159, 160
Dondon, Joan 213
Dondon, E. 63
Dondon, Jon 54
Dondon, Henry 47, 48
Dondon, Joan 219
Donegan, —, found guilty 157
Donegan's jury, NP empanelled on 157
Doneraile, Co. Cork 177
Donevan, D, pays for grazing 141
Donevan, Jon 52, 64, 89, 124, 136, 149, 157, 185
Donevan, Jon, his sister paid for weaving canvas 83
Donevan, Maurice 151
Donevan, Jon, paid money owed by Mrs Widenham 125
Donrel, Paddy, sold a strapper 210
Doolin, Darby paid in beans for cutting turf 82
Downs, —, NP dines with 142, 175, 176
Downs, Henry 179, 212, 213, 218, 228; dine with NP 180, 185, 187, 189, 200, 223; Mrs Downs 196; NP and Catherine dine with 197; NP and Catherine stay with 193, 222; NP dines with 189; NP visits 215; stays with NP 198
Downs, Jack, NP and brother dines there 181, 185
Downs, Jon, and his wife, dine with NP 189; christening 178; NP and Catherine dine with 202; stays with NP 197, 227
Doyle, Peter, tithe proctor 58, 118, 141, 143, 144, 156
Doyle, J., sold barley 135
Doyle, J., given a bridle and stirrup leathers 155
drink:
and a dram 153
beer: barm bought 139; price of 131; small 123, 124; small, bottled 123; water drawn for brewing 141

brandy 55, 105, 152, 158, 167, 169, 174, 178, 200, 220; brandy and lemon 83; dram 56, 64, 74, 104, 105, 108, 113, 125, 126, 128, 150, 153, 164, 189; dram and sugar 126; dram and turn 96
cider 81, 85, 92, 101, 102, 103, 105, 119, 120, 123, 124, 125, 126, 127, 133, 134, 139, 145, 149, 150, 151, 153, 167, 170, 172, 173, 175, 176, 183, 187, 198, 191, 193, 194, 195, 197, 215, 219, 223, 227; bottled 107, 114; drawn for Dublin 111; marked 127; racked 147; a ronlet of 121
hops 81, 89, 95, 104, 107, 116, 128, 140, 159, 174, 176, 178, 179, 180, 181, 184, 187, 189, 192, 193, 195, 198, 203, 207, 210, 212, 215, 216, 218, 220, 222, 224, hops, bought 81, 128, 176, Permont water 56
malt 95, 174, 176 177, 180, 193, 195, 198, 216, 200, 220, 224, 227; brewed 101, 108, 115, 119, 135, 141, 181, 198, 201, 204, 218; ground 83, 119, 104, 140, 141, 145, 158, 213, 224; Limerick 96; milled 171, 181, 185, 187, 191; sold 116; winnowed 133
punch served 134, 145, 164; made with rum, brandy and sugar 152; rum and brandy 152; sneaker 148
rum 47, 90, 105, 167, 169, 174, 183, 184, 185, 186, 194, 196, 197, 203, 209, 213, 215, 216, 218, 220, 223, 227, shrub 75, 80, 81, 180, 117, 139, 149, 180, 187, 204, 205, 212, 213, 214, 216, 217, 223; rum shrub 219; whiskey shrub 192, 196; whiskey shrub, broken 193
whiskey 125, 167, 183, 184, 185, 186, 187, 193, 204, 205, 209, 217, 219; ale bottled 91, 153; ankers of 117; bottles 187; bottles washed 91; cooper and son working 83; coopers paid 136; corks 87, 84, 196; metheglin 216; NP drank hard 176; NP drunk 116, 153; people come to 92; raspberry wine 124; sent to Dr Martin 108; sent to Limerick 122; spirits 145; stilling 122, 123, 129; stopped 115; tuned 108, 115, 174, 179, 181
wine 61, 68 104, 105, 124, 125, 150, 154, 156, 195, 203, 215; claret 181, 187, 204, 219; port 219, Lisbon 181, 183, 187, 203, 219, Lisbon, bought to patch up quarrel 138
drinking: NP spends day 151; NP spends at fair day 159; NP spends New Year's Eve 167
Drury, Tho 93
Dublin: glasses, cheese and carpet sent to 127; horses return from 149; NP goes to 113, 140; NP leaves 140; NP sent for urgently 108

Index 235

dye: dyeing frieze 81; dyeing threads 215; indigo 64, 83, 201, 203, 214; madder 185 190
dye stuff 65

Edigan, Martin 128, 146, 149
Edigan, Marg 177
Egan, Maurice of Charleville, Co. Cork 183
Elltown/Elton, Co. Limerick 109, 112
Enniscouse, Co. Limerick, NP visits 84, 94, 96, 99, 161
entertaining 47
Evans, Sam 99
Evans, Henry, saddler 73, 74, 77, 75, 78, 89, 91, 93, 127, 149, 150, 158, 169, 176, 195, 223, 224; serves subpoena on NP 226
Everitt, Jon 85

Faha, Co. Limerick 114
fairs 22, 60, 65, 87, 189; at Rathkeale 62; cattle to 95; hoggets sent to 81; Jon buys spreadtrees at 89; lambs, hogetts sold 82; wethers sent to 79
farm, NP takes possession of 50; barn broken open 161; cabin made for lambs 59, 93; cleaning barn 83; fencing 50, 78, 80, 82, 98, 119, 225; mice, a vast quantity in wheat 96; stable cleaned out 90, 157, 164
farm implements: axe 202; axletrees 180; baskets for drawing turf 84; bucket chain 50; churn 78; coop made 181; dray made 84; Dutch spades 50, 100; firkin 78; grind stone 213; hackles bought 113; harrow mended 89; hay knife 56; hoops 82, 174; hurdles bought 91; Irish plough sole 57, 59; Irish plough sole and irons lent 171; iron for truckles 82; plough irons 185; plough irons mended 79; pole axe 151; pounder at work 71; reap hook 97, 194, 203; riddle 50; rope 137; scythes 51, 108, 141; scythe stone 52, 43,; shovels bought 58, 186; shovel paid for in barley 187; sieves bought 107, 211; slane [spade] made 201; spade 74, 195, 197, 219; spoke shave 213; staves 83; sticks to make plough soll 183; swingle 70; truckles 64, 65, 82 179, 203; truckle car made 141; truckle mended 90; truckle size 78; truckle wheels shod 65, 83, 189; winnow sheet made 57, 100, 134
farming: cabbages 56
digging 59, 180; Court, digging for onions 154; digging for cabbage 107; digging for peas 222; digging for plants 222; second digging 138
dung: dunghills 70, 101,; spreading 75, 77, 80, 91, 155, 187, 212, 218;
fallow, 52, 56, 59, 69; fallow burned 51, 66; grazing 62, 82; fallow stubble 53
harrowing 61, 67, 75, 80, 155, 171

haymaking 83, 141, 177, 178, 180, 190, 204, 205, 215, 228; haystack 192, 215; iron salt 82
kiln 79; kiln roof collapses 151; kiln making lime 88
lime 65, 141, 146
parsnips 55, 56, 90; dug 91, 133, 156; sown 155
peas 187; plants bought 64; set 77, 155, 180
ploughing 48, 49, 51, 53, 54, 57, 59, 62, 66, 69, 70, 80, 88, 89, 93, 134, 156, 161, 171, 184, 198, 199, 206, 209, 211, 212, 219, 223; cross cut ploughing fallow land 80, 97, 184; fallow 89, 107; flax 224; ploughing and harrowing 60, 78; ploughing Portacacha 70; ploughing Sally bed 62; second ploughing 67
potatoes: Court, digging for potatoes 154; dug 55, 90, 97, 98, 103, 107, 131, 155, 171, 173, 179, 184, 187, 200, 206, 219, 222; earthed 62, 89; new trenched 213; planting 57, 60; ploughing for 209, 223; second digging 200; suffer from frost 73; trenching 62, 82, 172, 224
quicks bought for hedges 87, 93
reaping 53, 64, 178
reeds cut 54, 68, 127, 148, 179, 192
ricks destroyed by storm 88
sowing 48; bags bought 107; meads 54; parsnips 58
stones collected 64
stones, removing from land 86
sulphur 107
threshing 47, 65, 69, 71, 74, 75, 134, 150, 175, 222
weeding 51, 52, 101, 202
winnowing 47, 101
Farranan, Thos. 97, 99, 101
Farrell, Anthony 53, 57,
Farrell, Anthony, his widow 70
Farrell, James, paid money owed by Mrs Widenham 125; mowing 163
Farrell, Robert 75
Farrell, Torlo[?], paid for baskets 169
Feeneen, Darby, pays for grazing 129, 141
Feigh, M. 56, 71, 88, 129, 134, 149, 156, 174, 197, 198, 220
Feigh, Mick, paid for horse 163
Feigh, Mick, NP visits 179
ferry 120; fee 189; proposals for 134; profit from 142; rent from 142, 143, 144, 145
Filim, servant 204
Finch, Jon 136
Fisher, — 81, 86, 192, 195, 203, 213, 210; sells iron 188, 189
fitzGerald, Anstice 164, 176, 180, 196, 204
fitzGerald, David 112
fitzGerald, Edward 48, 50, 54, 57, 62, 84,

86, 96, 97, 108, 115, 127, 144, 145, 155, 156, 195; breaks pendulum of clock 105; pays rent of ferry 141; sends surrender of ferry 124; sent to Mrs Bastable 125; stock seized 131
fitzGerald, Honor 104, 109
fitzGerald, James 86, 162, 164, 185, 187; fixes grate 173; issues bill in favour of Tim Herlihy 164; mends oven 181; paid for building house walls 197, 215; paid for mending church wall 160; paid in part for new house 160; paid masons 191; stone colt delivered 162
fitzGerald, Jon, of Ardlahan, Kildimo 56, 63, 65, 66, 68, 74, 85, 87, 122, 109, 117, 131; NP asks for money from 92; pays for grazing 226; pays Mrs Widenham 85; promises to pay 67, 86; stock seized 57
fitzGerald, Mrs Howard, NP lends her money 104
fitzGerald, Margaret, paid earnest 207
fitzGerald, Martin 97, 100, 112, 114, 116, 117, 142, 145, 147, 148, 151, 153, 162, 163, 164, 165, 169, 171, 173, 175, 177, 187, 189, 217, 222; boy from 185; cuts turf 99; given bill 194; his brother 169; his horse 149, 211; his man trenching potatoes 98; NP and Catherine dine with 189; NP borrows from 164, 169, 174, 188; NP lends to 200; NP makes dairy agreement with 165; NP repays money 124; NP sends him cows 170; NP settles accounts with 118, 121, 129, 133, 154, 155, 183, 188, 197, 209; paid rent 182; pays for grazing 141, 148; pays horn money 206; pays mason 163; pays NP 213; returned guinea 212; sells butter 122, 176, 178, 180, 181, 192, 193, 194, 195, 201, 202, 203, 205, 207; sells NP calves 123; sold cows 159, 211, 212; sold sheep 184; tells NP his cow dead 156
fitzGerald, Mrs longs for sucking pig 75
fitzGerald, Mary spinner 77; dismissed 79
fitzGerald, Phill 86, 96; grazing paid in potatoes 97
fitzGerald, Richard 61, 205
fitzGerald, Thom 57
fitzGerald, Valentine 65
fitzGeralds, NP gives dinner to 197
Flahavan, Will 59, 73; affair of courtship 69
Flahive, James bleeds NP 65
Flanigan, R. 152
Flanigan, E. 201, 204; and wife, dine 215; NP and Catherine visit 190, 226; NP dines with 222; NP settles accounts with 226
flax 20; bags for seed 76; bogged 83; bought 56, 206; certificate for 76; dressed 119; dressing Mrs Hartstonges 118; dried 180; hackle mended 181; hackled 57, 73, 89, 166, 182, 207; hackler begins Mrs Hartonges 122; hackler finishes 154; hacklers paid 125, 205; pulling and binding 65, 66, 145, 146; seed, bought 211; seed cleaned 109, 155; seed ordered 111; seed sent to Limerick 76, 127; seed sold 79, 132, 157, 206; sent to Mrs Hartstonge 123; sold at fair 148; sold to Dr Martin 151; sown 63, 82, 199; sown for Mr Hartstonge 98; threshed 68; threshing 75, 85, 107; tongesing 85
Flyn, James 148
Fontenoy, news allied army beaten at 139
fodder: beans: dug 91, 171; bought 210; halm burned 82; price of 134; pulling 54, 66, 178, 191, 192; sent to Court 134, 185; sold 59, 133, 172; sowing 58, 49, 62, 78, 80, 90, 93, 105, 133, 138, 155, 171, 180, 184, 199, 219; sown in rush corcas 171; stored 55; threshing 48, 49, 58, 81, 90, 107, 154, 155, 184; to mill 82; traded for bed ticks 58; winnowing 78, 184, 154, hay 65, 68, 87, 217; ditch dug around yard 87; gathered in 87, 119, 123, 143, 145, 146, 163; mowing 52, 54, 218; NP sells 69, 135, 136, 137; rick made 218; rick thatched 151, 206; sold 119; yard cleaning 52, 66, 82
peas: cleaned 108; foul 154; harrowing pea field 112; pulling 66; sowing 60, 80, 105, 107, 154; threshing 69, 154; winnowing 69, 107, 154
food: apples 126, 147, 218, sent to town 123
asparagus 187
bacon 147; and beans 144; given to Sister McGan 80; hog salted 155; resalted 136; sent to town 123
bread 119, 120, 121, 124, 125, 126, 127, 131, 133, 134, 135, 137, 140, 142, 143, 144, 148, 149, 151, 152, 155, 156, 158, 159, 160, 161, 165, 169, 189, 190, 215, 220, 222, 225; a loaf of 155
beef 124, 125, 187, 198, 205; given to Sister McGan 80; salt 129, 209; sirloin of 217, 227
cheese 125, 196,
eels 150, 151; eelfry 210, 211, 135, 185, eggs 94, 149
fish 125, 175, 212
flour, English 217
fruit cake 91
geese boiled and roasted 147
goat 149; kid 118
ham 75
hare 129, 152, 165,
herrings 136, 198
honey 221
lamb 143, 158, 161, 170; beheaded by fox, for dinner 138; roast 140

lemons 186
liquorice 210, 223
lobsters 187
melons 215
milk 96
mutton 107, 108, 121, 145, 189, 214, 216, 218, 221; sent to town 104
oatmeal 93, 135, 145, 165, 201, 214
oil 54, 63, 70, 85, 95, 115, 121, 140, 141, 142, 179, 184, 194, 195, 206,
onions 196
oranges 183, 186,
oysters 104, 124, 155
parsnips 143
partridges 120
peas 220
pork: chine of 107; roasting, sent to Court 104; salted 75, 124, 132, 134, 165, 183, 209, 218, 221; sucking pig 71
potatoes 80, 119; new 143
rabbits 120, 183
raisins 195, 204, 210, 217, 220, 228
rice 186, 195, 203, 219, 222
salmon 137, 138, 147, 184, 201, 214, 224, 226
salt 58, 61, 70, 75, 85, 96, 98, 101, 105, 108, 109, 111, 121, 128, 132, 134, 139, 143, 151 157, 161, 165, 174, 176, 187, 192, 198, 201, 206, 209, 210, 213, 216, 217, 219, 220, 221, 227; Lisbon salt 137
shrimps, sent to Mrs Quin 141
spices: cinnamon 185, 216; cloves 217; food, ginger 210; nutmeg 185, 193, 195, 202, 211, 219; pepper 104, 135, 181, 195, 202, 210, 214, 217, 220
strawberries 101
sugar 90, 105, 135, 150, 158, 167, 169, 184, 186, 187, 189, 190, 194, 198, 202, 209, 211, 214, 215, 220, 203; jamaica sugar 181, 184, 186, 192, 204, 217, 222; powder sugar 181, 183, 186, 192, 204, 211, 217, 222; sugar candy 216, 220, 223, 224
tea 125, 184, 203, 211, 224
treacle 221
turkeys 51, 71; gift of 75; paid for 146; roast 108; sent to town 108
veal 199
vinegar 209, 212, 214, 217, 221, 228
forge. Irish, made 62; made at 131; plow iron made 59, 155; plow iron mended 51; spades, Dutch, made 63, 65
Fosbery family dine with Peacocks 175, 183
Fosbery, George, 48, 53, 54, 60, 62, 67, 68, 117, 120, 132, 138, 195, 200, 201, 213, 216, 218, 220, 222, 224, 227; dines at Court 144; dined with NP 137; paid for malt 207, 211; pays for grazing 163; stock grazed at 108

Fosbery, John 47, 49, 50, 52, 55, 60, 63; mends gun 81
Fosbery, John, dead 61
Fosbery, Mrs: lends harrow 78; lends thread to NP 93; pays church rates 63
Fosbery, William 58, 98, 174; agrees for grazing at Kilmoreen 158; breakfasts with NP 141; dines with NP after hunt 145; dispute about cattle 169; NP has difference with 162; pays for grazing 163
fox: kills geese 193; beheads lamb 138
Fox, Eleanor 164, 178, 179, 180, 200, 225
Fox, — 152, 209
Fox, Matt, NP repays money 155, 172, 210
Frost, Will, paid for cess 215, 223
funeral: coffin ordered 88, 200; gloves and hat band 119; Molly Quin's 95; Mrs Chapman, NP attends 209NP gets scarf and hat bands 103; NP hands out scarves and gloves 88; NP organises herse 88; NP pays bills for 88
Furnill, —, NP sells hay to 137
furniture: alphabet 99
bed 104, 141, 160, 179; bolster 160; headboards 71, 160, 179; moved down 180; NP lacks 111; put up 203; tester for 78, 160; vallance 160; wrought and blue 133
locks: for chest 70, 75, 90
chairs 104, 126, 144, 216; mended 90; NPO buys 99
chest of drawers 109
chests 58, 61, 73; sea-chest 109, 160
closet: mended 83
cradle: bought and fitted 187, 205
desk, bought at cant 131
drawers and desk 179
dresser 103
gardevine 109
Hartstonge furniture; sent to Cosgarys 109; sent to Shannongrove 109
looking glass 216; price of 92
sconce-glasses 109
scroles 109
suites of brought from Portacacha 126
table 81, 100, 126; breakfast 160; card 160; large 160; leaf made` 196; NP buys 99; tea 160
tongs and poker 116
trunk, leather-covered 109

gadds, 57, 155
gale, May paid 165
garden 226
beans planted 107, 154, 157, 200; threshed 61, 69, 90; harvested 88
cucumber plants for 158
digging 80, 114, 224; for cabbage 91; for parsnips 79, 98
ditch 221

fencing 60, 180, 185
finished south of the house 155
men working in 158, 215
NP in 159
oats 156
partridge peas, saved 79
peas: winnowed 90; threshed 90; harvested 88; sent to Dr Martin 108
ploughing 80; for peas 91, 95
potatoes 79, 80, 210; quality of 146
red wheat harvested 88
roses picked, to still 142
seeds sown: cabbage 66, 76, 77, 100, 222; carrots 61, 77, 79, 82; cauliflower seeds 212; garlic 198; hot bed, seeds sown in 157; kidney beans 79, 89, 158; leek 61, 112; lettuce 77; oats, 'naked' 80; onions 61, 112, 222; parsley 77, 158; parsnip seed sent by sister 92; peas 82, 154, 224, Short Hotspur planted 97, Long Hotspur planted 97, blue 79, 95, Glory of England 77, 79, 95, partridge 77, 95, set for eating 95, dwarf planted 97; radish 112; saffron 160; seeds 135, 185, 224; shallots set 61, 77, 91, 222; turnip seed 77, 158
sally bed 78
weeding in 140
west sown 211
Gardiner, Widow 215
garland, NP buys 129
garrott: wheat stripped by storm 110; thatched 110
Garrott, — 110
Geany, Phill, grazing paid for 230
Geary, John of Ardnavolla 149, 150, 164, 187, 190; christening 183; and his wife 169, 188; and his wife, dine with NP 179, 192, 203, 228
Geran, Maurice 184, 211
Gerin/Geran, John surveyor 61, 163, 181, 182; surveys Kilmoreen park 163
Gibon, Thomas 49, 60, 61, 62, 204, 212, 214; his boys 53, 54, 55; buys pig 105
Gibon, Maurice 211
Gillbreath, —, weaver 67, 193
gisement/gusmand 67
Gleesan, Thos 222
Gleesan, Thomas of Adare 220, 221, 222, 223, 224, 227
Glin, Nell 71
godfather: cost of being a 96, 124, 183; NP stands as 96
Gore, Revd. Francis, NP dines with 177
Gorman, Den 128, 174, 226
gouge, bought 95
Grady, Coghlen 98, 115
Grady, Coghlen 115
Grady, Standish 113, 114, 115, 131, 157; his horse 63; NP writes to 111, 112; seals

doors at Court 108; writes to NP 109, 110
Grady, Matt, pays for grazing 155
Graham, Mary see Widenham, Mary
grazing, money for 140, 141, 148, 155
Gready, Patt 52, 55
Gready, Mrs changes frieze 92
Greag, —, spinner 200, 203, 220; NP and Catherine dine with 225; NP and Catherine stay at 211
Greenfield, Co. Tipperary, NP stays at 177
Griffin (or Griffiths), paid church rates 211, 212
Griffy, Paul, sent plants 107
Griffy, James 98
Griffy, John of Adare 91, 140, 146
Grilla/Grillagh, Co. Limerick 147
Grimes, Betty: given a bushel of barley 138; NP pays her tithes 137; shirts to be made by 83, 143; son sent for money 132
groom, given tip for service of mare 180
guard, NP on 111
gun 56, 60, 63, 74, 225; gunpowder and shot, bought 93, 118, 152, 195; mended 180, 193; NP cleans 105; paid for with oats 80;

haberdashery: buttons 89, 90, 189, 195, 202, 221; moulds [for buttons?] 189; shirt 105; sleeve buttons 80
cadis 190, 210, 215, 216
combs 183, 215
lace 216
needles 51, 75, 90, 128, 133, 155, 172, 183, 206
pins 185, 190, 198, 202, 211, 219, 220, 223
quilting frames 116
ribbon 176, 181, 190, 195, 210
scissors 150, 181
steel hooks and eyes 131
tape 124, 198, 203, 204, 213, 215, 216, 221, 224
thimble 89
thread 62, 75, 185, 165, 180, 182, 187, 195, 206, 209, 210, 211, 220, 221, 224
hair, bought to make ropes 132
Hall, P. 81
Hall, B. 164, 177
Hall, Jon 81
Hallinan, Dea, tailor 55, 77
Halloran, Bridget, spinner 63, 70, 71, 73, 74, 75, 76, 77, 79, 80, 81, 83, 89, 93, 97, 117; brings tow thread 94; buys malt 100, 102; makes NP mittens 90; pays NP 101
Halloran, Jon 53
Halloran, Patt spinner 51, 52, 55, 59, 60, 75, 79, 80, 81, 97; wife knits stockings for NP 92
Halloran, Matt 58, 60, 64, 82, 83; mare ill with fever 84

Index

Halloran, Thomas 48, 49, 55, 58, 59, 60, 78, 81, 128; pays NP for grazing 132; to be married 67
Halpin, Mick, slate dresser; repairs house 125, 127, 141, 142, 149
Haly, Paul 48, 140, 141, 143; house burned down 117; mends harrow 110; NP borrows money from 110; NP pays for filly 141
Haly, Jon 115; paid for cow 60, 104; pays for grazing 155
Haly, Matt 140
Hanall, Mary 180
Handy, Mother, midwife 186, 203
Hannan, Jon 52
Hannan, Mary 55, 58, 59, 61,
Hannan, Mary paid 186, 190; paid and turned away 192
Hannan, Patt 213, 214
Hannan, —, paid for weaving canvas 195
Hanrahan, Tho 83, 89, 118, 121, 126, 128; buys lambs 213; goes to Ballylongford 68; goes to fair 121; Hartstonge furniture sent to 109; pays rent 187; repays money to NP 91; returns from Dublin 129; sells cow to NP 111; sent to Dublin with horses 119
Hanrahan, Patt 84, 86, 118, 119, 128, 135; pays for grazing 128, 178, 202, 205
Hanrahan, Matt 217; pays for grazing 192, 226, 228
Haragan, Mary 64
Harding, Sam 66, 86, 107, 116, 134; asks NP for money 131; claims compensation against NP 113; hires NP horses 87; invites NP to dinner 132; lends NP money 108, 115; NP hires his men 97; stays with NP 153; repays loan 131
Harding, Will, stays with NP 209
Harnedy, Richard 111, 117, 123, 165, 179,
Harnedy, John, apprenticed to NP 179
harness: bit and snaffle 159; bridle 117, 136, 149, 155; cavison 197; collars for horses 65; curry comb 90; finished 74; girth 63, 136, 150, 195; halters 165; made 73; martingale 197; pillion 184; saddle cloth 127; saddle paid for 89; side saddle bought 173, 175, 226; side saddle lent to Harry Downs 228; spurs 149, 212; spur rowells 198; straps and buckle 124; stirrups 89; stirrup leathers 155; whip 153, 173, 220, NP loses handle of 150, mended 122
Harney, John, refuses to take back light guineas 212
Harragan, Connor 73, 79, 81, 82, 86, 87, 190, 191, 194, 217, 218, 225; cuts thatch 88; goes to Limerick with barley, wool 99; goes to Limerick with wheat and oats 90; hire agreed 87; takes barley to Pallaskenry 99
Harragan, Edward, NP lends money to 216

Harragan, Darby 61, 75, 97; digging potatoes 99; fencing Caparoe 99; planting beans 108; threshing 99; threshing wheat 100; spreading dung 107
Harragan, Elenor, hired 177
Harragan, James 151, 224, 225, 226; and wife dine with NP 221; NP borrows from 203, 205, 207; NP repays debt 205; pays for grazing 194, 202, 204, 212, 213, 215, 216, 218
Harragan, Matt, paid 220
Harragan, Thomas 55, 67, 112, 116, 128, 151, 164, 171, 173, 174, 176, 178, 179, 199, 201, 215, 221, 224; and wife dine with NP 185, 221; given minute on Kilmoreen 198; ground measured 194; NP and Catherine dine with 225; NP borrows from 153, 169, 172, 178, 180, 181, 188, 189, 190, 198; NP buys cows from 184; NP repays debt 155, 159, 209; NP settles accounts with 142, 159, 198, 200; paid country charges 162; paid on account 188; pays for grazing 131, 149, 150, 218, 219, 202, 206, 213, 216, 226, 227; pays NP 182; pays rent 166; rent for Caparoe agreed 142; sells NP a cow 152; sells NP cider 194; takes Kilmoreen grazing 210
Harrold, Margaret, paid for her son 207
Harting, Mick 183
Hartney, Dennis, pays for grazing 128
Hartney, Jon 52, 108; dines at Court 144; NP dines with 112; pays NP for grazing 120
Hartney, Mick 156, 181, 187; funeral of 215; proposal for Court grazing land 134, 135, 136
Hartney, Patt, saddler 175; paid for side saddle 176, 179
Hartstonge, Alice 12, 48, 56, 57, 64, 79, 96, 114, 115, 116, 117, 121, 125, 129, 145, 169, 187, 226; Arthur Bastable pays 224; better 146; dines with NP 219; paid for sheep 173; gives NP expenses to go to Co. Cork 147; goes to Croagh 202; goes to Dublin 108, 181; goes to Limerick with NP 145; lends money to NP 183; letter from 135, 139, 173; NP borrows money from 186; NP explains abrupt departure to 173; NP lends money to 82, 87, 88, 100, 166, 178; NP settles up her accounts 123, 144, 204; NP writes to 117, 150, 131, 133, 153; NP visits 177; receives accounts for funeral 88; repays money 91; replies to proposals for Court land 136; returns to Limerick 120; sends for horses 119; sent for 203; sheep taken 183; stays with NP 123; taken by NP to Court 124, 144; unwell 145; with Catherine in labour 186
Hartstonge estate: beds sent to Corcamore 109; canted 113; canted goods at

Castletown 131; cattle seized 113; china dug up 127; coach geldings 116; Court and Coroheen advertised 110; cupboards at Court opened by sheriff 113; enquiry about children's cattle 113; goods seized and canted 108; goods taken to High Sheriff 114; grazing agreed 115; inventory sent to Mrs Hartstonge 110; land will not be set 112; legal advice on 110; Mrs Hartstonge buys goods canted 125; NP buys boards at cant 113; NP buys books from 113; NP buys slates at cant 113; NP writes to 117, 150, 131, 133, 153; pigs brought from Porte 127; ploughing stopped 110; spies out 109; stock listed 115; tenants 115; reversion served on Limerick tenants 119; wine sent to Dublin 118

Hartstonge family 12; come to see NP 171, 181; dine at Adare 162; go to Dublin 206, 148; go to Co. Cork 163; Hartstonge ladies invited by NP 183; in Limerick 103; NP meets on road 183

Hartstonge, Francis 12

Hartstonge, Henry 49, 68, 77, 127, 146, 147, 153, 158, 198; bill on 149; birthday celebrated 163, 177, 190; breakfast with NP 201; chosen as way warden for Kildimo 169; dines with NP 122, 123; dines with NP after hunt 145; goes to Bruff with NP 175; gives NP money 94, 99; his mare 110; health drunk 119; ill 121; letter from 132, 133, 134, 135; NP accounts for money to 99; NP asks for money from 91; NP buys heifer from 126; NP lends money 90, 147; NP sleeps in his bed 126; NP writes to 110, 133, 211; sends for NP 95; sends for NP 172; visits Adare 120

Hartstonge, archdeacon John 12

Hartstonge, Lucy 145

Hartstonge, Pryce (1692–1743) 12, 52, 63; death of 108; NP lends money to 102; very ill 108

Hartstonge, Ralph 12

Hartstonge, Standish (1627–1701) 12

Hartstonge, Standish (–1751), death of 227

health: balsam 184, 196, 221; brimstone 220, 221; Catherine bled 205; Catherine has toothache 203; Catherine Peacock ill 180, 188, 205; Catherine refuses to have tooth drawn 205; doctor visits Court 145; doctor visits NP 102; manna 189, 195, 222, 204; measles 209; medicine paid for 194; NP bled 189; NP buys physic 82, 107; NP falls off horse 92, 173; NP has cold 105, 146; NP has earache 183; NP has pleurisy 102, 179; NP has toothache 149; NP ill 52, 61, 65, 82, 84, 90, 107, 117, 126, 144, 145, 148, 156, 166, 169, 185, 186, 210; NP ill with ague 141; NP has hangover 117; NP ill and takes physic 89; NP loses tooth 107; NP has earache 183; NP and Catherine bled 175; NP takes physic 69; NP takes puke 186; oil, of marshmallows 148; Pryce bled 223; pills and ointment 178; Pryce ill 194; puke; bought 198; tar surfit water 219, 221; tooth drawn 175; vomits bought 83

Hearlyhy [Herlihy], Tim 124

hearth money paid 50, 82, 98, 137, 173, 185, 224; not paid 67; paid for Jon Bryen 225

hearth money collector, paid with bill on the Westropps 122; paid by NP to seize glasses 113; stays with NP 158

Heas, Bridget 61, 62, 75

Heas, Jon lays foundation of the stable 85

Heas, Jon 55, 64, 71, 87, 88, 133; breaks stones to fill kiln 84

Heas, Dan 47, 53, 206

Heas, Dan, his son trenching potatoes 98

hemp 61, 66, 190, 204; bogged 191, 205; breaking 201; dressed 184, 205; hackled 70; hemp brake formed 191; kiln to dry 69; pulled 190, 191, 194; seed sown 186, 224, 225; steel 134

Henesy, Morton, paid for hire 177, 180

Henry, Simon 152; and wife dine with NP 179, 181; herdsmen, at Killmoreen 143

Herlihy, Tim 143, 149, 164, 167, 176, 178; paid church rates of Kilmoreen 198; pays for Terence Cenedy 133

Hewson, Robert of Ballyengland, Co. Limerick 112, 115, 122, 149, 222; and wife dine with NP 185; gives NP a sow 122; May rent demanded by 148; May rent paid 146; NP and Catherine stay with 178; NP asks for money 123, 127; pays for grazing 126, 128; rent paid 135; refuses to accept part of rent 141; sent money for Mr Thornhill 163; sent wooden ware 134

Hewson, Mrs, godmother 217

Hewson, George 193; his cant 223

Hewson, Mrs, her boy 217

Hewson, James paid for service of mares 207

Hickey, Mick 52, 55, 65

Hickey, Ned pays earnest 86

Hickey, J., pays for grazing 152

Hickey, Patt, wages paid 161, 167, 170, 172, 174, 183, 190

Hickey, Will 68, 133, 179, 219; makes breeches for boys 118; pays for grazing 204

Hines, Eleanor 61, 68, 69, 70, 82, 82, 87, 135; ill 69; spoils NP's frieze 119; wages for year 87

Hodge, Sarah 74

Hogan, Jon, pays for grazing 124, 134; made churchwarden 157; mowing 163

Hogan, Peter, horse rider 165, 170, 197; given bill on Thomas Harragan 171

Index

Hogan, Patt 217
Hogan, Patt 53, 64, 217, 218; hogsheads, bought 124, 125, 140
Hohehan, Mary spinner 71, 76, 91, 94, 98,
Holehan, W., thresher 100
Holly Park, Co. Limerick 56, 146; Catherine visits 228; NP draws timber from wood 88
holy days 52, 63, 66, 84, 94, 145, 157, 179, 213, 219
hoops 213
horse rider 69, 164, 165; doctors colt 82; takes mare in hand 68; trains bay colt 84
house: bedroom cleaned 73; cleaned 124; glazier leads window 179; new window 179; NP cleans his room 144; NP fits out his room 179; NP settles the 174, 217; room enlarged 179; size of 201; slating agreed 185
household linen: baskets 169, 179, 180, 185; bed ticks 58; blankets, silk and coarse 141; blanketting 187; brass inkhorn 213; brooms 179; cadow, for the Palatine 104; curtains 75, 191, 193; curtains, NP makes 75; glasses 105, 113, 216; grate 173; knife blade 175; linen drawers 170; locks 83, 124; meal tub 83; padlock 70, 110; pillow cases, NP makes 128, 131, 132; powder 105; powdring tub 83; quilt 69; sheets, coarse 143; sheets, made of new bandle cloth 143; sheets, new put on bed 104; sheets, NP makes 103, 115; tablecloths 68; tablecloths, diaper linen 67; ticking 124, 128, 152; tin 'things' 201; towels 67, 68; tumbler 192, 205; whiteing 219
Hoyle, Susan 12
Hungary water 203
Hunt, Robert of Askeaton NP dines with 101
Hunt, Jon, 224; dined with NP 192, 227; godfather 217; NP dines with 227
Hunt, Mr, NP and Catherine dine with 216
Hunt, Vere of Friarstown, Co. Limerick 103, 184, 192
hunting: NP goes and fails to find 125; with Henry Hartstonge 123.145, 187, 193
huntsman, tipped by NP 123
Hure, Patt, horse rider 70, 84, 89, 124, 140, 143; breaks horse 96; pays for horse 100; rides gelding 98, 99

ink 110, 163
Ireland, Coursey 73, 87; funeral of 94
Ireland, Elizabeth, see Attkins, Elizabeth
Ireland, Norse 126; removes his mother's goods 94; repays loan 105
Ireland, Sister: gives NP woollen gloves 86; gives receipt 211; goes home 209; moves to Rathkeale 97; NP divides her effects 112; NP lends money to 133; NP writes to 210; paid for beans 102; sends shirts 107; sent firkin of butter 100; stays with NP and Catherine 204; visits NP 96; wants her things moved 94
iron and steel 49, 58, 65, 84, 87, 89, 92, 98, 99, 102, 119, 137, 139, 150, 154, 155, 159 160, 161, 165, 170, 174, 175, 181, 185, 188, 194, 198, 201, 202, 211, 214, 215, 219, 222, 223; Irish iron 70, 192; piggins, iron 213; steel 49, 65, 105, 172, 180
Ivers, —, paid for light 214

Jack 48, 53
Jackson, Tho, chandler 51, 71, 83, 94, 112, 143, 144, 145; advised about Mr Hartstonge's deer 104; advises on tenants' quit rent 127; delivers beds 133; dines at Court and divides goods 125; gives up pictures to Mr Mellsop 111; NP buys peas from 80; receives Hartstonge cattle and pictures 109; tells NP of betrayal 109
Jackson, Joseph, attorney, Limerick 154
Jackson, Robert 172, 197; pays for grazing at Court 157
James, Jon stacks wheat 86
Jermy, Elizabeth 12
Johnston, Mrs of Limerick, maltster 92, 95, 154
Johnston, John, 80, 112, 116, 117; and son, breakfast with NP 142; breakfasts with NP 155; his son, eats all NP's cheese 138; NP stays with 176; NP visits 188; NP writes to 111; visits Court 138
Johnstown, Co. Tipperary 113
Joynte, Dudley, dines with NP 223
Judy, — 112, 171, 194; dines with NP 179; stays with NP 193
juries 25
jury: discharged 157; NP discharged from 174; NP on Grand Jury 169; NP serves on at trial of Hugh Massy 145

Keating, Valentine 109
Kelly, — 190
Kelly, —, servant 81
Kenny, Patt 112
Kent, Simon 59, 187
Kilbreedy, Co. Limerick, NP visits 86, 100, 129
Kilbreedy, Man, pays for grazing 132
Kilceed/Kilkeedy church, NP attends 171
Kilcorly, Co. Limerick 47, 52, 63, 151
Kildimo, Co. Limerick 51, 85, 157, 158, 183; Kildimo green 69; new house at 158; new house at, foundations marked out 160; new house at, scaffolding made of wattles 163; new house at, walls started 161; NP dines at 119, 159
Kildorrery, Co. Cork, fair 121, 228
Kilfinney, Co. Limerick 87, 90

Kilgabon/Kilgobbin, Co. Limerick 64, 65, 104, 108, 143, 161
Kilknockan, Co. Limerick fair 119
Killinan, Co. Limerick 202
Kilmallock, Co. Limerick: NP leases land at 87; NP sleeps at 150; NP visits 115
Kilmoreen, Co. Limerick: 50 acres to let 133; boundaries mended 117; boundaries to be settled 143; grazing agreement 201; grazing at paid for 125, 140, 141, 150, 151, 152, 182; grazing of 210; house at 212; men sent to 210; NP and Catherine visit 177; NP leases land at 111; NP lets west part 172; NP views calves 123; NP views cattle at 183; NP visits 54, 111, 128, 138, 139, 144, 148, 158, 160, 162, 163, 165, 169, 189, 190, 195, 199, 200, 210, 202, 211, 212; park surveyed 163, 182; rent of 194; shows M. fitzGerald land 154; stock sent to 160, 174, 176; terms of letting settled 184; Thomas Harragan given possession of west park 174; trees planted at 132, 166; viewed stock at 152
kiln, repaired 214
kiln, made 180
Kiltenan, Co. Limerick 165, 194, 206, 210, 211; NP and Catherine dine at 184, 191, 212; NP visits 222
kitchen; basting ladle 160; brass cock 83; bottles 116; chafing dish 160; cheese vat 104; colander 160; copper ladle 160; crock 75, 210, 215; fish kettle 160; fleshfork 132; grater 212; kettle and cover 160; knife 56, 213; lamp 160; masher 83; moulds 131; padlock 51; pot hangers 69; potting pans 195; salt box 104; sauce pan 173, 181, 187, 220; scales 69, 212; scissors 89; sieves 83; snuffers 89; steel 220; stewpans 160; stool for sieve 83; strainer 202; tartpans 195; wooden bowl 96; wooden ware 103
knitting 195

labourers: cleaned yard 123; cows pounded 160; hay making 146; hired 162, 219; mowers paid 68, 120, 133; NP accounts with 138; numbers of pulling flax 120; paid with beans 80, 87; slates gathered up 127; women cleaning barley 135
lamp black 70
land agent, duties of 17
land: bog surveyed 159; conditions for letting meadows 136; in Co. Cork, very bad 147; leased at Portacacha 49; lease drawn 153; meadow sold 122; meadows divided 144; NP measures meadows 101; NP leases in Co. Cork 153; surveyed 61
landlord and tenant law 17
laundry: blankets washed 141; ironing 144;

linen washed 78, 90, 107, 160, 115, 128, 129, 134, 141, 160, 176, 180; list of 170, 172; NP sends clothes to 164; soap 74, 75, 84, 98, 111, 113, 115, 121, 128, 137, 140, 148, 159, 164, 170, 174, 175, 181, 183, 188, 192, 198, 202, 206, 212, 216, 223, 226; starch 125, 176195, 213, 219, 226; stone blue 82, 137, 176, 180, 183, 188, 195, 212, 217, 190, 195, 198, 206, 223, 226; washerwoman paid 126
Leamy, Will pays tithes on turf 84
Leary, Tim 165
Leary, the pedlar 94
leather: bought 71; buckskin 93; calf skin 104, 183, 165, 166; cow hide sent to NPs sister 70; cow hide sold 56, 90, 105, 151, 165, 166, 185, 196, 198, 199, 201, 215, 209, 211, 220, 221; dressing 48; horse hide, sold 70; sheepskin 198, 219; skins 108
legal, fees: for affidavit 157; for attorney and clerk 153
legal processes 53, 57; administration sworn 133, 161; affidavit 158; fiery facies 270; NP gets administration 136; NP gets case sealed and signed 133; NP served with injunction 140; justicia 165; master in Chancery 170; process bought 81; process served 86, 157; replevin 153; subpoena and execution served on Henry Studdert 170; subpoena served 152; subpoena served on NP 226; warrant 62; warrant signed 62
letters: Edward McGan pays for 129; paid for 127, 132, 137, 138, 139, 140, 143, 163, 176, 181.191, 19 3.207, 227; NP receives 160; NP writes110; wafers bought for 64
Lewellin, — 203
liquorice 177
lights, repaired 214
lime kiln, burned and emptied 84
Limerick City 52, 79; assizes, NP goes to 157; cost of room in for assizes 157; gates opened 121, 126; gates shut at night 110; lodgings 126; NP breakfasts at 113; NP dines in 122, 124, 126; NP goes to 60, 64, 74, 81, 88, 94, 108, 110, 113, 127, 140, 159, 174, 190, 216, 228; NP goes to with Mr Hartstonge 95
Limerick, fair 79, 83, 91; barley and wheat sold at 99; barley sold at 91, 92, 99; beans sold 133; oats to 75, 102; peas sold at 93; wheat sold at 99, 101, 133
limestone, for kiln 143
linen: bandle cloth 58, 59, 80, 82, 143, 164, 179; bandle cloth, lost at fair 206; bandle cloth to weaver 96; bed tick sowed 170; bleaching 95, 103, 203, 219; bleachyard

101; bought for winnowing sheet 57; bought for shirts 81; coarse and fine sheets and pillow case bought 123; diaper 66; intended for breeches 82; Irish linen wheel 75; Linen Board, Limerick 76; Linerd, Batt 150; Lisdogan, Co. Cork 174, 182, 210, 223; Lohort Castle, near Castlemagner, Co. Cork 177; Longane, Darby, weaver 48, 51, 79, 89; NP and Catherine return from 209; NP and Catherine visit 188, 200, 205, 206, 213, 220, 224; NP visits 153, 173, 175, 177, 191, 197, 205, 202, 207, 209, 211, 227; spools for spinning 62; taken to Court 120; thread 82, 90, 96, 103, 121, 134; towels made by MP 110; ticking 170, 180, 181
Longane, Jon 104, 117, 152; given sucking pig 94; pays NP 92
lossid 56
Lowry, Alice, paid by Catherine 204
Lowry, Joan 73, 83, 87, 164, 196, 204
Lowry, Mick 157
Lowry, Nicholas 60, 63, 79, 91, 152, 214
Lowry, Tim of Adare 47, 49, 65, 68, 84, 85, 214, 218, 220; gelds lambs and pit 82; paid for sheep 221; pays for wheat 87; sends NP mutton 216
luck money 116
Lynch, Jon 52
Lynch, C., paid for picking potatoes 155
Lynch, Tho. 59
Lynches, NP at the 126
Lynes, Dennis: NP pays tithes 203; paid by note 192
Lynes, Will pays NP 87

Maddin, Connor, pays for grazing 161
Maddin, Den, spinner, paid for tow 205
Maddin, James, cooper 75, 93, 117, 118; cleaving wood 93; cutting wattles 93; delivers ferkins 101; delivers peck barrels 98; given coop 75; given hoops 114; hewing timber 89; makes barrells 95, 97; making boards 98; making staves 94; paid in oats 92, 101; paid in oats and peas 98; paid in wheat 90; pays for grazing 141; paid for 6 chairs in oats 92; pays NP 86; sells NP a churn 98; sells truckle 78; slitting hoops 93; staves a barrel 96
Madigan, Den 59
Madigan, Jon 50
Mahony, Tim 211; asks for rent 124, 129, 169, 176; distrains NP 163; paid May gale 178; paid November gale 195; paid September gale 198, 203; rent paid to 129, 163, 176, 182, 188, 192, 193, 207; rent paid to for deer park 139; rent paid to for deer park 159, 161; settles accounts 117
maids: at frieze 143; binding 191; examined about pork, wool, flax and oats 156; paid 104; spreading hemp in the bogs 191
Maigue river, overflows 170, 223
Malony, Timothy, measures turf cutting 84
man, Mr Thornhill's, issues note 162
Mane, Jon, pays for grazing 117, 178
Maney, Ms 116
Mangadan, Mary picks potatoes 131
Margett 180, 185, 187, 188
Margett, old, paid for spinning 192, 194, 195
marriage: dowry, inadequate 52; locket ring bought 173; negotiations 13, 52, 69; NP and Catherine are marriage 175; NP and Chapmans all agree to 173; NP cannot marry without a licence 174; NP's wedding clothes, made 173; wedding ring bought 173
Martin, Dr John 51, 69, 75, 91, 113, 158, 161, 190; buys wheat 81, 128; offers to buy Supple's horse 122; owes NP for peas 94; visits NP 109
Martin, Dr John jnr. 147, 159
Maryborough, Queen's County, NP at 113, 140
masons, given money to drink 162
Matty, paid 126, 149; given an old shirt 131; given money to buy sand and soap 140
Maunsell, Aphra 12
Maunsell, John 134, 135, 146
Maunsell, Alderman Richard 212, 219, 226
McCoghlin, James 115
McDaniell, Michael, saddler 157, 216
McGan, Edward 52, 55, 60, 69, 121, 126, 127, 128, 150, 154, 156, 157, 158, 159, 176, 179, 186, 189, 193, 195, 203, 205, 209, 228; and Sister, visit NP 112; and wife stay with NP 209; bill on issued 183; dispute with NP 132, 133; drinks with NP 116; lent black mare 171, 172; met at Clampitts 148; NP borrows from 158, 215; NP borrows his mare 110; NP lends money to 155, 213, 178; NP visits 188; NP writes to 191, 210; paid 181; paid 150, 180, 190; pays for mare 118; pays some of Studderts money 154; stays with NP 186, 204
McGan, Harry 80
McGan, Nick 102, 108
McGan, Sister 49, 51, 52, 62, 74, 75, 99, 105, 108, 126, 128, 134, 152, 157, 159; death of 180; funeral of 180; presses and dyes frieze 81; gives present of oysters 81; NP pays for tobacco from 92; in an uproar 138; pays for butter 132; visits NP 80
Mcmahon, James 149
Mcmahon, Bryan 53, 13; deals with Studdert execution 157, 162, 170
McMahon, Den 54, 68
McMahon, James 128

McMahon, Jon 150
Mcmahon, Terence 149
McMahon, Tim 177
McMahon, Torly 193
McMahon, Tom 219
Mcnemara, Honor 162, 173
Mcnemera, Matthew 47
Mcquier, Jon 61, 69, 74, 80; buys cow 70; owes money to NP 79; served with process 81
McTamos, James 129, 141, 149, 189
McTomas, Mick 128, 142, 149
McTomas, Robin 80
Mead, Mick 54, 57
Mead, Jon 85
Melan/Mellon, Co. Limerick 94, 112; church 125
Melane, John 109, 115
Melane, James pays for mare with cash, a cow, and yearling 111
Mellsop, George, visits NP 110, 111
Mellsop, John 78, 102, 103, 117, 136, 144, 149 158,
Mellsop, John, NP writes to 150
Mellsop, John, sold cow 190
Mellsop, John, sent oats 170
Mellsop, John, paid for cow 191
Mellsop, Mr and Mrs dine with NP 112, 210, 216
Mellsop, John 149
Mellsop, Mrs paid 138, 154
menus, dinner 131
Meny, Patt 50, 52
Merowny, Edward 124, 137, 165
Michael, James, paid 202
militia array 150
mill 48, 98; barley 63, 66, 67, 165, 213; beans 82; corn 74; malt sent to 171, 179 181, 185, 187, 191; NP goes to 214; oats 50, 71, 74, 93, 102, 166, 176, 180, 183, 189, 192, 201, 206; price of 187; wheat 124, 191, 192, 198, 199, 216
miller, paid 183
miller, paid for tucking 187
Miller, Peter 70, 71, 73, 75, 77, 83
Miller, Peter, his wife 76
Mills, —, given visitation fees 148
minister, asked to christen Price 187
minute, NP writes for Brettridge Badham 111
Molcheen, Jon, 217, 131; NP buys cow from 129; NP lends money to 218
Molcheen, Mick 219; pays for grazing 129, 133
Molcheen, Thos, hired 227
Monasterevan, Co. Kildare, NP sleeps at 113, 140
Moneen, Jon 151
money 146; 4/0d pieces 224; bank bills sent to Mr Thornhill 142; guinea would not pass 212; light guineas given 'if he can pass them' 213; luck money 116; moidore 200; Mrs Hartstonge's taken to Limerick 165; NP finds 1d on road 134; NP gives children 118; NP pays with bills 112; NP tries to borrow 163; pistoles 143, 172, 173; rate of exchange 143, 167; token given for a guinea 124; tries to get payment of 89; use of foreign 25, 69
Moor, Luis 96
Moran, John, pays for grazing 122
Morihy, Darby, grazing 139
Morony, Edmund 60
Morony, Father Nicholas 124
Morphey, John, promised £12 on behalf of NP's brother 138
Morphey, John, NP buys hair from 213
Morphey, John 150
Morphy, Mary 97, 98; pays for malt 101
Morroh, —, boatman 49; his sloop 137; paid for hire 125, 140
Morroh, —, boatman, paid for hire 140
mower, paid for mowing walks 123
Moynham, Councillor Matthew 61, 110, 157, 223
Mulcheer, D. 52
music, Jew's harp 94

Naas, Co. Kildare, NP dines at 140
Naghtin, Edward 57, 101, 118, 129, 133, 142
Naghtin, Fran hires out horses 85
Nagle, James, slater paid 201, 202
Nancy, Miss: dines with NP 219; NP lends money to 148; paid 126
Nantinan, Co. Limerick, fair 218
Nash, Ed 215
Nash, John 55, 77, 87, 98, 116, 128; and his brother reaping dutch barley 86; and his brother reaping wheat 86; assaults John Bryen 115; digging potatoes 99; pays for grazing 151, 205, 227; threshing 99
Nash, James 150, 186, 190, 191
Nash, Mick 201; paid for hire 211; sells a pig to NP 102; threshing 100
Nash, Patt 226; pays for grazing 178, 180
neighbours, disputed lamb 91
Nelan, David 47, 62, 64, 68, 81
Nele, Jon 86, 97, 98, 116; digging potatoes 99; given oats 101; NP accounts with 147; pays for grazing 132; sold bed of potatoes for wool 103; sold potatoes 103; threshing 100, 116, 117
Nele, Tim paid money owed by Mrs Widenham 125; paid in oats 90
Nele, Owen, hired 227
Nell, NP borrows from 190
Nell, wages paid 195
Nenagh, Co. Tipperary, NP dines at 113

Index

Neve, Jon 97
New Year's Eve celebrated 152
Newman, Jon 47, 52, 59, 63, 73, 74, 77, 102, 116; lent money 82; pays for grazing 178, 204; wife's funeral 62
Newman, Patrick 59
Newman, Will reaping 84; his boy 53, 54; shovelling potato trenches 83
newspapers bought 110, 111, 112, 115, 119, 120, 121, 122, 124, 131, 132, 134, 135, 138, 139, 140, 144, 148, 149, 151, 152, 154, 155, 190; paid for year's 172
night, unsatisfactory 172
Nihill, Lawrence 119; bill on by Wyndham Quin refused payment 92
Nolan, D. 52
Norse 151, 216; dines with NP 179; NP gives barley to 140; NP sells barley 139; pays NP debt 139; son Harry 138; son Jack 138

O Bryen, Dennis 4, 91, 157; NP lends money to 150
O Daniel, Mick 195
Odle/Odell, Thomas 151, 186
Oliver, — 116
orchard: apples 103; apples picked 127; apple trees 105, 107, 133, 180; ditch 93; fallow 69; plums sold 217
Ormond, Robert, sells trees to NP 104
ostler, NP tips 126

Page, Henry of Askeaton 149; asks NP for rent 142, 146; draws up NP's case 137; NP pays rent to 47, 122, 159, 188, 213
Palatine, Moll 122
Palatine, the 70
Palatine, Phil 132; buys potatoes 139; makes bags 109; sent to Limerick 104
Pallaskenry, Co. Limerick 95, 193; bleach-yard 216
Pallaskenry market: beans sold at 171; frieze bought at 132; oats sold at 101
Palliser, — 149
paper 70, 81, 89, 101, 110, 111, 113, 119, 122, 126, 131, 132, 135, 172, 187, 189, 198, 203, 226
parish, taxes 161; rates raised 84
Parker, —, NP dines with 162; his barn burned 154
Patrick's Pot, money given to servants 79, 136
Peacock family 12
Peacock, —, brother 100; brother asked for money 101; brother, asks NP for £12 142; brother at Enniscouse 84, 96; brother at Rathkeale 107; brother has a son 96' brother, letter from about lawsuit 136; brother NP borrows money from 109; brother NP visits 103; brother offers to pay in a year 135; brother refuses to lend money to NP 116; brother threatens court action 117; brother writes to NP 110; brother's baby dies 97; brother's child's christening 96; taken by John Morphey for debt to NP 138
Peacock, Catherine, returns home 178; agrees to marry Nicholas Peacock 173; given money 181; has tooth drawn unsuccessfully 175; her mother's legacy 210; hires maid 213; ill 180; negotiations for her hand 172; neighbours visit 216; NP writes to 174; pays maid 213; stays at Court 175, 201; stays at Shannongrove 180; stays away to wean baby 228; writes to Mrs Hartstonge 209
Peacock, George of York 12
Peacock, George (1749–): born 203; has measles 209
Peacock, James 12
Peacock, Pryce (1748–): bled 194; born 186; christened 187; ill 194; left with Mr and Mrs Geary 188; very ill 223
Peacock, Richard of Graigue 12
Peacock, Robert, hearth money collector 67, 82, 163, 199, 220; lives in Limerick 92; and wife 196; debt repaid 193; gives present to Pryce 197; NP borrows from 189; NP dines with 157, 158, 174, 176; NP repays debt 205; paid 219; pays quit rent 207; repays debt 159; sent for 196; stays with NP 193
Peacock, Nicholas gets cornet's commission 56
Peacock, NP's sister, letter about law suit 136
Peacock, Nicholas, travels part way to Dublin 148
Peacock, Nicholas, patches up quarrel with sister 138
Peacock, Nicholas and Catherine return home 175
Peacock, William 12
Peacock, William (1750–): born 216; weaned 228
Peacock, William, bids for Lisdogan 197
pedlar: handkerchiefs from 202; paid 194
pens, NP makes 111, 142
Peppard, Patrick 50
permit, fee for 189
Pery, Mrs of Limerick takes in plate 85
Pery, Mrs Jane 116
Pery, rev. William Cecil 125; and Mrs Pery visit NP 190
Pery, rev. William Cecil 102
pets, linnets in a cage 112
pictures, sent to Tim Bork 109
piggins, iron 75
Pike, Ann 13
piper, the, paid 225; pays for grazing 133

pledge given 73
Pordon/Purdon, Jon 150; godfather 217
Pordon/Purdon, Nell, hired 225
Portacacha, Co. Limerick 49, 50, 110, 120, 121, 122, 134, 149, 159, 161
corcas 75, 77; little corcas finished ploughing 171; rent of 194.212
Porter, Will, vails for 98
porters, paid 127, 128, 189
pot, mended 197
poultry: chickens 96, 225, 226; drake 59; drake sent to Court 128; goslings hatched 97; geese laid eggs 93; fowl laid eggs 93; turkey cock found dead 127; turkey and ducks sent to Sister 167; turkeys, sent to NP's brother 9; turkeys sold 207
pound: cows sent to 62; NP's stock put in 112; sheep to 63
powder and shot 74, 186
pregnancy and confinement 30
Price, Anne 12
prices: barley 58, 74, 91, 92, 135; beans 59; bull 175; butter 87, 143, 147, 148.153, 162, 163, 164, 165, 175, 176, 178, 179, 181, 186, 188, 192, 193, 194, 195, 196, 199, 201, 202, 203, 204, 205, 206, 207, 210, 214, 215, 220, 227, 228; butter did not get right 183; cheese 196; coat 137; comparison of 24; grazing 158; hay 135; hemp 205; hide 165; markets have fallen 133; milling blanketting 187; nails 140; oats 75, 102, 135, 162; service of two grey mares 142; sheep dressing 181; shoe for black mare 165; slating 202; weaving 181; wethers 175; wheat 102, 133, 210, 221; wool 163; yearlings 175
priest: NP makes towels for 110; NP sells horse to 121; paid 61; stays the night with NP 227; visits and receives Christmas money 105
prints, NP buys at cant 100
privy ['little house'], cleaned 143
proclamation against roman catholic clergy 110
Purcell, Cyprian 160, 163, 190, 191, 198, 215, 216; advises on illness of child 194; asked to draw tooth 205; asks for money 170; and mason at work 213; bleeds Catherine 180; buys wood for sashes at Court 132; dines with NP 223; draws tooth unsuccessfully 175; given malt 181; given red breeches 216; hangs wallpaper 103; ill 181; makes case for glasses 127; makes leaf of table 196; makes proposals for ferry 134; measures house 201; mends beds 141; NP borrows from 189; paid 142, 162, 163, 165, 192, 211; paid in oats 214; paid on account 171, 173, 175, 203; plans of new house and orchard 159; puts up sashes in hall at Court 138; sells trees to NP 132; sells trees to NP 165; swaps cow with NP 138; tries to mend clock 116

Pursill, Garratt, pays NP 86, 123; paid for trench 138; paid money 128; paid money owed by Mrs Widenham 125; sells yearling 134
Pursill, James 81
Pursill, Jon 218; binding 217; brings melons 215; pays for hay 125; sent to buy plants at Limerick 135; sent to Standish Grady 108; watches turf 105
Pursill, Mick 128, 132
Pursill, Patt 183, 189; buys land 144; paid christening money 198, 211
Pursill, Will, paid in barley 139
Puttney, — 196, 200

Quain, Martin pays for grazing 63, 191
Quain, John, weaver 111, 118, 124, 128, 151, 155, 156, 218, 221, 227, 228; NP buys heifers from 112, 131; NP settles accounts with 142; paid for weaving 204, 216; paid in oats 171; pays NP 177; pays NP a guinea 190; sells NP a cow 152
Quain, Martin 57; sells NP calves 122
Quain, Mrs, paid on account 155
quarry, NP searches 226
Quarter Sessions 169
Quelly, Connor 109, 139, 149; account made up 161; paid his wages 123, 126
Quelly, Law 53, 55
Quin, Connor 56
Quin, Andrew 165, 173
Quin family, met at Court 197; NP dines with 162
Quin, George of Kiltenan 125, 206; lent black mare 204
Quin, Mrs, breakfast with 136
Quin, Henry: death of 145; dies 137; funeral of 145
Quin, Jon: pays NP 86; sells NP two cows 129
Quin, Mary (1682–1776) 12, 13, 50, 53, 63; brought home by NP 144; NP dines with 211; NP goes to see 69; NP pays 210; NP talks to about Betty Ireland 210; paid for tithes 81
Quin, Molly, NP's sister, died 95 NP's sister, dies 95
Quin, Mr and Mrs 128
Quin, Mrs, sent a pig 134
Quin, Thady 13
Quin, Valentine (1678–1744) 13, 51; death of 119; funeral of 119; refuses to pay rates 62; NP dines with 65; NP dines with 84; NP visits 61; NP visits 104; NP visits 110; very ill 110; very weak 119
Quin, Wyndham (1717–1789) 55, 125; invited to stand as godfather 187; lends

money to NP 92; sent last year's tithes 143;
    sent tithes 203, 215
Quin, Widower 82; asks NP for money 101
Quinlan, Will 226, 227
Quinleven, Margaret, servant 97; dismissed
    148; has flax 98
Quirk, pays NP 86
quit rent: collector 111; drivers fees 127, 190,
    194, 206, 209, 216; drivers, take NP's
    stock 112; NP borrows to pay 128; NP
    offers to pay but is refused 138; NP pays
    51, 56, 64, 83, 89, 103, 112, 125, 149,
    165, 180, 207, 217, 225; NP pays his own
    and Mr Hartstonges 99; paid for Kildimo
    183, 194, 209, 213; paid for Portacacha
    194; paid not to drive NP 103, 111;
    receipts 112, 171, 187, 195, 218,

Rahily, Man, tithes 118
Rahily, Paul 52, 53, 65, 146; pays for grazing
    128; receipt 79; views NP's tithes 102
Rahily, 'Young' asks for tithes 109
Rathkeale, Co. Limerick 73, 100, 148, 174,
    213; barley sold at 101; fair 87, 120; NP
    visits 112; oats fail to sell at 101
razors, mended 206
receipt, printed bought 76; used 189
reed, cutting 217
Richardson, William 60
road; money for 178; use of agreed 115
robbery, butter, meal and salt from barn 161
Roche, Philip, Limerick draper in Main Street
    88
rock lime 160
Rockfield, Liscarroll, Co. Cork 153
Rohan, Jack, has clothes made 96
Rohan, James pays for grazing 145, 156
Rohan, Jon 180, 183, 191, 194, 220; sent to
    Lisdogan 189, 209, 210; paid for hat 178;
    paid on account 225; paid on account 214;
    sent to forge 221; sent to Limerick 189;
    sent to Limerick 211, 223; sweeps chimney
    220; NP borrows from 221
Ronan, Will 114
Roscrea, Co. Tipperary, NP at 113, 140
Rose, Richard of Rathkeale, Co. Limerick
    137
Rose, George jnr, served with subpoena 158
rosewater, stilling 143
Rourke, John 60, 61, 87, 172, 206, 207, 219,
    227; given money for Mrs Thornhill 218,
    225, 227; his man 211; paid for Richard
    Thornhill 176, 187, 195, 202; stays with
    NP 202; visits NP 177, 213, 218
Royce, Annabella, Mrs 226
Ruddle, David 59, 60
Russell, Mrs Elizabeth 58
Russell, George I 180; pays for grazing 133
Russell, James, pays for grazing 142

Russell, Hew, paid 213, 216
Russill, James 57
Ryan, A. 56, 64
Ryan, Mick: buys oats 98; rents meadow 102
Ryan, Jon 55, 188, 227

sally bed 92, 216, 222
saltpeter 135, 195, 209
Sammon, Mary 63
Sawn, Martin 82, 84, 122; buys malt 98
Sawn, Mary, knitter 122, 143
Sawn, Mary, her daughter 145
Sawn, Matt, pays for grazing 133
servants: at mass 77; boys' breeches 132;
    boys' hats 142; boys' shoe soles 138;
    brogues for 80, 132, 156; cooper 81; dog
    boy 117; given Christmas box 196; given
    clothes 77; given flannel 75; given leave to
    go to Kildimo 85; given money and punch
    on Christmas Day 152; given money for
    drink 60, 68, 82, 88, 148, 171, 199, 211,;
    given money for Patrick's pot 199, 211,; go
    to funeral 62; go to Kildimo 51, 85; go to
    Pattern 145, 171; hired paid 87; money for
    Pattern 65; paid 103
Shannongrove, Co. Limerick 85, 122; com-
    pany at 144; NP and Hartstonge family at
    92, 98, 125; NP invited to 101, 180; NP
    visits 138, 154, 156, 161, 199; salmon and
    trout sent to 138
Shea, Dan cobbler 48, 50, 65, 71, 112, 116,
    132, 217, 219, 222; makes new shoes 96,
    136; marriage of his daughter 224; new
    slippers 191; NP accounts with 91; NP
    buys new pumps from 174; offered land at
    Kilmoreen 133; pays for grazing 226; Shea,
    Nick, his boy 207; sold barley 135
Sheehan, E.: NP pays church rates to 139
Sheehan, Patt, 152, 162; makes proposal for
    grazing at Court 134; NP borrows money
    from 122; pays for grazing 156; proposal
    for grazing sent to Mrs Hartstonge 134
Sheehee, Ned 118, 138; pays for grazing 131;
    sells NP calves 127
Sheehy, —, made churchwarden 157
Sheehy, Ed 175; NP borrows from 171
shepherd, paid 193, 200
sheriff 110, 112; given executions against
    Henry Studdert 172; his court 154; in
    town 171
sheriff, sub- 131; NP delivers wigs and buck-
    les to 140; NP writes to 174; seizes
    Hartstonge stock 112, 113; sub-, asked to
    open boxes 125
sheriff bailiff 223; paid 226
Shilbred, James 49, 54, 55, 64; paid for turf
    68
shoes and brogues: boots 65, 155; boys
    brogues 116; brogues 48; brogues and

stockings for little boy 165; brogues and stockings for Matty 124; brogues for John Rohan 102, 191; brogues for servants 80; hob nails 213; pumps 114, 174, 222; shoes 49, 136; shoes and buckles 140; shoes for Catherine 182; soles 80, 102, 121, 169, 183, 220
Sillver, Mrs 146
silver: cup presented to NP by Mrs Hartstonge 140; punch ladles 187
Silvermines, Co. Limerick 113, 126, 140, 148
sister-in-law, visits 185
Slattery, Charles 138
Slattery, Charles 115, 118, 141, 170, 173; mowing 163; NP settles accounts with 174; paid for dressing hemp 149, 205; paid in oats 171; pays for grazing 125
Slattery, Patt 71
Slattery, James, pays for grazing 150
Smith, —, slater 194; ran away 198
snuff 81, 94, 99, 100, 104, 105, 108, 111, 112, 115, 118, 126, 128, 132, 135, 136, 137, 143, 147, 152.154, 155, 158, 163, 165, 166, 169, 172, 175, 176, 179, 181, 183, 184, 188, 187, 189, 191, 193, 198, 199, 200, 201, 203, 204, 205, 206, 207, 209, 210, 211, 212, 213, 214, 215, 216, 217, 219, 220, 221, 222, 223, 224, 225, 226, 227, 228; bought for Mrs Hartstonge 119, 121, 123; Havana 126; snuff box 112, 192, 202
Southwell, Lord, militia troop not at array 150
spark, sent to Limerick for a 191
spickets and fossels 199
spinners, paid 81, 177, 185, 186, 195, 206, 207
spinning, tow 71, 73; cost of 91; flannel 69, 70; finished 185; hemp 73; linen yarn 48, 191; reel 71; thread 62; thread washed 82; wheels and reel 193; wheel flyers 73; woollen yarn 191
St Patrick's Day 79, 136, 171, 199, 211
St Leger, William 75
Staff, W, pays NP 123
Stanford, Ms 146
stock: cattle: bull sold 211, 214; bullocks 64; bullock and cow sent to fair 193, 195; bullock killed 165; bullock price of 96, 103; bullock sold 202; bullocks sent to graze 96; calf 215; calf dies 104, 165, 166; calves and yearlings gelded 82; calves taken in 135; Court cattle drenched 180; cattle did not sell 174, 188; cows 213; cattle sold 174, 175; cows bought 56, 58, 61, 64, 103, 111, 112, 151, 152, 155, 156, 187, 211; cow calved 59, 78, 80, 94, 155, 156, 165, 219; cow, did not sell 165; cow dies 104, 166, 199, 201; cow dead disputed by NP 156; cow drowned 60, 157; cow's carcase taken by NP's servants 156; cow, fat 145; cow hipshot killed 199; cow slaughtered 70, 105, 128, 151, 161, 164, 220; cow sold 141, 183, 188, 95, 190, 201, 204, 207, 209, 211, 212, 219, 220; cow sold at fair 228; cow skinned 185, 211; cow swapped for corn 58; cows given to M. fitzGerald 154; cows in wheat 62; cows sent to Coroheen 186; cows to fair 175, 200; Hartstonge cattle guarded 109; Hartstonge cattle sent to Elltown 109; Hartstonge heifer sold 117, 120, 142, 200; heifers 56, 225; heifer bought 111, 112, 128, 151, 155, 160, 213; NP pays for calves 138; strapper 170; strapper swapped for in calf heifer 197; veal killed 199; veal sold 178
horses 64: ball 61; bay garron shod 213; bay mare shod 210, 213, 217; bay mare and colt sold 228; big gelding and mare sold 118; black colt 109, 218; black mare shod 176, 202; black mare sold to Henry Hartstonge 104; bought 52; branded 63; broken 96; colt and filly sent home 145; colt gelded 82, 138; colt harnessed 78; colt sold 228; colts 134, 157, 165; colts shod 165; compensation 116; Court horses shod 117; dies 137; docked 62; farrier sees black mare 227; filly 165; filly shod 165; filly sold 210; for Dublin chosen at Adare 140; for sale 100; gelded 50, 63; gelding and black mare shod 174; gelding bought 161; gelding sent to NP's brother 89; gelding harnessed to truckle 107; gelding lame 162; geldings fail to sell 117; geldings shod 100, 117; gelding sold at Lisdogan 177; grazing at Court 128; grazing at Kilmoreen 141; grey colt shod 117; harnessed 78; Hartstonge to be given up to sheriff 110; hired and paid for with money and drink 85; horn branding 88; horn money paid 151; horse bled 62, 63, 64; horse 'Pet' shod 100; horse, 'Phebe' shod 173; horse sold 120, 219; horses with fever 64; mare shod 155; mares shod 165, 211; 'Mcgan' sold 96; 'Meloge' killed 221; NP tries to sett 90; pad 123; 'Pet' schooled 100; physic for 227; pillion hired 119; 'Red Robin' 128; sent to Mrs Hartstonge 123; sent to Shannongrove for sale 90; shod 57, 83, 84, 88, 90, 101; stone colt bought 161; stone horse 161; taken in 134; traded for beans 59
mares 139; bay mare 171; bay mare refuses horse 124; black mare 162, 192, 217; blacksmith treats mare 114; Dublin mare takes horse 138, 160; grey mare foaled 141, 160; grey mare's foal sick 143; grey mare refused horse 115; grey mare refused

Index

horse 143, 158; grey mare took the horse 117; horse, 'Jet' 148; little mare foaled 225; mare 128, 139, 158; mare bought 138; mare foaled 63, 113, 98'; mare harnessed to truckle 101; mare, 'Harriott Blew' foaled 142; mare, 'Harriott Blew' took the horse 142; mare, 'Harriott Blew' refused the horse 143; mare has cough 75; mare ill 153; mare took horse 115; mare took James Hewson's horse 158; price of service of two grey mares 142; rowelled 50; white mare 111, 222; white mare sick 220; young mare serviced 180

yearlings: 'Richball' died 73: yearling 117; yearling bought 134; yearling bought for Mrs Hartstonge as present 125; yearling bought with cash and corn 111; yearling filly bought 141; yearlings bought 140; yearlings gelt 138; yearlings sold 95, 114, 118, 141, 142, 156, 200, 225, 227'

pigs 171, 219; hoggets bought 114; hogget drowned 185; hoggets sold 95, 211; pig bought 51, 165; pig slaughtered 58, 74, 132, 134, 157, 165, 166, 181, 184, 181, 209, 210; pig sold 105; piglet died 91; pigs gelded 159; sow farrowed 91; sow gelded 63; sow pigged 169; sow slaughtered 107, 224; sow sold 197

sheep 105, 135; at Kildimo 188; counted 139, 160; dagging 139; dies 133, 158; dressed 181; drowned 198; from Studdert 173; from Kilmoreen 170; grazed 55, 121, 156; Hartstonge, sold 120; killed 165, 180, 187, 219; lambs 62, 138; lambs born 78, 79; lamb dies 128; lambs gelded 158; lambs from Kilmoreen 170; lamb killed 96, 170, 201, 225; lambs marked 225; lambs shorn 65, 85; lambs sold 188, 213; lamb stolen 170; lost 67; marked 176; ram 51; ram borrowed 218; ram bought 62; ram skin put on horse 64; sent to Court 115, 188; shearing 51, 63, 83, 96, 99, 115, 141, 160, 176, 188, 200, 213, 225; shearing lambs, price of 143; sheep bought 213; sold 104, 105, 122, 166, 173, 174, 175, 178, 186, 201, 220, 2 21, 226; trespassed 63; washed 62, 82, 99, 114, 140, 159, 161, 188, 200; wethers 152, 172; wethers bought 64; wether dies 156; wether killed 104, 181; wethers sold 81, 188

stock: dogs killed for killing stock 113; driven 57; driving threatened 122; NP lists 128, 114, 174; viewed at Portacacha 152

Studdert, Henry 148, 151; fiery facies against 170; legal fees 171; letter from 172; NP has him arrested 174; subpoena and execution served on 170

Studdert, Mr, sends money 155

Studdert, Thomas, of Ardlaman, Co. Limerick 139, 147, 148, 151, 172; asked for money 155; his bond 137; serves injunction on NP 140

Studdert, — 161, 210; NP settles with 209; NP writes to 210; pays money to NP 161

Studderts 150; NP goes to Limerick to meet 153; NP visits 158; sell sheep to raise money 154

Sulivan, Jon: paid in oats 93; pays debt 144; pays for grazing 128

Supple, Geary/Gerry 152

Supple, Henry 58, 97, 139, 105, 114, 118, 120, 129, 169; death of wife 197; given barrel of oats 134; invites NP to dinner 131; paid with cash, oats and barley 136; salary accounted for 139, 156, 163; served with bill on High Sheriff 122

sword, given to Mrs Hartstonge 144

tailor, NP pays 128, 151, 180, 197; making coat 157, 180; making coat and waistcoat for NP 207; making coat for little Jack 145; making servant's clothes 204; paid for boys clothes 190; paid for Jack's clothes 172; paid for making coat 213; paid for making surtout 210; sent for to make NP clothes 195

tallow 83; rendered 92, 128, 165, 167

Taylor, Mrs Ann, funeral of 103

Taylor, Col. Edward, 49, 110, 117, 149, 186, 199, 219, 226; asks NP for quit rent 101; barley for 75; buys lamb 77, 93; gives leave to cut turf 98; his men 77; letter from 64, 69; NP gets bill on 133, 135; NP gets commission from 56; NP pays rent to 56; NP pays rent to 103, 174, 176, 178, 193, 200, 205; oats for 59, 75, 76

Taylor, William 56, 146; money collected from 69; rents grazing 136; stock leaves grazing 141

Thornhill, Richard 142, 163; asks for money 162, 165; asks for rent 170, 172; his mare 182; letter from 127, 129, 133, 139; lent £40 in bill on Studdert 139; NP dines with 177; NP meets 174; NP pays quit rent to 176; NP sends money to 149; NP writes to 129, 133; visits NP 111

Thornhill, Mrs, paid quit rent 213; NP writes to 228

tinker, paid 64, 205

Tipperary, Co., NP plans trip to 177

tithe proctor 65, 74

tithe proctor, his fees 85, 118, 120, 184, 191, 192, 203, 214, 227, 228

tithe proctor, paid last year's tithes 165

tithe proctor, Revd. Bucknor's 177

tithes paid 50, 53, 74, 63, 65, 81, 85, 137, 141, 146, 163, 183, 197, 203, 205, 227, 228; agreed 203, 215; in wool 160; levied

on turf 54, 65, 84; Kilmoreen, paid 144, 178; Mr Quins 192; NP did not agree Court 144; NP gives Martin fitzGerald a share of 170; NP pays his and Mr Hartstonges 118, 119, 136; NP pays late 102, 142; of Court, advertised 144; set 86
tobacco 54, 55, 56, 58, 60, 62, 63, 64, 65, 67, 69, 70, 79, 80, 81, 87, 89, 90, 85, 98, 101, 118, 129, 138, 139, 148, 150, 154, 155, 161, 177, 182, 213; pipes 81, 213
Tobin, Watt 117; his daughter spinner 73, 74, 76, 77, 79, 96
Tobin, —, instructed to summons Cleary and partner 100
Tomson, Andrew 219, 222, 225,
Tooreen, Co. Limerick 160
Tories, feared in Crataloe wood, Co. Limerick 85
Tough, Co. Limerick, NP goes to 48, 57, 58, 59, 60, 61, 63, 64, 65, 66, 67, 74, 84, 85, 92, 108, 109, 114, 117, 119, 122, 126, 128, 135, 140; cattle driven at 67
Townsend, —, dines with NP after hunt 145
transport, by river 26, 49; boat aground 135; boat carries turf 104; boat hired 108; boat, mended 157; boat to Court 122; cost of, by boat, 103, 127; cider to Dublin by boat 127, 137; horse hired 219; timber by boat from Limerick 189
trespass, 141; NP pays for 128; sheep 132
Tullamore, King's County 113
Tullow, Co. Tipperary 126, 148
turf 126, 161, 217; cut 49, 64, 81, 141, 187, 201, 203, 214, 219, 225; bought 53; bought 52; drawn 64, 68, 84, 85, 86, 98, 116, 149, 164, 165, 177, 179, 189, 190, 191, 203, 204, 214, 217; footing 143, 188, 189, 202, 226, 227; house 124, 149; kishes bought 55; rick of 83; truckles and harness ready to draw the 83
turnpike fee 103, 104, 105, 107, 109, 124, 126, 127, 128, 132, 144, 148, 150, 152, 165, 173, 176, 179, 180, 181, 185, 187, 191, 192, 193, 194, 196, 197, 199, 201, 204, 205, 207, 211, 213, 216, 217, 220, 221, 223, 224, 226
Tutthill, George of Kilmore and Faha, Co. Limerick 119
Tutthill, 'Young', buys timber 127

Vahan, Ed 86, 128, 178
Vahan, N., pays NP 101; N, paid in oats 174
vails: NP gives to cook and housemaid 102; NP gives to cook, boy, groom, man 153; NP gives to maid and ostler 126; to Bastable servants 182; to boy 164, 213, 217, 222; to Chapman servants 173, 182; to Court groom 175; to Gore servants 177; to groom 161, 162; to huntsmen 187; to Hartstonge maid 183, 184, 186; to Hartstonge servants 181; to Hewson servants 178; to Johnston servants 176; to Lisdogan servants 177; to maid 211, 213, 218, 221, 224; to servants 177, 190, 199, 205, 223; to Thornhill servants 177
Varah, Tige, paid in oats 222; paid for hire 189
Vincent, Thomas, clerk to Linen Board 76

wages: in bandles of flannel 129, 151; in barley 134, 135; in barley and cash 135; in beans 80, 82, 84, 87, 94, 161, 201; in brogues 183; in brogues and soles and cash 179; in cash and corn 103; in malt 97; in oats 83, 86, 91, 92, 96, 98, 101, 115, 165, 196, 207, 213, 214, 223; in oats and cash 215; in tallow and cash 184; in wheat 90, 91, 118; in wool 214, 225; oats and beans 83; oats and peas 98; servants 177; with pig 171
Walcot/Wollcutt, Councillor John Minchin 223, 226
Wall, Will 51, 74
Wallis, John, Dublin 153, 157, 158; asked for fiery facies 170; consulted on Studdert case 165, 166; gives NP copy of bill 228; issues subpoena 151; issues writ 140; sends bill of costs 171; sends execution 155
Wallis, Mrs, and NP dined at Court 187; goes home 187
Wallis, Thos. 218, 220, 221, 226, 227
Wallis, sister, at NPs 186
Warner, Mrs in a huff about grazing 116
Warner, Rev. Simon, tithe paid 120, 136
watch, pocket 127; found in NP's breeches 145; mended 164; NP buys 140; NP thinks he has lost his 145; sent to be mended 186; sent to Robert Peacock 220
wax, stick of sealing 108, 111, 152, 214
way warden, NP chosen as for Kildimo 169
weather:
  frost 56, bitterly cold 110; frost 91, 150; frost continues 78; frost too hard for plough 155, 184; great frost 181; ground frosted 62; prevents ploughing 71; terrible frost 71; too cold to shear sheep 63; very cold 141; thaw 73, 156
  snow 134; heavy snow 135; snow and frost hinders work 78; snow, sun melts 134; snow thawed 78; snow too deep to go out 135; too wet to make hay 143;
  weather, too wet to plough 148; wet 54, 59, 78, 83, 84, 93, 98, 123, 126, 161, 170, 187, 203, 204; wet hinders sowing 97; very wet 156, 228
  storms: at night 110, 121; violent 75, 88, 110, 121, 131, 165, 217, 219; violent, strips houses 127, 222

Index

weavers: paid 139, 148, 193, 228; thread sent to 86
weaving: canvas 48, 78, 79, 132, 206, 216; bandle cloth 128, 191, 206; flannel 48, 74, 90, 181, 204; frieze 69, 195, 206; ticking 167; yarn 213
weir: constructed 136; near wing set down 156; problem with foundation 157; rebuilding of 139, 140; swept away 139; timber painted for 136
Welch, Bridget 73, 77
Welch, David 52, 54, 55, 70, 73, 75, 190, 146, 214, 215; hires horses from NP 80; paid in tallow and cash 186; pays for grazing 178, 192, 218
Welch, Ed 194
Welch, Maurice 194
Welch, May 164, 172, 180
Welch, Patt 82, 84; cuts thatch 88; paid for hurdles 92; paid in oats 84, 90, 98; threshing 95, 97
Welch, R. 71
Welch, Thomas, printer in Limerick 172; paid 173
Welch, William, 128, 189, 191, 203, 204, 207, 211, 214, 218, 219, 220, 221, 222; paid in bandle cloth 217; paid in beans 201; paid in wool 214; paid on account 213; pays for cow hide 211
Welch, Thomas221; NP pays rent to 226
well, cleaned 143, 146
Welldon, Tho 194, 216
Wellens, James 80
West park, water for 62
Westropp, Revd. Cecil 118, 119, 121; asked for money 122, 123, 127; pays NP 133
Westropp, Mountifort, pays hearth collector's bill 123; gives NP grazing for mash 136
Whiddon, Mrs, paid for tartpans 195
Whistler, —, cutler in Limerick 153, 154, 176
White, Thomas 81
Widenham, Alice see Hartstonge, Alice
Widenham, Mary 13, 47, 48, 51, 53, 54, 56, 57, 58, 61, 63, 64, 66, 81, 87, 124, 142; burial of 88; buys barley 69; death of 88; her sheep shorn 83; her silver plate taken to Limerick 85; ill 53; NP makes up accounts 65, 74, 139; NP sells barley to 80; pays mowers 68; very ill 81; yearlings marked 62
Widenham, Mrs 219
Widenham, Mary see Quin, Mary
wife, NP muses on need for a 156
Wight, archdeacon Richard 74, 85119,
will, NP writes his 151
Williams, — 92
Williams, Jon: paid for combing worsted 86; pays for grazing 150
Williams, Mick 173
Williams, Mrs 217, 222
Williams, Mr 225
Willson, John paid 184
woodman, paid 84
wool: fleece of 63, 160; for flannel 181; lambs 164; lambs wool 69; price of 121; sold 83, 87, 115, 119, 148, 149, 176, 188, 201, 226; sorted 63

Yeamons, Frank, Limerick merchant 136, 145, 152, 163, 183, 185, 186, 196, 198, 212, 215, 216, 217, 219; NP borrows from 158; NP pays debt 159; NP pays Hartstonge account 150; NP pays old account 195